Handbook of Microscopic Anatomy

Continuation of Handbuch der mikroskopischen Anatomie des Menschen
Founded by Wilhelm von Möllendorff
Continued by Wolfgang Bargmann

Edited by A. Oksche and L. Vollrath

Enrico D. Canale   Gordon R. Campbell
Joseph J. Smolich   Julie H. Campbell

Cardiac Muscle

With 101 Figures

Springer-Verlag Berlin Heidelberg NewYork Tokyo

Handbook of Microscopic Anatomy
Volume II/7: Cardiac Muscle

Dr. E.D. Canale   Dr. G.R. Campbell

Cardiovascular Research Unit, Department of Anatomy, University of Melbourne,
Parkville 3052, Victoria, Australia

Dr. J.J. Smolich   Dr. J.H. Campbell

Baker Medical Research Institute, Commercial Road, Prahran 3181, Victoria, Australia

Professor Dr. Drs. h.c. A. Oksche

Institut für Anatomie und Zytobiologie der Justus-Liebig-Universität, Aulweg 123, D-6300 Giessen

Professor Dr. L. Vollrath

Anatomisches Institut der Johannes Gutenberg-Universität, Saarstraße 19–21, D-6500 Mainz

ISBN 978-3-642-50117-3   ISBN 978-3-642-50115-9 (eBook)
DOI 10.1007/978-3-642-50115-9

Library of Congress Cataloging-in-Publication Data. Cardiac muscle. (Handbook of microscopic anatomy; vol. II/7) Bibliography: p.   Includes indexes. 1. Heart – Muscle. 2. Heart – Muscle – Anatomy. I. Canale, E.D. (Enrico D.) 1958– . II. Series. [DNLM: 1. Heart – physiology. 2. Myocardium – ultrastructure.   QS 504 H236 Bd. 2 T.7] QP113.2.C369 1986 596′.047 86-3784
ISBN 978-3-642-50117-3

2122/3130-543210

Preface

In the ever-expanding field of heart research the needs of established researchers, students and general readers can vary considerably, making it difficult therefore to cater for all types of audience within a single volume.

The aim of this book has been to provide a comprehensive and up-to-date review of the structure of the heart, including its cell biology. The ultrastructure of the working myocardium and all portions of the conduction system, together with their development, is covered in detail. Also included are chapters on the morphometry of cardiac muscle, the innervation of the heart, cardiac hypertrophy and regeneration, and the development of the coronary circulation. A detailed review of cardiac muscle in cell culture is also provided.

It is to be hoped that readers, whatever their background, will find the information contained herein useful for their needs.

This work was supported by a grant from the National Heart Foundation of Australia. The authors wish to gratefully acknowledge the following people for their invaluable assistance in preparation of the manuscript: Professor Yasuo Uehara, Dr. Takashi Fujiwara, Dr. Peter Baluk, Dr. Seiji Matsuda and Bill Kaegi for providing unpublished micrographs; Fabian Bowers, Patricia Murphy and Janet Bennett for typing; and Lucy Popadynec, Nella Puglisi, Maggie Mackie, Mary Delafield and Liana Butera for assistance with references and figure preparation.

THE AUTHORS

Contents

A. General Introduction

The mammalian heart is a compact muscular organ consisting of four main chambers, a right and left atrium and a right and left ventricle (Fig. 1). The left and right sides of the heart pump blood respectively into the systemic and pulmonary circuits of the vascular system. Venous blood enters the right atrium through the superior and inferior venae cavae, passes to the right ventricle via the tricuspid orifice and is then pumped through the pulmonary arteries to the lungs where gaseous exchange occurs. This oxygenated blood returns to the left atrium via the pulmonary veins, then passes to the left ventricle through the mitral orifice for distribution throughout the body by the aorta and its branches.

The wall of the heart consists of three layers: the inner layer, or endocardium; the middle layer, or myocardium; and the outer layer, or epicardium. The heart also has a fibrous skeleton which serves as a support and firm anchorage for the origin and insertion of the atrial and ventricular musculature as well as the valvular tissue.

The *endocardium* completely lines the atrial and ventricular cavities and covers all the structures that project into the lumen of the chambers (valves, chordae tendineae and papillary muscles). It is lined by endothelium directly continuous with that of blood vessels entering and leaving the heart. These endothelial cells rest on a thin subendothelial layer of delicate connective tissue. Between this subendothelial layer and the underlying myocardium is the subendocardial layer of connective tissue containing blood vessels, nerves and branches of the conducting system of the heart.

The *myocardium* is composed largely of cardiac muscle cells. They comprise more than 70% of the heart wall volume, but make up less than 30% of the total cell number (see NAG 1980; GEVERS 1984) (Table 1). Cardiac muscle is responsible for the rhythmic, unceasing activity of the heart from early embryonic life to death. In man, some cardiac muscle is found in the walls of the great veins at their point of entry into the heart; however, in smaller vertebrates such as rats and mice cardiac muscle is present in veins some distance from the heart. The myocardium forms a complete coat around the heart except at the atrioventricular junction, where fibrous rings encircling the mitral and tricuspid valves separate the atrial musculature from that of the ventricles. It varies considerably in thickness in different parts of the heart, being thinnest in the atria and thickest in the left ventricle. The architecture of cardiac musculature in the myocardium is very complex, particularly in the ventricles (FREEMAN et al. 1985). A transverse section through the central region of the ventricles shows an oblique muscle fibre orientation in the superficial layer near the epicardium, transverse orientation in the middle (Fig. 2) and a vertical orientation

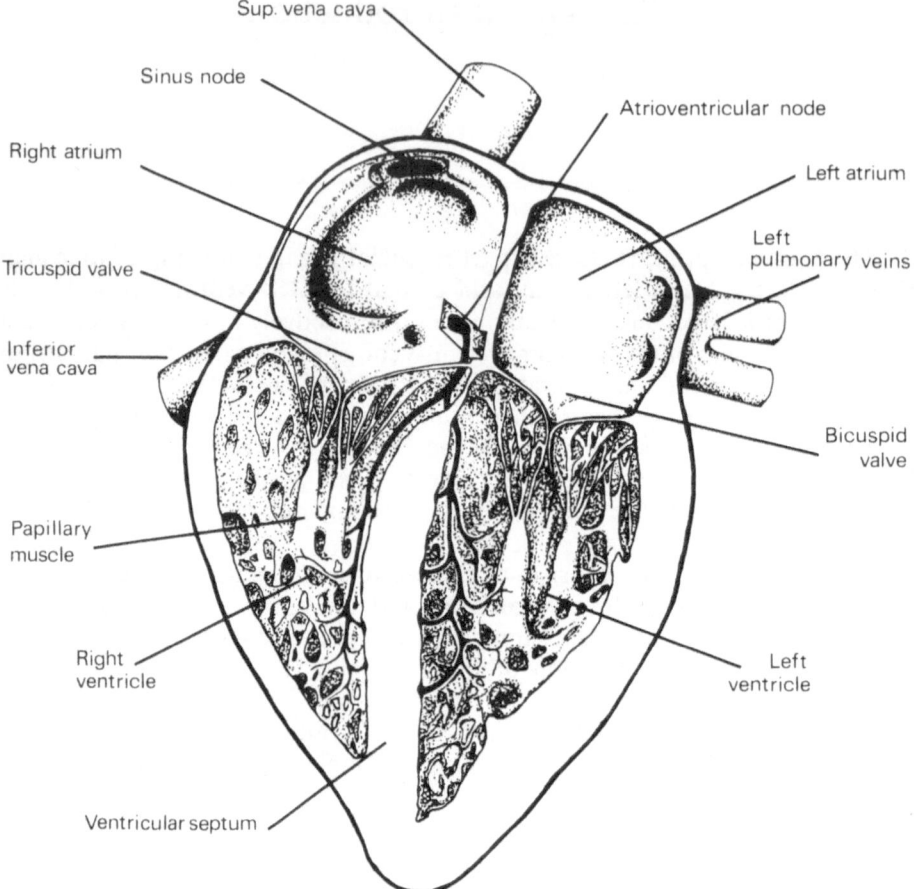

Fig. 1. Cutaway schematic view of heart demonstrating the rear portion. The nodes and specialized conduction pathways of the ventricles have been superimposed in the correct position in black

close to the endocardium (NAVARATNAM 1980). This appears to be due to the orientation of muscle fibres along figure-of-eight pathways. In the left ventricle, for instance, the pattern is characterized by a train of muscle cells adjacent to the epicardium that spirals both down and into the wall, past midwall to the endocardial side, and then spirals up on the inside past the equator, where a similar route past midwall returns the fibre train back to the epicardial side. This completes the figure-of-eight by starting another figure-of-eight without actually joining it (see STREETER 1979).

The *epicardium* is covered externally by a single layer of mesothelial cells supported by a thin layer of connective tissue. A subepicardial layer containing blood vessels, many neural elements and adipose tissue attaches the epicardium to the myocardium.

The heart possesses a system of specialized cardiac muscle cells whose function it is to coordinate the heart beat by regulating the contractions of the

Table 1. Major cell types of the heart (see NAG 1980; GEVERS 1984)

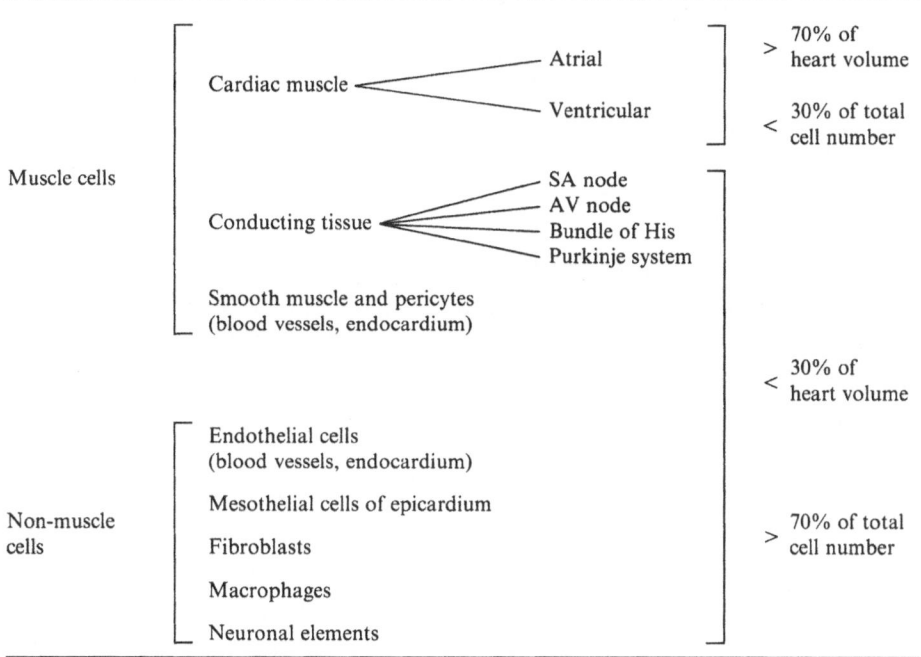

atria and ventricles. Contraction (systole) of the atria is followed by contraction of the ventricles. The initiating impulse begins in the sinoatrial (SA) node, passes through the atrium to the atrioventricular (AV) node (no specialized fibres connect the two nodes of the human heart, although preferential conduction pathways exist – see D.I.8), the AV bundle, its branches, then finally to the cardiac muscle cells via the Purkinje system and the transitional cells. All areas of the specialized conduction system of the heart, as well as the cardiac muscle cells (under certain conditions), are capable of spontaneous discharge. However, the SA node normally discharges at a faster rate, and the wave of depolarization spreads rapidly from it to the other regions before they discharge. The SA node is therefore the cardiac pacemaker, with its rate of discharge determining the rate at which the heart beats.

The SA node is located in the upper part of the crista terminalis at the insertion of the superior vena cava (SVC) into the right atrium (Fig. 1). The AV node lies in the atrial septum, anterior to the valve of the coronary sinus. The AV bundle arises from the AV node and passes through the right trigone of the fibrous skeleton of the heart near the mitral valve to the ventricles, thereby forming a muscular connection between the atria and ventricles. At the anterior edge of the muscular septum the AV bundle divides into left and right branches descending on both sides of the muscular ventricular septum beneath the endocardium. These major branches further separate into smaller branches finally passing to all parts of the ventricles (Fig. 1).

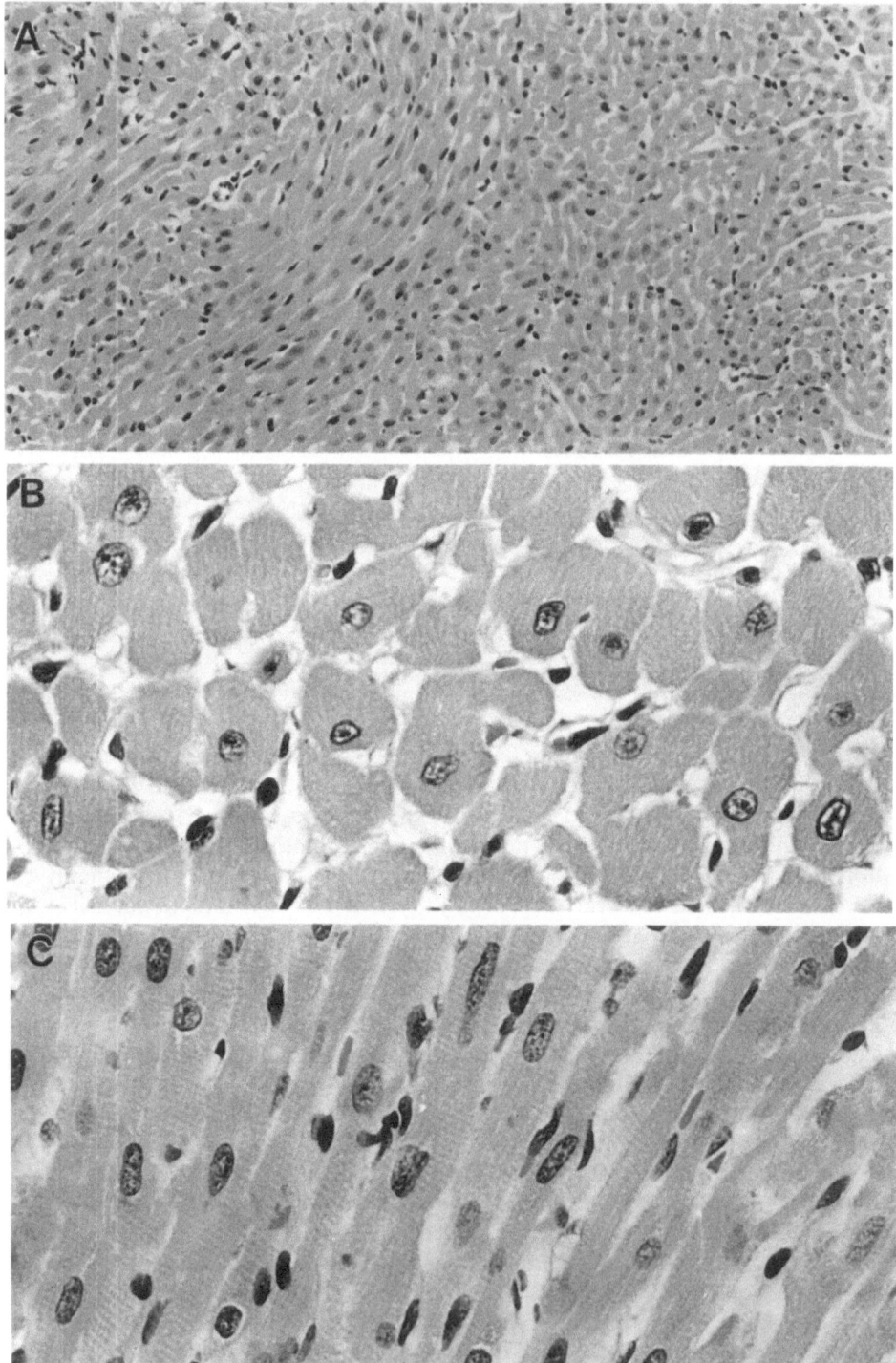

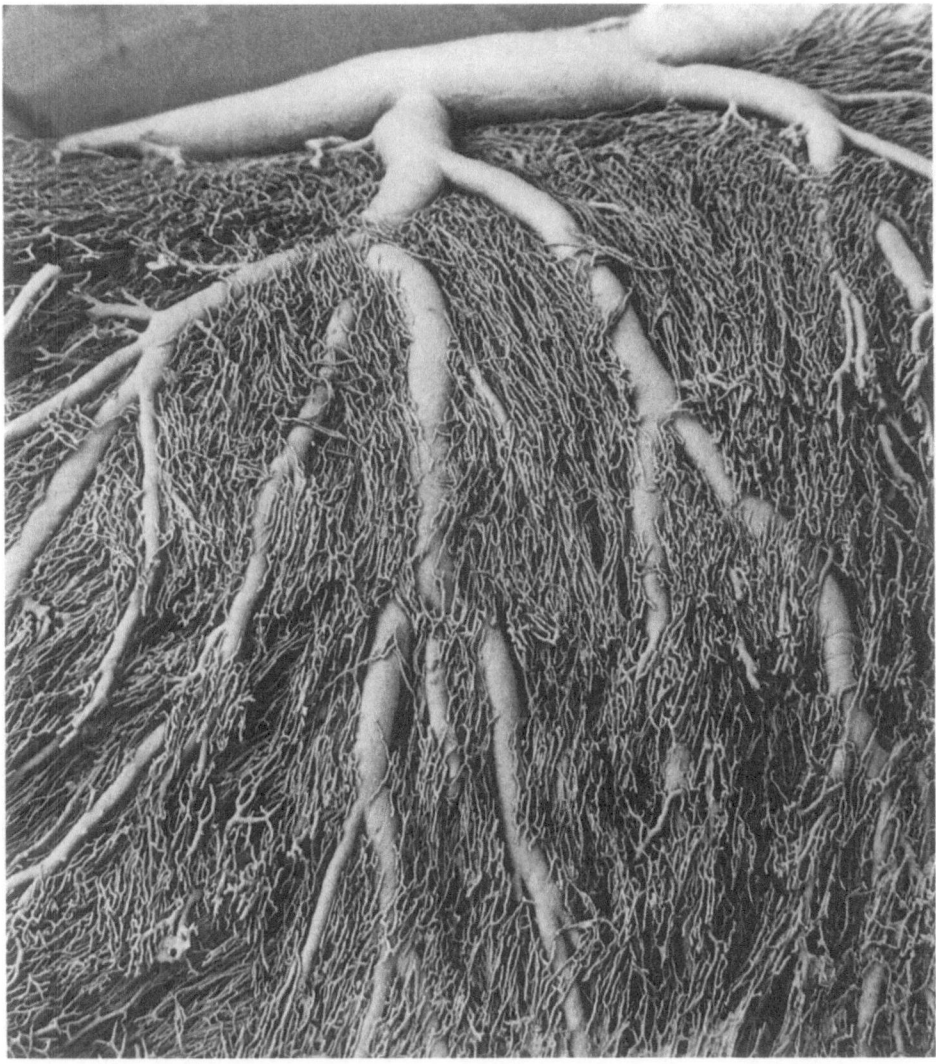

Fig. 3. Scanning electron micrograph of methacrylate cast of coronary arteries of rat ventricle. Large coronary arteries run in the epicardium giving off smaller branches to the myocardium. These further branch and ramify as a fine capillary network between the myocytes. × 25. Micrograph courtesy of Dr. PETER MOSSE

Fig. 2 A–C. Light micrographs of immersion-fixed rat ventricular myocardium. **A** Bundles of myocytes are cut both in transverse and longitudinal section demonstrating the complex arrangement of fibre pathways. × 187. **B** Myocytes in transverse section. Cells tend to be rounded with centrally located nuclei. The endomysium between the cells contains many capillaries whose endothelial nuclei stain more intensely than those of the muscle cells. × 562. **C** Myocytes in longitudinal section. Nuclei are centrally located and elongated or cigar-shaped. Cross striations can just be detected. Note how the fibres appear to branch and anastomose creating interstices in which endothelial cells are found. × 562. Paraffin embedded, H and E stain. Micrographs courtesy of Mr. BILL KAEGI

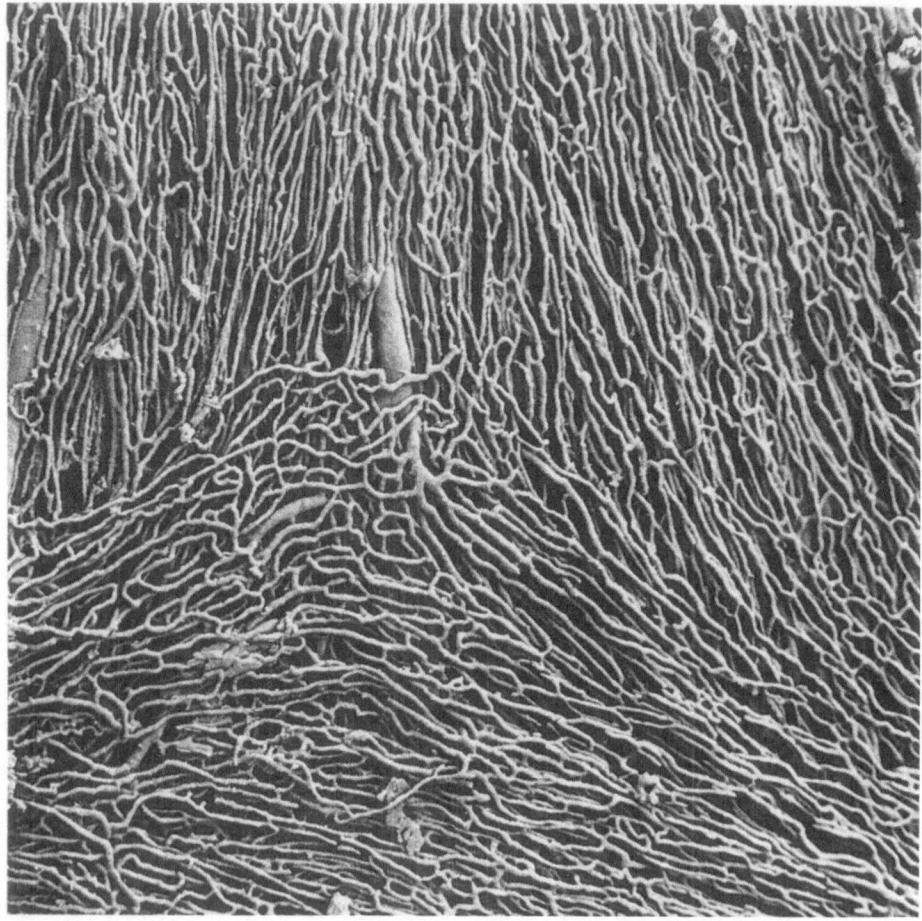

Fig. 4. Scanning electron micrograph of methacrylate cast of capillary network of rat ventricle. Capillaries branch and interdigitate, but closely surround the myocytes parallel to their long axes. This micrograph, like Fig. 2A, also shows how neighbouring bundles of myocytes run in different directions. × 60. Micrograph courtesy of Dr. PETER MOSSE

The arterial supply to the heart is almost totally met by two coronary arteries (the left and right coronary arteries), which arise from the aorta at the sinuses of Valsalva behind the anterior and posterior cusps of the aortic valve. The two arteries descend subepicardially across the heart giving off branches (Fig. 3) and smaller perforating vessels which break up in the myocardium into a rich capillary plexus, which runs parallel to the longitudinal axis of the muscle cells and shows frequent anastomoses (see IZUMI et al. 1984) (Fig. 4). A similar capillary density is found in the myocardium of most mammals, with the ratio of capillaries to muscle cells approximately 1:1. This suggests that, on the average, adjacent capillaries are separated by a single muscle cell (see BERNE and

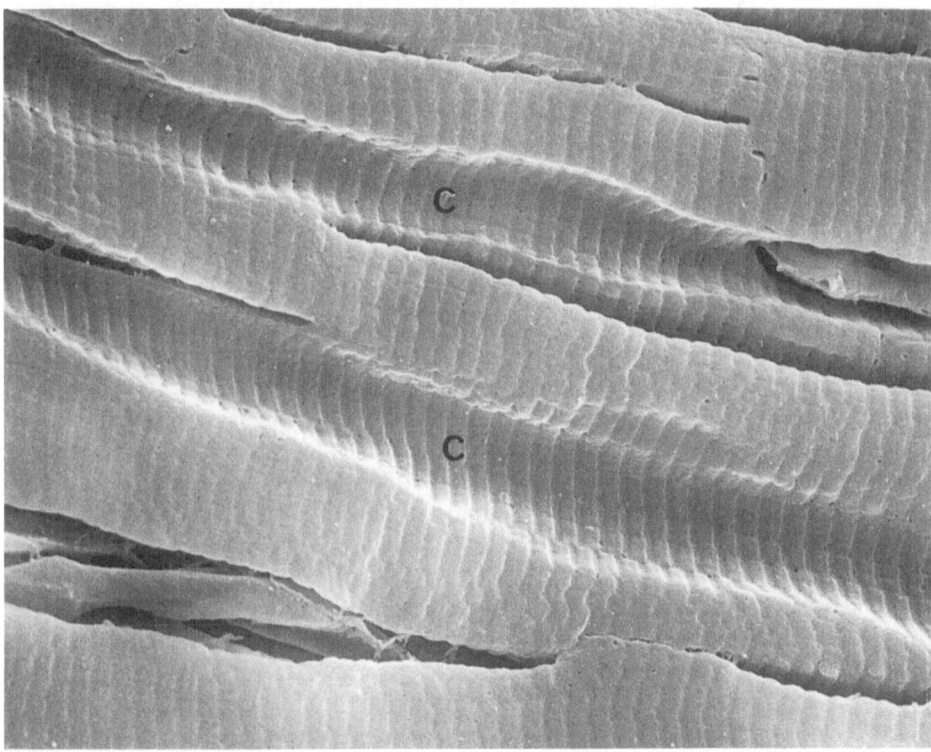

Fig. 5. Scanning electron micrograph of myocyte surfaces from rat ventricle after fixation and removal of surrounding connective tissue by hydrolysis in 8N HCl. The close association of capillaries and myocytes is demonstrated here by the two parallel depressions on the surface of the myocytes (*C*). These indentations represent the sites of capillaries before hydrolysis. × 1300. Micrograph courtesy of Dr. TAKASHI FUJIWARA

RUBIO 1979) (Figs. 5, 6, 7). The capillary plexus drains into cardiac veins which empty mostly by way of the coronary sinus into the right atrium. A small proportion of coronary blood flow is also returned directly to the cardiac chambers.

B. Morphology of Cardiac Muscle

I. Light Microscopy of Cardiac Muscle

Cardiac muscle is made up of cross-striated fibres, quasi-cylindrical in shape, which bifurcate and connect with adjacent fibres to form a complex three-dimensional network (Figs. 2, 6, 8). Each fibre is a linear unit composed of several cardiac muscle cells joined end-to-end or end-to-side, in an interdigitating rectangular step-like manner, by specialized junctional complexes called intercalated discs (Figs. 6, 8, 12). Since individual muscle cells (myocytes) rarely branch, the bifurcation of fibres is mainly due to end-to-side interdigitations (see SOMMER 1982).

When measured under the light microscope the ventricular myocyte of the mammalian heart is approximately 80 µm long (range 35–130 µm) and 10–25 µm wide (MUIR 1965; TRUEX 1972; HIRAKOW and GOTOH 1975). Atrial cells have a smaller diameter. For example, cat ventricular muscle cells have an average diameter of 10–12 µm (FAWCETT and McNUTT 1969) while cat atrial cells have an average diameter of 5–6 µm (McNutt and FAWCETT 1969). Measurements of isolated rat cardiac myocytes demonstrate that there is an increase in both width and length during postnatal growth (see F.VI.2), the average cell being 15.7×88.2 µm at 60 days and 21.5×110.3 µm at 250 days (from BISHOP and DRUMMOND 1979).

The nucleus of myocytes is usually cigar-shaped or fusiform, and orientated along the long axis of the cell (Figs. 2, 9, 27). Its location is usually central, although it may be eccentric in position, and close to the sarcolemma. Myocytes with more than one nucleus are a common finding in the adult and their presence appears age-related (BISHOP and DRUMMOND 1979). The nuclear chromatin is usually dispersed with scattered clumping. With the electron microscope it can be seen that the nucleus is surrounded by a nuclear envelope consisting of two membranes separated by a narrow perinuclear cisterna. This nuclear envelope is interrupted by nuclear pores and is continuous with the sarcoplasmic reticulum. The cytoplasmic side of the nuclear envelope may be associated with ribosomes. The perinuclear region at each pole often contains a prominent Golgi complex, mitochondria, glycogen, a few lipid droplets, lysosomes, and, with age, increasing deposits of lipofuscin granules (Fig. 27).

Mitochondria are present either in the perinuclear region or in longitudinally displaced single rows. These rows of mitochondria separate myofibrils (Figs. 6, 7, 9) and give an appearance of longitudinal striations. Myofibrils are cross-striated with A, I, M, H bands and Z discs identical to that of skeletal muscle (Figs. 2, 6, 9). Sarcomere lengths are generally uniform, the largest reported

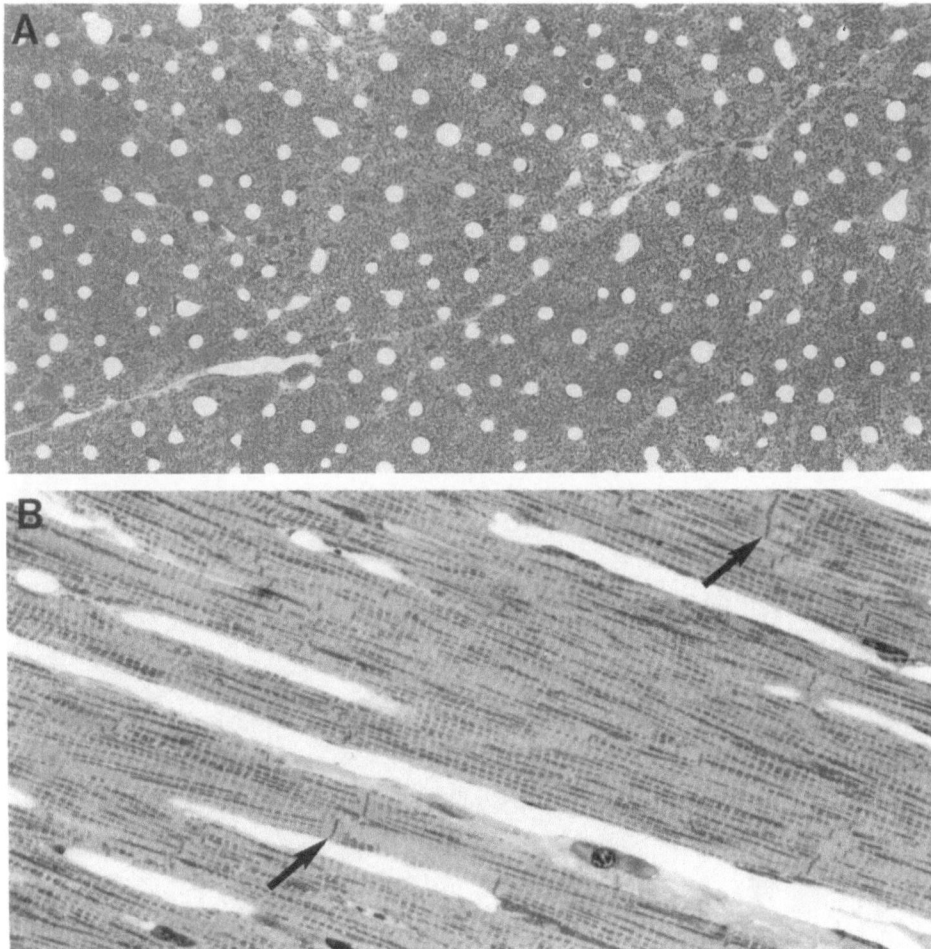

Fig. 6A, B. Light micrographs of perfusion-fixed myocardium of sheep heart. **A** With this method of fixation cells are tightly packed making it difficult to delineate individual myocytes, although mitochondria can be distinguished as dark spots within them. Capillaries which have been cleared during fixation are evenly distributed through the tissue and run parallel to the long axes of the myocytes. × 363. **B** Longitudinal section clearly showing cross-striated nature of cells and intercalated discs (*arrows*). × 537. Plastic embedded, Toluidine blue

range in a single tissue being 1.6–2.0 μm, in cat ventricle (HOYLE 1983). This range is probably a reflection of the contractile state at the time of fixation (CANALE et al. 1983b). The average sarcomere length is 2.2 μm, the same as in vertebrate skeletal muscle although there is some variation between species. Finch ventricular myofibrils have a mean sarcomere length of 1.3 μm (BOSSEN et al. 1978), guinea-pig papillary muscle, 2.3 μm (SIMPSON et al. 1973) and rat ventricle, 1.5 μm (PAGE and FOZZARD 1973).

Intercalated discs appear as dark lines which extend transversely across the muscle fibre in an irregular, zig-zag or step-like manner (Figs. 6, 12). They

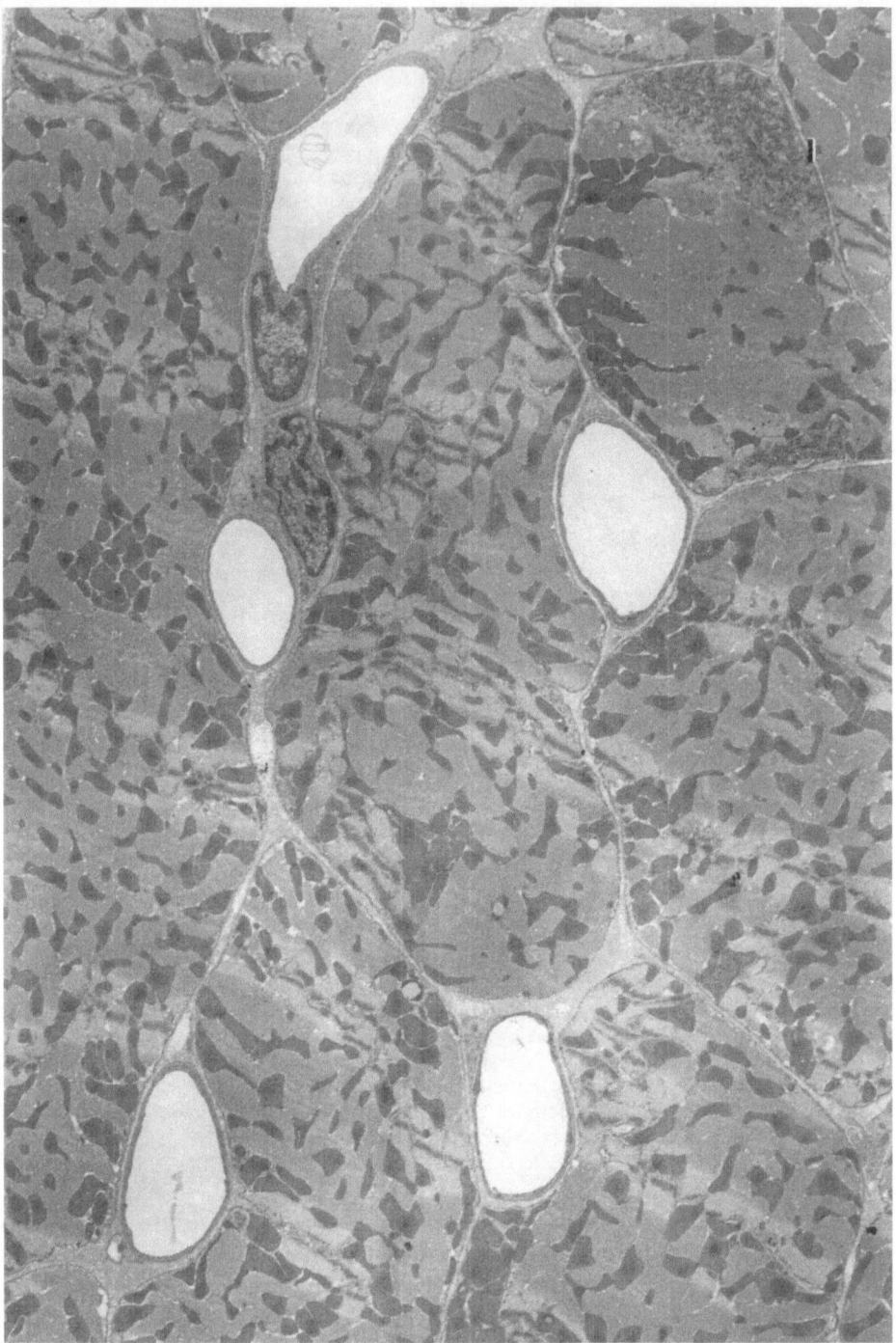

Fig. 7. Transverse section of myocytes from perfusion-fixed rat ventricle. Note close association of cleared capillaries to myocytes and the relatively large proportion of the cross sectional area that is occupied by mitochondria. These mitochondria vary considerably in size and shape. Complex intercalated disc region (*I*). ×4000. Micrograph courtesy of Dr. SEIJI MATSUDA and Professor YASUO UEHARA

always cross fibres at the level of the Z disc, and although relatively inconspicuous when stained with haematoxylin and eosin, they can be clearly demonstrated in iron-haematoxylin or phosphotungstic acid-haematoxylin preparations.

II. Ultrastructure of Cardiac Muscle

The fine structure of mammalian myocardium has been studied extensively and the reader is encouraged to consult a number of excellent reviews for additional structural and functional information (see CHALLICE and VIRÁGH 1973; PAGE and FOZZARD 1973; LEGATO 1973; McNUTT and FAWCETT 1974; SIMPSON et al. 1974; ISHIKAWA and YAMADA 1976; SHIBATA 1977; SOMMER and JOHNSON 1979; BOURNE 1980; DE MELLO 1982; SOMMER 1982; LANGER et al. 1982; WINEGRAD 1982; FABIATO 1983; FORBES and SPERELAKIS 1983, 1984; HOYLE 1983; ABE et al. 1984; LEGATO 1985).

1. Sarcolemma

The term sarcolemma is derived from early light microscopy studies of muscle, and applies to the boundary between the cellular cytoplasm and the extracellular space. Sarcolemma of muscle is in fact a synonym for plasmalemma and is approximately 9 nm in diameter with the typical trilaminar appearance of other cell types (see ROBERTSON 1957, 1958). The sarcolemma has an undulant or scalloped topography, with regular depressions and intervening elevations. The depressions form long grooves (Z grooves), which run perpendicular to the long axis of the muscle cell and overlie Z discs (Fig. 8). The grooves and intervening elevations alter according to the sarcomere length, with scalloping more pronounced in contracted muscle (HAGOPIAN and NÚÑEZ 1972). This suggests a firm attachment of the Z disc to the sarcolemma, a process possibly mediated by intermediate filaments (LAZARIDES et al. 1982).

Freeze-fracture studies of cardiac muscle sarcolemma demonstrate at low magnification the regularly-spaced openings of the transverse tubules (T tubules) in the Z grooves (RAYNS et al. 1967) and caveolar necks distributed apparently randomly over the surface of the cell (LEVIN and PAGE 1980; LANGER et al. 1982). At higher magnification intramembranous particles approximately 8–10 nm in diameter can be discerned. These particles apparently represent membrane proteins intercalated within the lipid bilayer and considerably more are found on the cleaved face adjacent to the cytoplasm (P face) than on the cleaved face adjacent to the extracellular space (E face) (RAYNS et al. 1968; McNUTT and WEINSTEIN 1970; McNUTT 1975; ASHRAF and HALVERSON 1977; FRANK et al. 1980; LANGER et al. 1982). Muscle cells from rabbit ventricle have 2300 intramembranous particles/μm^2 on the P face and 332 particles/μm^2 on the E face (FRANK et al. 1980, 1984). Adult rat ventricle has 1373 particles/μm^2 on the P face and 226 particles/μm^2 on the E face (GROS et al. 1980). Recent

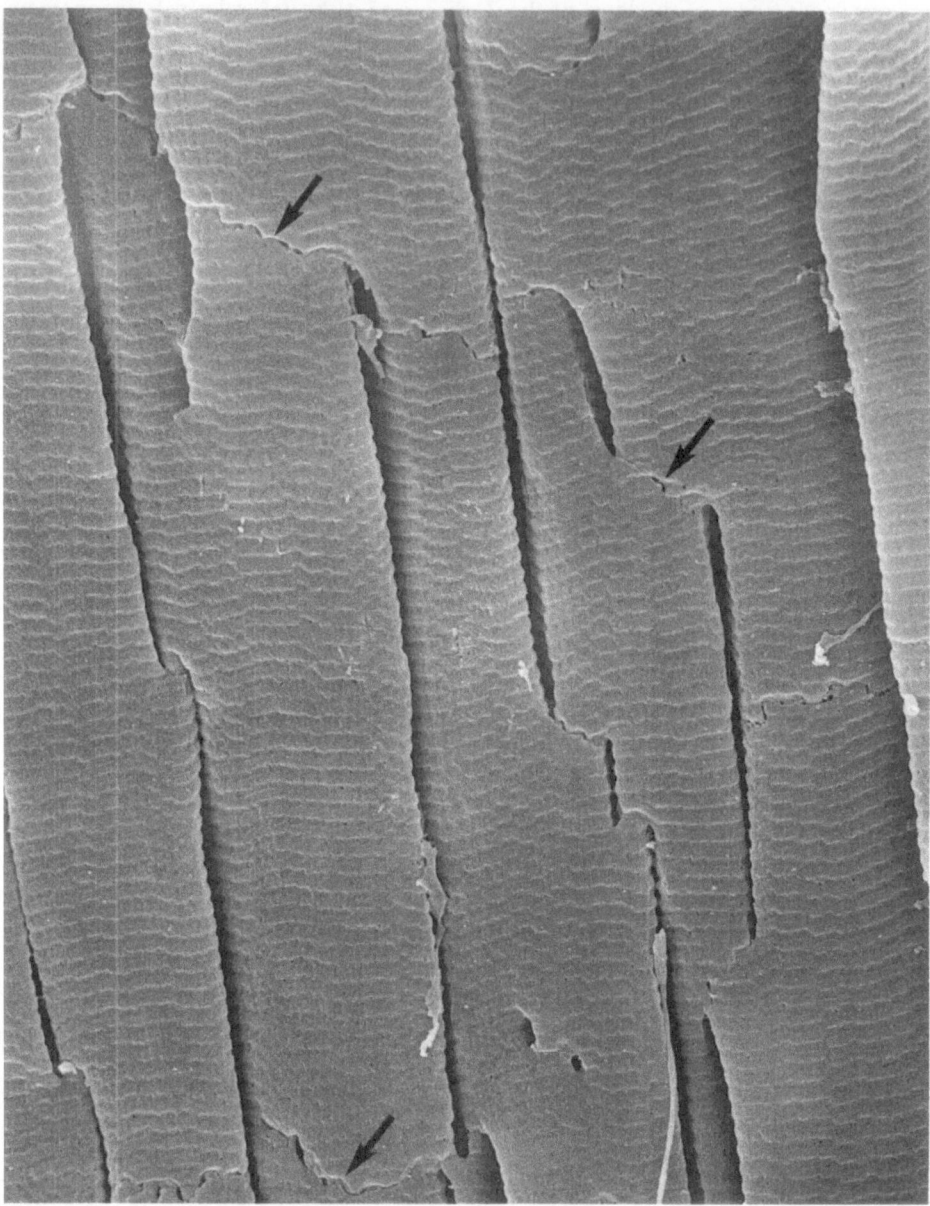

Fig. 8. Scanning electron micrograph of myocyte surfaces from rat ventricle after fixation and removal of surrounding connective tissue by hydrolysis in 8 N HCl. This treatment exposes the true surface of the muscle fibre, showing a smooth surface with a cross-striated pattern that conforms to the underlying myofibrils. Grooves occur in the region of the Z disc. Note how cells join end-to-end in a step-wise fashion at the intercalated disc (*arrows*). × 1100. Micrograph courtesy of Dr. TAKASHI FUJIWARA

studies using stereoimaged replicas tilted with a goniometer stage suggest these values may be only half the real values. For ventricular muscle cells, KORDYLEWSKI et al. (1983) found intramembranous particle values on the P face of 4525 (rat), 4799 (rabbit), 4122 (8-day chick), 4281 (adult chicken) and $5848/\mu m^2$ (frog). Consistent with ultrastructural studies which show gaps or holes in the sarcolemma following prolonged anoxia (GANOTE et al. 1975; HEARSE et al. 1975; FEUVRAY and DE LEIRIS 1975), FRANK et al. (1980) found aggregations and a significant decrease in the number of intramembranous particles on both the P and E faces.

External to the sarcolemma is the basal lamina (or glycocalyx), which is approximately 50 nm thick consisting of an inner less dense component 20 nm thick and an outer more dense component 30 nm thick (FRANK et al. 1977). This surface coat is similar in appearance to the basal lamina of a number of other different cell types (MARTINEZ-PALOMO 1970) and is probably composed of similar elements. Components described in the basal lamina of other cell types include type IV collagen, laminin, fibronectin and a heparan sulphate proteoglycan (KEFALIDES et al. 1979; TIMPL et al. 1979). Some of the heparan sulphate proteoglycans have their core proteins embedded in the lipid bilayer of the cell membrane (KJELLEN et al. 1981). Polycationic stains which bind to the negatively-charged surface groups of the proteoglycans have been used to study the basal lamina of cardiac muscle (HOWSE et al. 1970; FRANK et al. 1977; LANGER et al. 1982). Perfusion of the adult heart with Ca-free media results in separation of the basal lamina at the junction of the inner and outer components, as well as reorientation of intramembranous particles. These changes can be prevented by hypothermia and cadmium substitution (FRANK et al. 1982), and do not occur in young animals (FRANK and RICH 1983). These and other experiments in which removal of components of the basal lamina of cultured heart cells by enzymes markedly increased the influx of Ca into the cell (LANGER et al. 1976), prompted these authors to suggest that the basal lamina (glycocalyx) plays a role in the regulation of cell permeability to Ca. Subsequent studies have shown that only approximately 10% of the Ca bound to the cell membrane is in the basal lamina, but more than 80% is bound to anionic phospholipid sites within the sarcolemma itself (PHILIPSON et al. 1980). Since this bound Ca is 20 times greater than the amount required by the cell to activate contraction to 90% maximum, LANGER suggests that the sarcolemma is a source of Ca for the initiation of contraction during excitation (LANGER et al. 1982; LANGER 1984).

2. Caveolae

These are flask-shaped static invaginations of the sarcolemma, 50–80 nm in diameter which open to the extracellular space via a narrow neck (Fig. 26). The neck can be discerned in freeze-fractured material and the caveolae appear to be randomly distributed over the surface of the muscle cells with a density of between 4–6 necks/μm^2 (GABELLA 1978; LEVIN and PAGE 1980). Transmission electron microscopy reveals that sometimes two or three caveolae open into

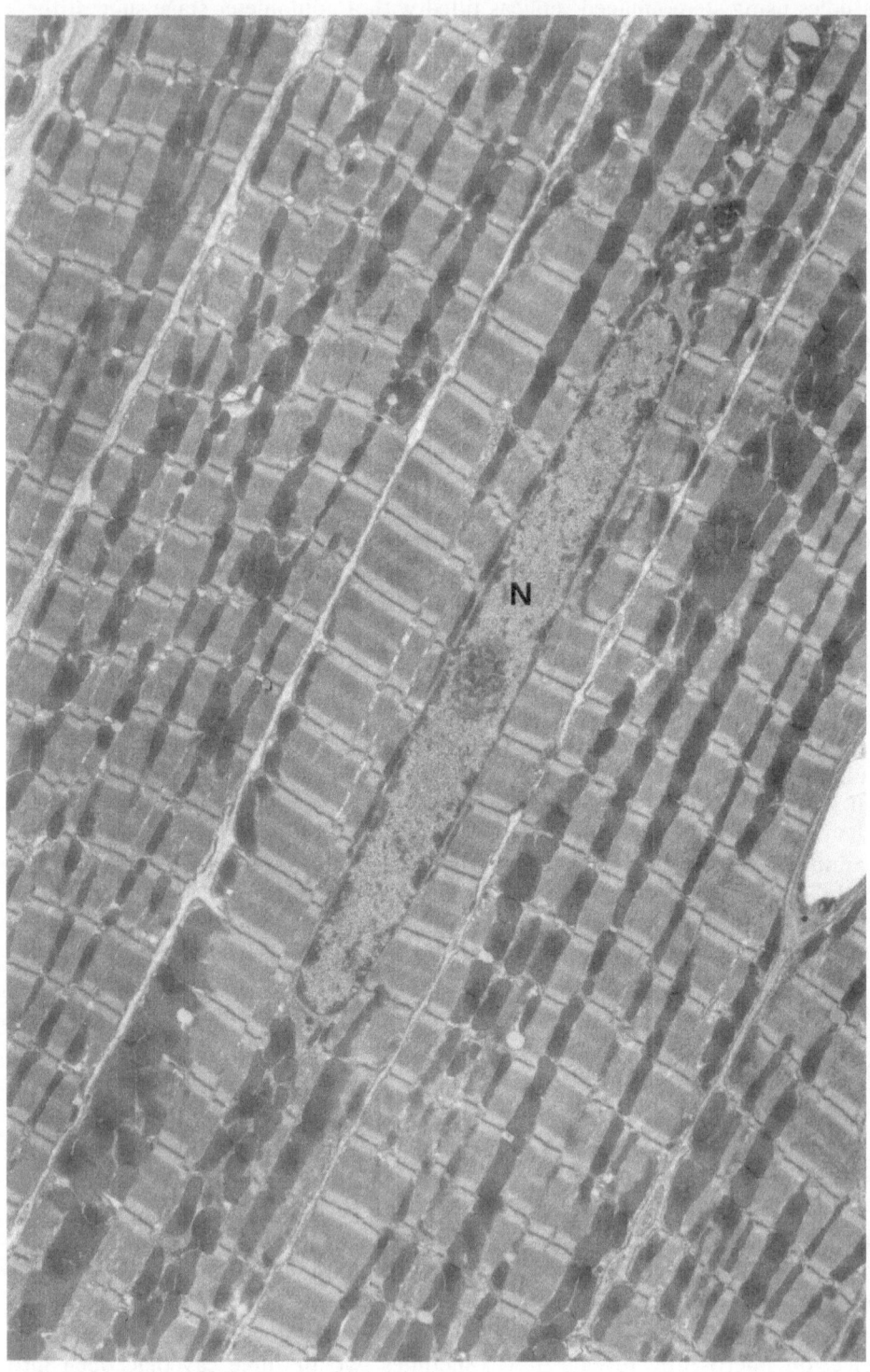

the same neck (Fig. 26). They are similar in appearance to the surface invaginations of smooth and skeletal muscle and endothelium (GABELLA 1973; DUL-
HUNTY and FRANZINI-ARMSTRONG 1975; BRUNS and PALADE 1968; SIMIONESCU
et al. 1978). In line with the other muscle types, but unlike endothelial cells,
these vesicles do not play a role in pinocytosis (see Fig. 31). Their function
in cardiac muscle remains obscure. Although they significantly increase the
cell surface area they do not function as a reservoir of membrane to be recruited
when the cell is stretched. LEVIN and PAGE (1980) calculated, assuming 2–3
caveolae per neck, that caveolae increased the cell surface area by 21–32%.
One function therefore suggested is the maintenance of a high surface-to-volume
ratio in muscle cells. Caveolae are particularly numerous in cells of the conducting system of mammalian hearts (SOMMER and JOHNSON 1979; MASSON-PÉVET
et al. 1980), and in working myocardial cells of vertebrates which lack T tubules
such as the lizard and fish (FORBES and SPERELAKIS 1971; BREISCH et al. 1983).
Severe anoxia significantly decreases the number of caveolae present on the
sarcolemma (FRANK et al. 1980). Caveolae appear to play a role in the formation
of T tubules of cardiac muscle by joining together to form a long tube or
network (ISHIKAWA and YAMADA 1976) (see F. V.2e). This mechanism appears
similar to the formation of T tubules in skeletal muscle (EZERMAN and ISHIKAWA
1967; ISHIKAWA 1968).

3. T System

The T system is a network or a series of networks of tubular invaginations
(T tubules) of the sarcolemma and comprises in the adult rat ventricle 27–36%
of the total sarcolemmal area (PAGE 1978). The continuity of the T tubule
lumen with the extracellular space has been confirmed using marker compounds
(GIRARDIER and POLLET 1964; FORSSMANN and GIRARDIER 1966, 1970; SOMMER
and JOHNSON 1968a; RUBIO and SPERELAKIS 1971; SIMPSON et al. 1973; SYBERS
and GANN 1975; SOMMER and WAUGH 1976; FORBES et al. 1977; FORBES and
SPERELAKIS 1980; MYKLEBUST et al. 1978a; LEESON 1978, 1980), and their ostia
are always found at the level of the Z disc in the Z grooves (RAYNS et al.
1967) (Fig. 10). The use of extracellular markers in thick sections examined
with high voltage electron microscopy has been used, as in skeletal muscle
(see PEACHEY and FRANZINI-ARMSTRONG 1983), to demonstrate that while the
T tubules generally course in a transverse direction encircling myofibrils at
the level of the Z disc, many adjacent tubules are interconnected by axial tubules
thereby forming a three-dimensional network (FORBES and SPERELAKIS 1983).
   The tubules of the T system vary in size and shape (Figs. 10, 11) with axial
branches having relatively small and uniform diameters as opposed to the large

Fig. 9. Longitudinal section of myocytes from rat ventricle. Note close packing of the cells. Nucleus
(*N*) is cigar shaped and centrally located. Cross-striated myofibrils run parallel to the longitudinal
axis of the cells and are separated by rows of mitochondria. These mitochondria vary in size but
often approximate a sarcomere in length. × 4400. Micrograph courtesy of Dr. SEIJI MATSUDA and
Professor YASUO UEHARA

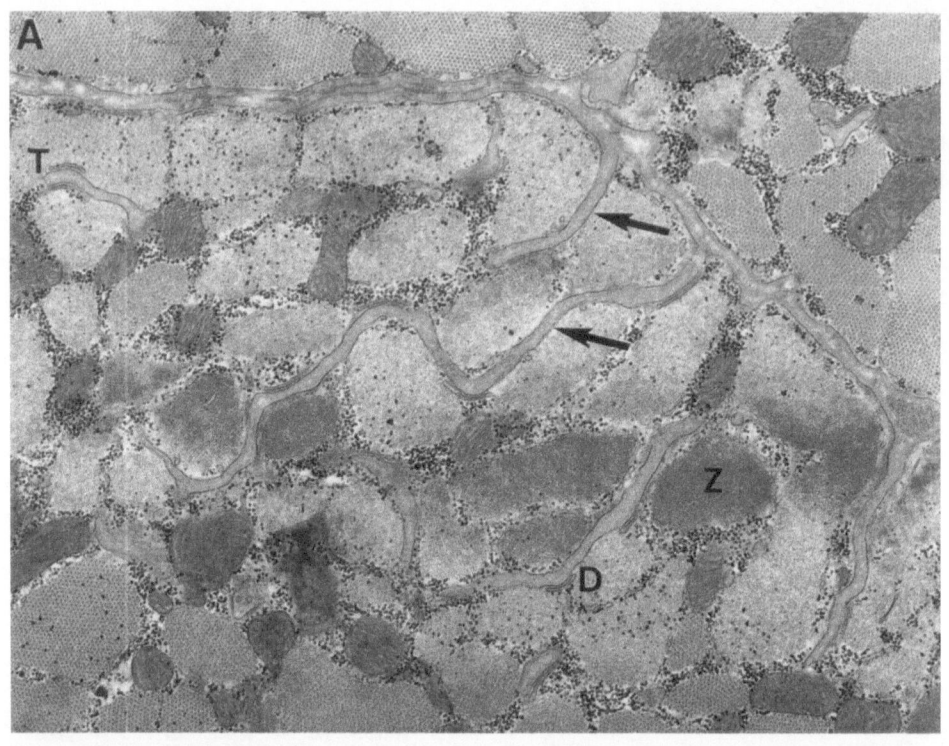

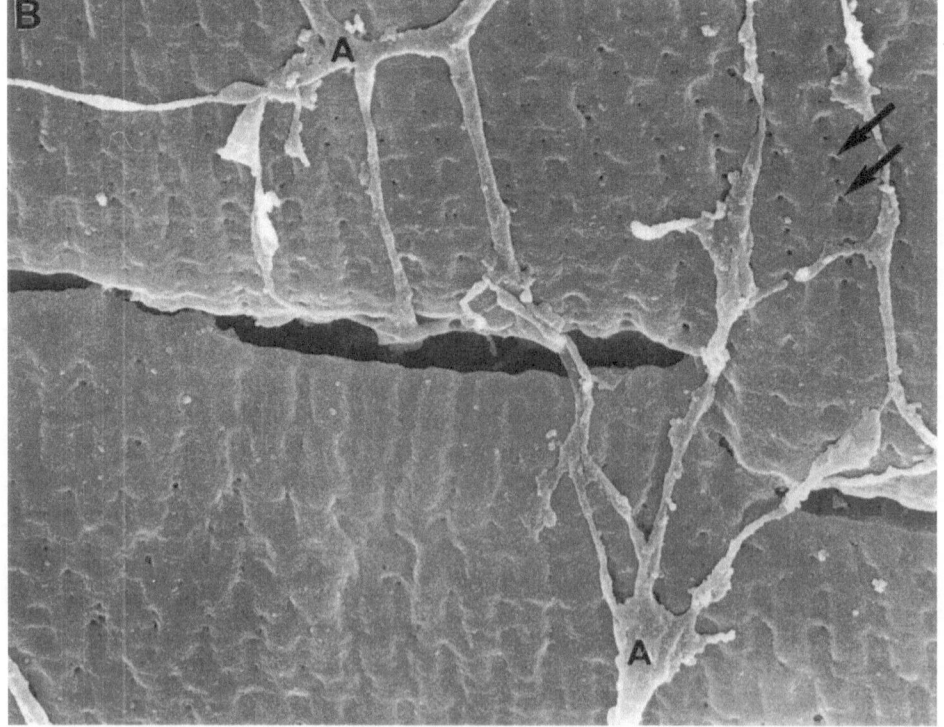

and sometimes irregular profiles of transverse tubules (SOMMER and JOHNSON 1979). T-tubule diameter can be altered according to the ionic composition of the extracellular space (LEGATO et al. 1968) and also varies considerably depending on the species. The smallest diameter T tubules reported are 50–80 nm in rat (FORSSMANN and GIRARDIER 1970), ferret (SIMPSON and RAYNS 1968) and mouse ventricle (FORBES and SPERELAKIS 1973), the largest 450–500 nm in the golden hamster and grey seal ventricles (AYETTEY and NAVARATNAM 1980, 1981), with the majority of other species having a diameter range of 100–350 nm (FAWCETT and McNUTT 1969; McNUTT and FAWCETT 1969; RUBIO and SPERELAKIS 1971; RAYNS et al. 1975; MYKLEBUST et al. 1975, 1978a; HIRAKOW and KRAUSE 1980).

The lumina of transverse and axial tubules over 100 nm in diameter usually contain basal-lamina material (Figs. 11, 19) and sarcolemmal caveolae (FORBES and SPERELAKIS 1983). These are major differences from the T tubules of skeletal muscle which have a smaller diameter (20–50 nm) and do not contain basal lamina material or caveolae (PEACHEY and FRANZINI-ARMSTRONG 1983).

There is evidence to suggest that the occurrence of T tubules is related to the diameter of the myocardial cell, with small diameter fibres lacking them (HIRAKOW 1970). Cat ventricular muscle fibres have a diameter of 10–12 µm with prominent T tubules, but cat atrial muscle cells with a diameter of 5–6 µm have strikingly fewer T tubules (McNUTT and FAWCETT 1969). HIBBS and FERRANS (1968) reported an absence of T tubules in rat atrial muscle fibres although they were present in ventricular fibres. More recent studies by LEESON (1978, 1980) using tannic acid showed that T tubules are only absent in the smallest fibres. As well, nonmammalian vertebrate cardiac muscle cells are generally smaller than their mammalian counterparts and lack T tubules (see SPERELAKIS et al. 1974; MARTINEZ-PALOMO and ALANIS 1980; LEESON 1981; McDONNELL and OBERPRILLER 1983a; BREISCH et al. 1983). Purkinje cells of large mammals with a diameter of 30–50 µm also lack a T system, however here the myofibrils are confined to a thin layer subjacent to the sarcolemma, the central region of the cell being essentially devoid of organized contractile filaments (SOMMER and JOHNSON 1969; THORNELL and ERIKSSON 1981).

Freeze-fracture studies indicate the same arrangement of intramembranous particles on the P and E faces of the T tubules as seen on the peripheral sarcolemma (RAYNS et al. 1968) suggesting a similar function. The major function of the T system is therefore to provide a direct and close contact between the extracellular environment and the deeper regions of the cell. By analogy with skeletal muscle, it has been suggested that cardiac muscle T tubules facilitate the transmission of the action potential into the cell and are involved

Fig. 10. A Transverse section of a sheep heart myocyte demonstrating how T tubules (*arrows*), which are clearly invaginations of the sarcolemma, run in a tortuous path across the cell at the level of the Z disc (*Z*). Note the T tubule is filled with similar material to the basal lamina of the cell. T tubules have interior couplings forming triads (*T*) or dyads (*D*). × 12 500. **B** Scanning electron micrograph of myocyte surfaces from rat ventricle after fixation and removal of all surrounding connective tissue elements except nerves (*A*) by hydrolysis in 8 N HCl. Openings for T tubules (*arrows*) can be seen on the surface in a groove. × 3500. Micrograph courtesy of Dr. TAKASHI FUJIWARA

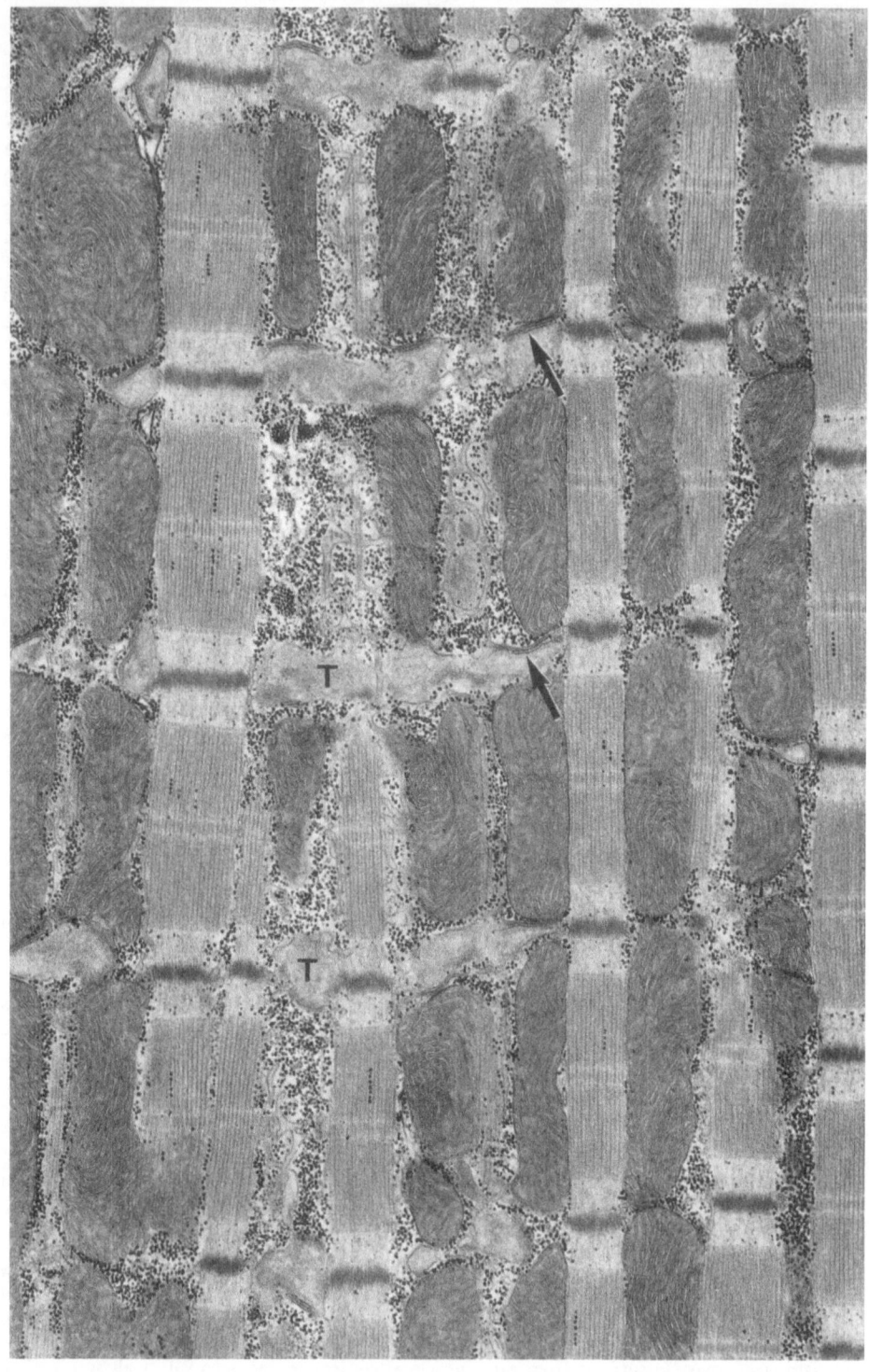

Fig. 11. Longitudinal section, sheep heart. Myofibrils are separated by rows of mitochondria and glycogen. T tubules (*T*) appear regularly at the level of the Z disc and form couplings with the sarcoplasmic reticulum (*arrows*). ×16000

Fig. 12. Intercalated disc (*I*) between two cardiac muscle cells, guinea-pig ventricle. Note that the intercalated disc occurs at the region of a Z disc so that I band filaments appear to attach to it. Thus myofibrils appear at low magnification to continue from one cell to another. × 9000. Micrograph courtesy of Professor YASUO UEHARA

in ionic exchange with the interstitium as well as excitation-contraction coupling (SOMMER and JOHNSON 1979).

4. Intercellular Junctions

The concept of early light microscopists that vertebrate cardiac muscle formed an anatomical syncytium (HEIDENHAIN 1901; GODLEWSKY 1902) was dispelled by the ultrastructural studies of SJÖSTRAND and colleagues (SJÖSTRAND et al. 1958) who demonstrated that heart muscle is made up of individual cells joined end-to-end (Fig. 8) at the intercalated disc by three specialized types of junction (Figs. 6, 12, 13). These are now called: (1) the nexus (gap junction), (2) desmosome (macula adherens) and (3) fascia adherens (intermediate junction). The intercalated disc ranges in complexity from a simple apposition (Figs. 12, 13) to a complex network (Fig. 14).

a) Nexus

In mammalian cardiac muscle, nexuses are usually located along the longitudinal segment of the intercalated disc parallel to the myofibrils, although they can occur independent of the intercalated disc. Morphometric analyses indicate that the nexus occupies approximately 10% of the intercalated disc area (SPIRA 1971; see DEWEY 1969; PAGE and McCALLISTER 1973b), which is 1% ($0.0047\ \mu m^2\ \mu m^{-3}$) and 0.7% ($0.0042\ \mu m^2\ \mu m^{-3}$) of the total sarcolemmal area in rat and rabbit ventricle respectively (see PAGE and SHIBATA 1981). It should be noted that considerable size variations exist. SHIBATA and YAMAMOTO (1979) showed that gap junctions in the rat heart varied in size from 0.0014 to 7.7 μm^2.

The nexus of cardiac muscle is similar in many respects to those of other tissues and is characterized by the close and parallel apposition of the membranes of two adjoining cells (Fig. 15). No filaments are associated with these junctions. The 2 nm gap between the adjoining cells is readily penetrated by extracellular tracers and in longitudinal section they clearly delineate an array of hexagonal subunits bridging the gap with a centre-to-centre spacing of approximately 10 nm (REVEL and KARNOVSKY 1967). Freeze-cleaving studies indicate a regular hexagonal lattice of pits (centre-to-centre spacing of 9 nm) on the E face with complementary particles in register on the P face (KREUTZIGER 1968; CHALCROFT and BULLIVANT 1970; GOODENOUGH and REVEL 1970; McNUTT and WEINSTEIN 1970; PERACCHIA 1973, 1980; GREEN and SEVERS 1984) (Fig. 16).

Isolated gap junctions from rat and mouse liver (GOODENOUGH and STOECKENIUS 1972; GOODENOUGH and GILULA 1974; GOODENOUGH 1976) have resulted in the formation of a model for the nexus (CASPAR et al. 1977; see HERTZBERG et al. 1981). The apposed lipid bilayers of the cell membranes are penetrated by protein configurations called connexions, each of which is thought to be composed of six subunits. These subunits are arranged such that an axial channel with a diameter of approximately 2 nm is formed in the centre. Two connexions join in the centre of the intercellular gap to form a hydrophilic

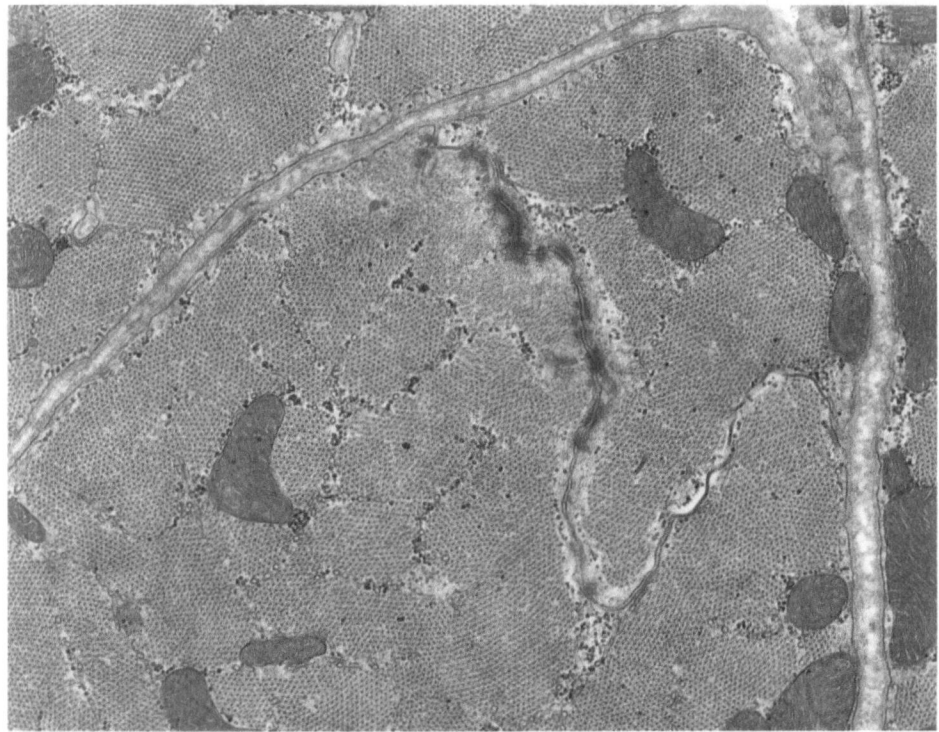

Fig. 13. Transverse section through myocardial cells of sheep heart. The intercalated disc in this instance is uncomplicated. Note the myofibrils are of the Felderstruktur type. That is, the contractile elements are arranged in a more or less continuous mass, only partially broken up into bundles of varying size by SR and mitochondria. The sarcolemma of each cell is surrounded by an electron-dense basal lamina. × 13000

channel connecting the two cells. These channels allow the exchange of chemical and electrical messages between apposing cells giving the nexus its role as the physiological low-resistance pathway in electrical coupling between myocytes. Biochemical analysis suggests that the rat heart gap junction protein consists of an intramembrane component of ~29500 daltons and a second 14500–17500 dalton component which extends out of the lipid bilayer onto the cytoplasmic surface of the membrane (MANJUNATH et al. 1984).

Nexuses from a variety of tissues are permeable to small positive ions and molecules with a nominal diameter of up to 1.2 nm (i.e. molecular weights up to 1000). Sugars and negatively-charged polypeptides can cross the junction while proteins, nucleic acids and other macromolecules cannot (BENNETT and GOODENOUGH 1978). Intercellular channels of the nexus in heart have also been shown to be permeable to substances such as 42 K (WEIDMANN 1966), Procion Yellow (IMANAGA 1974; NÚÑEZ-DÚRAN et al. 1983), Lucifer Yellow CH (DE MELLO et al. 1983), tetraethylammonium (WEINGART 1974), cyclic AMP (TSIEN and WEINGART 1976) and fluorescein (POLLACK 1976; DE MELLO 1979a, b) demonstrating their highly permeable hydrophilic nature.

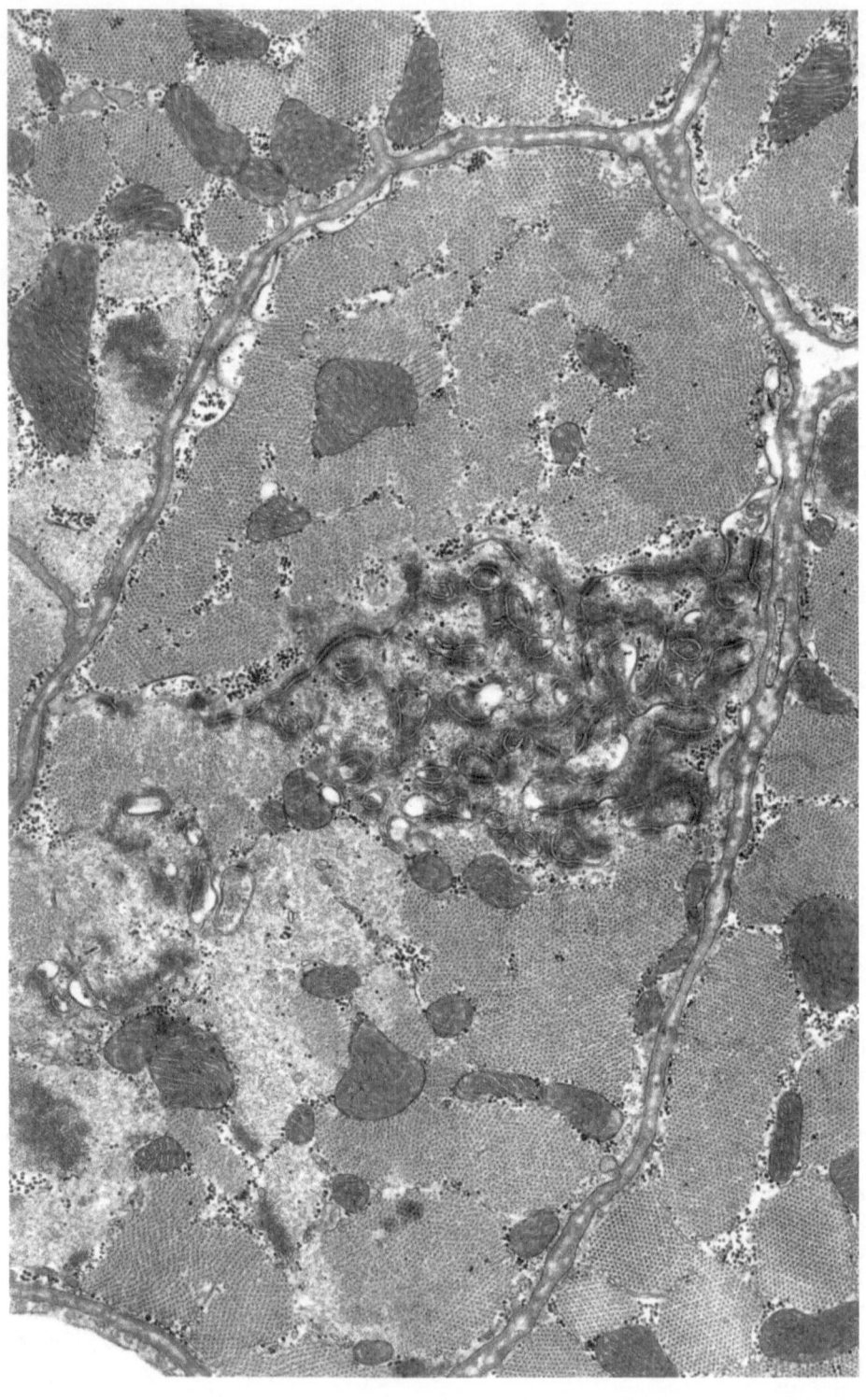

The permeability of nexuses between cardiac muscle cells can be altered. When the muscle is injured, damaged and undamaged cells uncouple. This phenomenon, known as "healing over", requires the presence of Ca^{++} (DE MELLO et al. 1969, 1982a, b; DELÉZE 1970). This process was clearly demonstrated by DE MELLO who injected Ca^{++} into a normal cardiac muscle cell causing decoupling, then immediately injecting EDTA iontophoretically into the decoupled cell and re-establishing 70–80% of the control coupling (DE MELLO 1972, 1975, 1979a, b). Freeze-fracture studies demonstrate differences in intramembrane particle distribution between coupled and uncoupled junctions of cardiac muscle (REVEL and KARNOVSKY 1967; ASHRAF and HALVERSON 1978; BALDWIN 1979; see PAGE and SHIBATA 1981; PAGE et al. 1983; SHIBATA and PAGE 1981; BURT et al. 1982).

Morphological and biochemical differences between nexuses from different tissues have been detected. Myocardial junctions in isolated preparations have a greater width (18.5–19 nm) (GOODENOUGH et al. 1978; MCNUTT and WEINSTEIN 1970) than those from liver (GOODENOUGH and REVEL 1970; GOODENOUGH and STOECKENIUS 1972). This appears due to a thickened electron density along the cytoplasmic surfaces of the junctions (KENSLER and GOODENOUGH 1980). In negatively-stained preparations the tightness and orderliness of the packing of the connexion units appears intermediate (KENSLER and GOODENOUGH 1980) between the highly disorganized packing in nexuses of lens fibres (GOODENOUGH et al. 1978; GOODENOUGH 1979) and the ordered packing seen in nexuses of liver (CASPAR et al. 1977; HENDERSON et al. 1979; MAKOWSKI et al. 1977). As well, polypeptides associated with myocardial nexuses appear to be of different molecular weight from those of liver (KENSLER and GOODENOUGH 1980). Whether these differences reflect functional differences has not been determined.

A number of earlier reports suggested that gap junctions were absent in a variety of nonmammalian vertebrates, such as frog (NAYLER and MERRILLEES 1964; TRILLO and BENCOSME 1965; STALEY and BENSON 1968; SOMMER and JOHNSON 1969; SPERELAKIS et al. 1970), snake (YAMAMOTO 1965; LEAK 1967) and bird (MIZUHIRA et al. 1967; SOMMER and JOHNSON 1969). Since all these muscle cells exhibited syncytial behaviour, this presented fundamental problems for physiologists (see PAGE and FOZZARD 1973; MARTINEZ-PALOMO and ALANIS 1980 for discussion). However more recent studies including freeze-fracture have demonstrated the presence of nexuses in these animals, although they are usually smaller, sparser, and may have a different structure (MAZET and CARTAUD 1976; SHIBATA 1977; MAZET 1977; SHIBATA and YAMAMOTO 1979; see MARTI-NEZ-PALOMO and ALANIS 1980; LEESON 1981). The early difficulty in finding nexuses in lower vertebrates may be (as with T tubules) related to the size of the cell, for in all vertebrate cardiac muscle a correlation can generally be made between the size of the cells and the size of the nexus joining them (SOMMER and JOHNSON 1979).

Fig. 14. Transverse section through myocardial cells of sheep ventricle showing complex arrangement of intercalated disc. Round profiles represent interdigitations of the one cell into the other, each connected by elements of the disc. × 14500

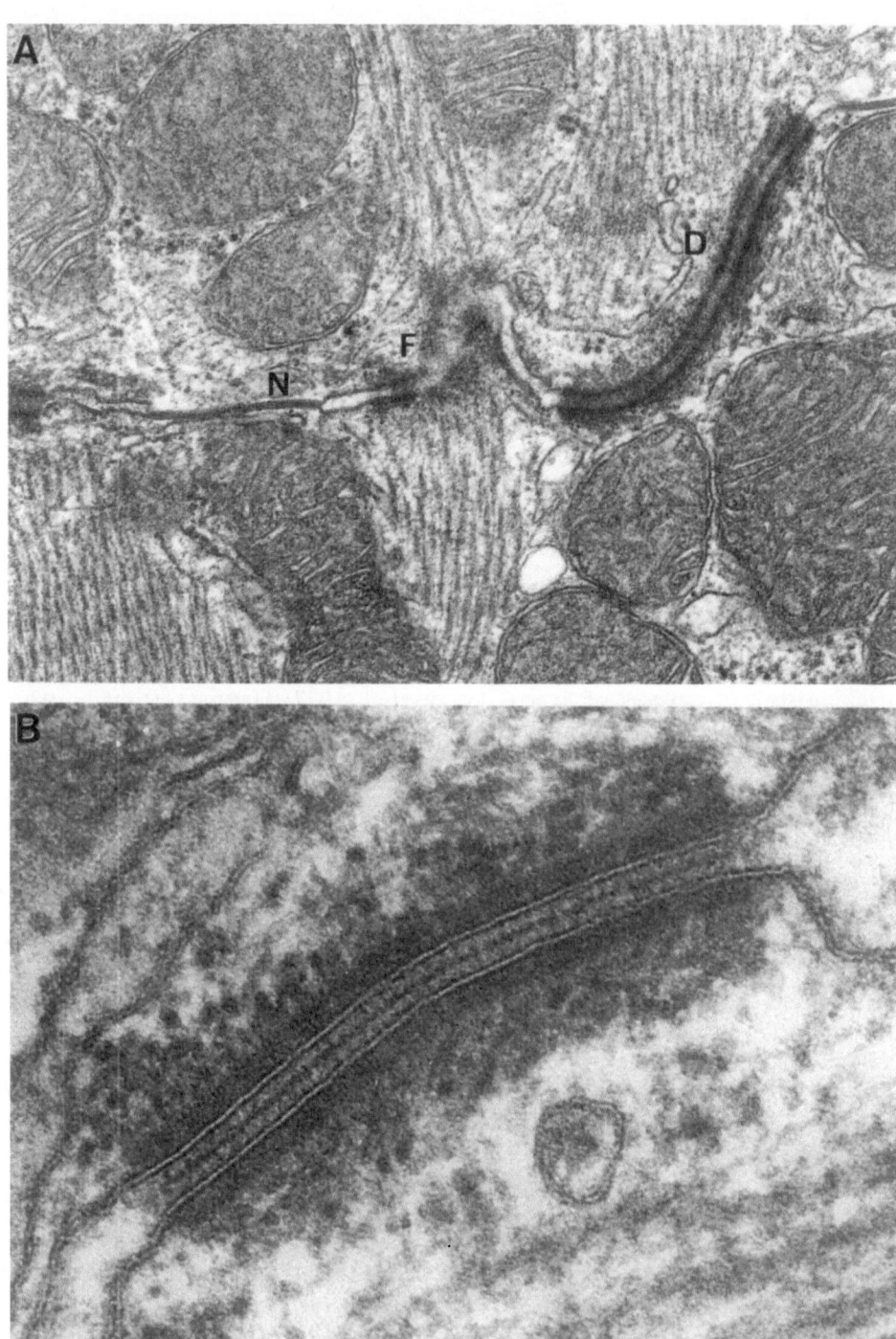

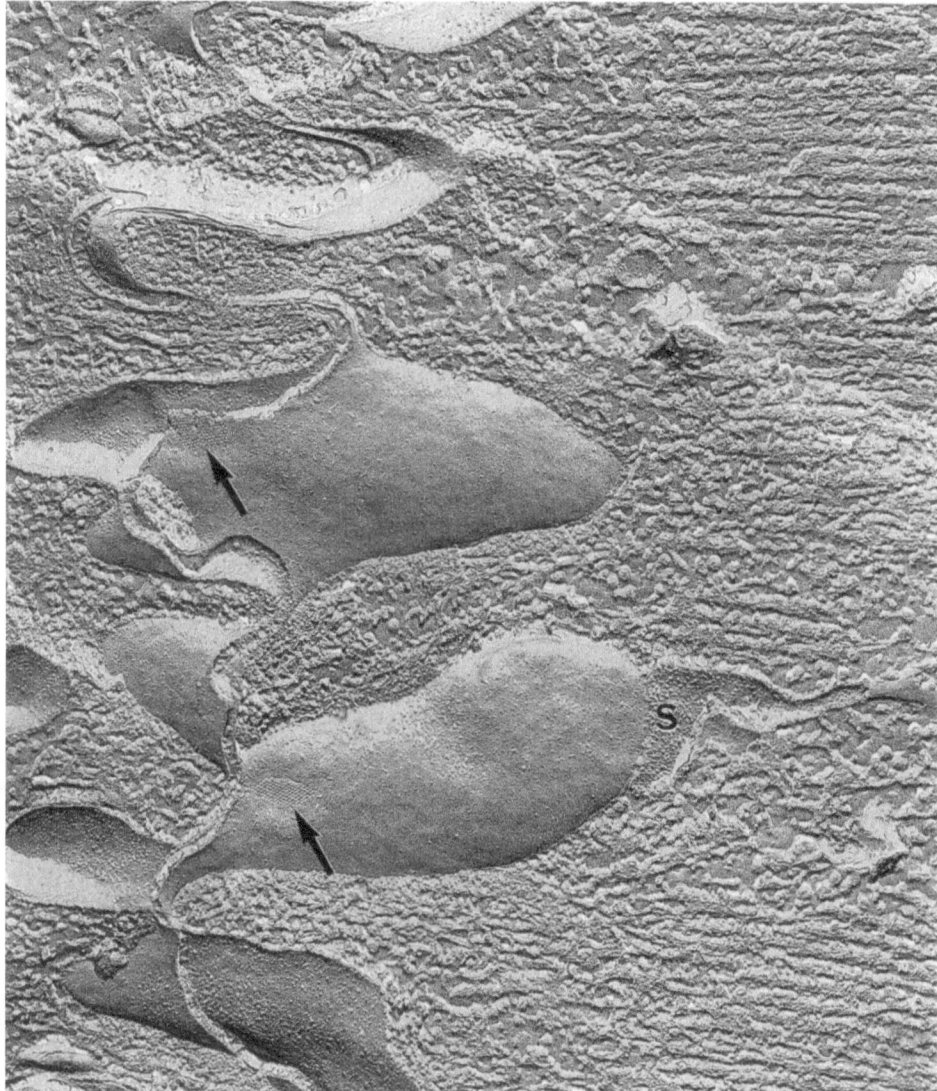

Fig. 16. Freeze fracture of dog ventricle showing gap junctions (*arrows*) at the intercalated disc. A broadened aspect of SR (*S*) is seen in contact with a process of the intercalated disc near a gap junction. × 46000. Micrograph courtesy of Dr. DONALD J. SCALES and THOMAS YASUMURA. (From J Mol Cell Cardiol 13:373–380, 1981)

Fig. 15. A Intercalated disc between cardiac muscle cells of guinea-pig ventricle. It consists of three components: the nexus (*N*), the desmosome (*D*) and the fascia adherens (*F*). Myofilaments attach only to the fascia adherens. × 42000. **B** Desmosome between cardiac muscle cells of guinea-pig ventricle. Note the parallel membranes of apposing cells with the extracellular material in the interspace organized into an electron-dense line midway between. The cytoplasmic side of the two sarcolemmas is covered by a dense fibrillar mat into which insert intermediate filaments. × 190000. Micrographs courtesy of Professor YASUO UEHARA

Nexuses have been described between membranes of the same cell in normal myocardium (MYKLEBUST and JENSEN 1978), in Purkinje cells (CANALE et al. 1983c), in cultures of myocytes (KELLY and CHACKO 1977) and in abnormal myocardium (BUJA et al. 1974). A similar feature has been reported in Schwann cells (SANDRI et al. 1977; ROSENBLUTH 1978) and smooth muscle (IWAYAMA 1971; CAMPBELL et al. 1971). The significance of these findings is unknown.

A close association between mitochondria and nexuses has been reported in myocytes of several mammals (FORBES and SPERELAKIS 1982; BAKEEVA et al. 1983). In the mouse ventricular wall, for instance, over 40% of the length of the nexus is juxtaposed to mitochondria (FORBES and SPERELAKIS 1982). Again this association appears common to a number of different cell types (see BAKEEVA et al. 1983).

b) Desmosome

Desmosomes are round, spot-like modifications of the sarcolemma, 30–300 nm in diameter, located on both the longitudinal and transverse portions of the intercalated disc, and involved in intercellular adhesion. They are composed of parallel membranes of apposing cells separated by a constant intercellular space of 20–30 nm. The space contains extracellular material organized into an electron-dense line midway between the two sarcolemmas (KAWAMURA and JAMES 1971; SOMMER and JOHNSON 1979; FERRANS and THIEDEMANN 1983). The cytoplasmic side of the sarcolemmas is covered by electron-dense material 10–15 nm in thickness into which insert intermediate filaments (FRANKE et al. 1982; KARTENBECK et al. 1983) (Fig. 15). Myofilaments do not insert into desmosomes.

Studies of cardiac muscle desmosomes with lanthanum reveal a staggered set of projections and pits on either side of the central extracellular electron-dense line (RAYNS et al. 1969). Recent studies suggest that desmosomes of myocytes of higher vertebrates have a very similar structure to those of epithelial cells (FRANKE et al. 1982; COWIN et al. 1984). Lower vertebrate heart desmosomes contain some, but not all of the proteins found in epithelial cells and myocytes of higher vertebrates. Neither desmosomes nor any of their proteins isolated thus far are found in skeletal or smooth muscle (COWIN et al. 1984).

c) Fascia Adherens

The fascia adherens makes up most of the transverse portion of the interdigitating face of the intercalated disc and is therefore much larger than the desmosome. These junctions are readily stainable in histological section and therefore largely responsible for early light microscope recognition of the intercalated disc. They consist of two parallel apposed sarcolemmas, separated by an extracellular space of 20 nm containing a central electron-dense line (Fig. 15). This line, however, is much less distinct than that in desmosomes. Along the cytoplasmic side of the sarcolemma is an electron-dense "filamentous mat" (McNUTT 1975) which is similar in appearance to, and indeed sometimes continuous with, the Z disc (SIMPSON et al. 1973; McNUTT 1975). Thin (actin-containing) filaments in the sarcomeres at the terminal ends of myofibrils within the cells are embedded into the filamentous mat. Since the filamentous mats of apposing cells occur usually where a Z disc would be anticipated in a sarcomere, myofibrils often appear to pass from one cell to another scarcely interrupted. The fascia adherens has therefore been considered, in essence, to be a bisected Z disc providing mechanical coupling of the force-generating myofibrils of adjacent cells (see KAWAMURA and JAMES 1971; FORBES and SPERELAKIS 1980; FERRANS and THIEDEMANN 1983). The filamentous mat and the Z disc have biochemical similarities since both are extracted by immersion in solutions of urea (RASH et al. 1968). Freeze-etch studies have demonstrated differences in the number of intramembranous particles found in the fascia adherens compared with non-junctional sarcolemma (McNUTT 1970).

In lower vertebrates the main junctional complex appears as a composite of the typical mammalian fascia adherens and desmosomes and often appears as a continuation of Z disc material (FAWCETT and SELBY 1958b; SOMMER and JOHNSON 1969; STALEY and BENSON 1968; McDONNELL and OBERPRILLER 1983a).

5. Sarcoplasmic Reticulum

The sarcoplasmic reticulum of striated muscle is a completely internal reticulum of membranous tubules not in continuity with the sarcolemma. The sarcoplasmic reticulum is the muscle cell's equivalent to the smooth-surfaced endoplasmic reticulum of other cell types and is often continuous with the nuclear envelope (PEACHEY 1965). However, in muscle it plays a major role in regulating the concentration of calcium in the sarcoplasm and hence the contractile state of the cell (see MARTINOSI 1984, and later discussion).

Although the general architecture of the sarcoplasmic reticulum of cardiac and skeletal muscle is similar there are a number of significant structural differences which must be reflected functionally. These structural variations, as well as species differences, have led to a somewhat confusing array of terminology. The terms chosen for this chapter are suggested by the excellent reviews of SOMMER and WAUGH (1976) and FORBES and SPERELAKIS (1983).

Cardiac muscle sarcoplasmic reticulum (SR) forms a "cylindrical" mesh of branching and anastomosing tubules intimately surrounding the myofibrils and displaying a repetitive, though by no means congruent, organization from sarcomere to sarcomere (Figs. 17, 18). It can be divided into two broad components, junctional SR of couplings and network (free) SR.

a) Junctional SR

The junctional SR is analogous to the terminal cisternae of skeletal muscle (PORTER and PALADE 1957) and in cardiac muscle may be found in close association with the peripheral sarcolemma to form a peripheral coupling (Fig. 26), or with a T tubule to form an interior coupling (Figs. 19, 20). The word coupling is derived from studies on skeletal muscle and implies a site involved in excita-

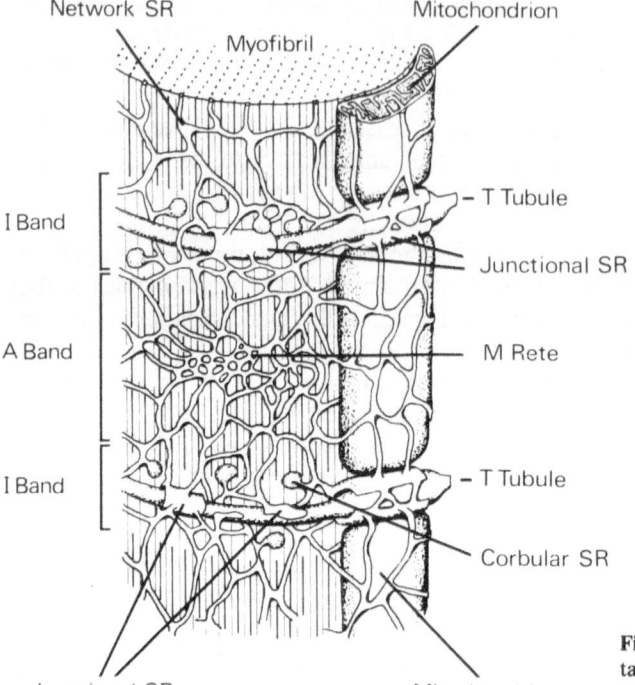

Fig. 17. Diagrammatic representation of the three-dimensional organization of SR and T tubules

tion-contraction coupling (see FORBES and SPERELAKIS 1977; SOMMER and JOHN-SON 1979; PEACHEY and FRANZINI-ARMSTRONG 1983). In Purkinje fibres, or animals without T tubules such as birds, only peripheral couplings occur (JEWETT et al. 1971; THORNELL and ERIKSSON 1981; SOMMER 1982). In the region of the coupling the network SR expands into a flattened saccule and is closely apposed to the peripheral sarcolemma. The coupling is characterized by the presence of two morphological features: 1) Junctional processes, which are regularly-spaced linear quasi-membranous bodies (also called pillars) that appear fused with the unit membranes of the junctional SR and the sarcolemma (SOMMER and JOHNSON 1970; FORBES and SPERELAKIS 1982b). Similar junctional processes have now been observed in couplings of skeletal, cardiac and smooth muscle (FORBES and SPERELAKIS 1983). These junctional processes or pillars have been suggested in skeletal muscle to function as electromechanical connections between the apposed membranes of the junctional SR and the T tubule (SOMLYO 1979; EISENBERG and GILAI 1979; FRANZINI-ARMSTRONG 1980; see MARTINOSI 1984). SCHNEIDER (1981) further proposed that each pillar represents an assembly capable of blocking and unblocking Ca channels.

Fig. 18. Longitudinal section of sheep myocyte demonstrating the complex arrangement of network (free) SR around the myofibrils. Note the T tubules (*T*) at the level of the Z disc and the fenestrations in the SR at the M rete (*M*). × 22000

2) Junctional granules consisting of electron-dense particles that form a dotted central line bisecting the lumen of the junctional cistern (SOMMER and JOHNSON 1968; FORBES and SPERELAKIS 1977). The junctional granules have been alternatively called the central membrane or density (WALKER et al. 1970, 1971).

In most mammalian cardiac muscle the junctional SR is associated with T tubules (both transverse and axial) in the form of a triad, i.e. two elements of junctional SR form couplings on opposite sides of the tubule (KELLY 1969; SOMMER and JOHNSON 1979). However, often one junctional SR element is missing forming a dyad which is sometimes wrapped around the T tubule (see FORBES and SPERELAKIS 1983) (Figs. 17, 19, 20).

b) Network (Free) SR

This is defined as comprising all sarcoplasmic reticulum excluding the junctional SR of couplings and consists of a number of specialized and apparently unspecialized regions (Figs. 17, 20). When stained with the Golgi method the network SR appears as an extensive continuous reticulum extending throughout the whole of the cell. It surrounds not only the myofibrils but also the mitochondria (Figs. 21, 22) (SCALES 1981, 1983). Those elements associated with the myofibrils are approximately 20–35 nm in diameter and form a cylindrical network around them.

α) Central Region of Sarcomere

The network encircling this region consists of two distinct elements. At the M band, like in skeletal muscle, the tubules form a lacework called the M rete (VAN WINKLE 1977). Longitudinally-orientated branching tubules connect those of the M rete to the majority of other network elements which are found in the region of the Z disc (see FORBES and SPERELAKIS 1980, 1983; SCALES 1983) (Figs. 17, 20, 21).

β) Z Disc Region

The major components of the network SR found in the Z disc region are Z tubules, corbular SR and cisternal SR (FORBES and SPERELAKIS 1983). These together with other elements form what is termed the Z rete (SOMMER and JOHNSON 1979) (Figs. 17, 23).

Z Tubules. These consist of elements of network SR which encircle myofibrils at the level of the Z disc following its course closely. Strands of material extend between the substance of the Z disc and the SR (EDGE and WALKER 1970; FORBES and SPERELAKIS 1980) suggesting an adhesive contact between the network SR and the myofibrils. Although Z tubules have now been observed in a wide range of species such as dog (EDGE and WALKER 1970), mouse (SOMMER and WAUGH 1976), ferret (SIMPSON and RAYNS 1968) and guinea pig (SIMPSON et al. 1973), for a number of years controversy existed as to their presence (see McNUTT and FAWCETT 1974).

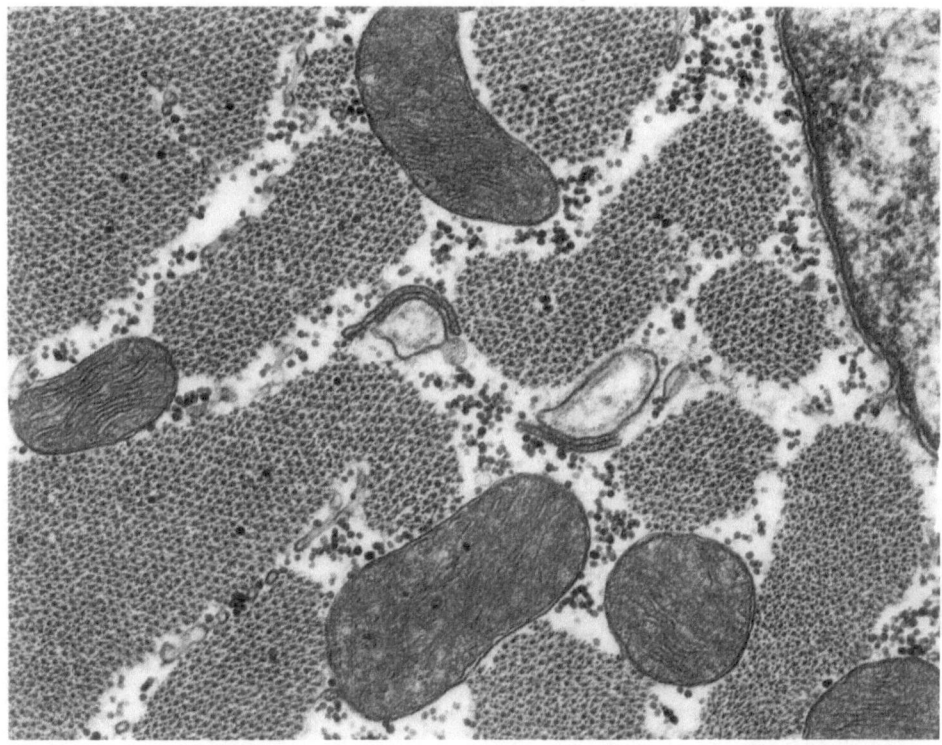

Fig. 19. Transverse section of sheep myocyte illustrating two T tubules and SR in close apposition. Note the basal-lamina material lining the T tubule. × 27000

Corbular SR. These consist of round or oval vesicles, about 70 nm in diameter, which are in fact outpouchings of elements of the network SR (Fig. 20). They have electron-dense granules within their core and a coat consisting of junctional processes and lattice fibres (SOMMER and WAUGH 1976). Corbular SR is always found at the level of the Z disc (Fig. 17). FORBES and SPERELAKIS (1983) suggest that corbular SR is a specialized form of SR called extended junctional SR which is commonly seen in animals whose muscle cells lack T tubules. Here the SR component morphologically resembles junctional SR but does not contact T tubules or the peripheral sarcolemma (JEWETT et al. 1971, 1973; SAETERSDAL and MYKLEBUST 1975). Extended junctional SR is also found in atrial myocytes of human heart (THIEDEMANN and FERRANS 1976, 1977). DOLBER and SOMMER (1984) have recently confirmed by thin sections and freeze-fracture that structural homology exists between corbular SR and junctional SR in general, and the extended junctional SR in particular.

Cisternal SR. These have only recently been described and consist of flat extended regions of the network SR in the region of the Z disc (DOLBER and SOMMER 1980; SCALES 1981, 1983).

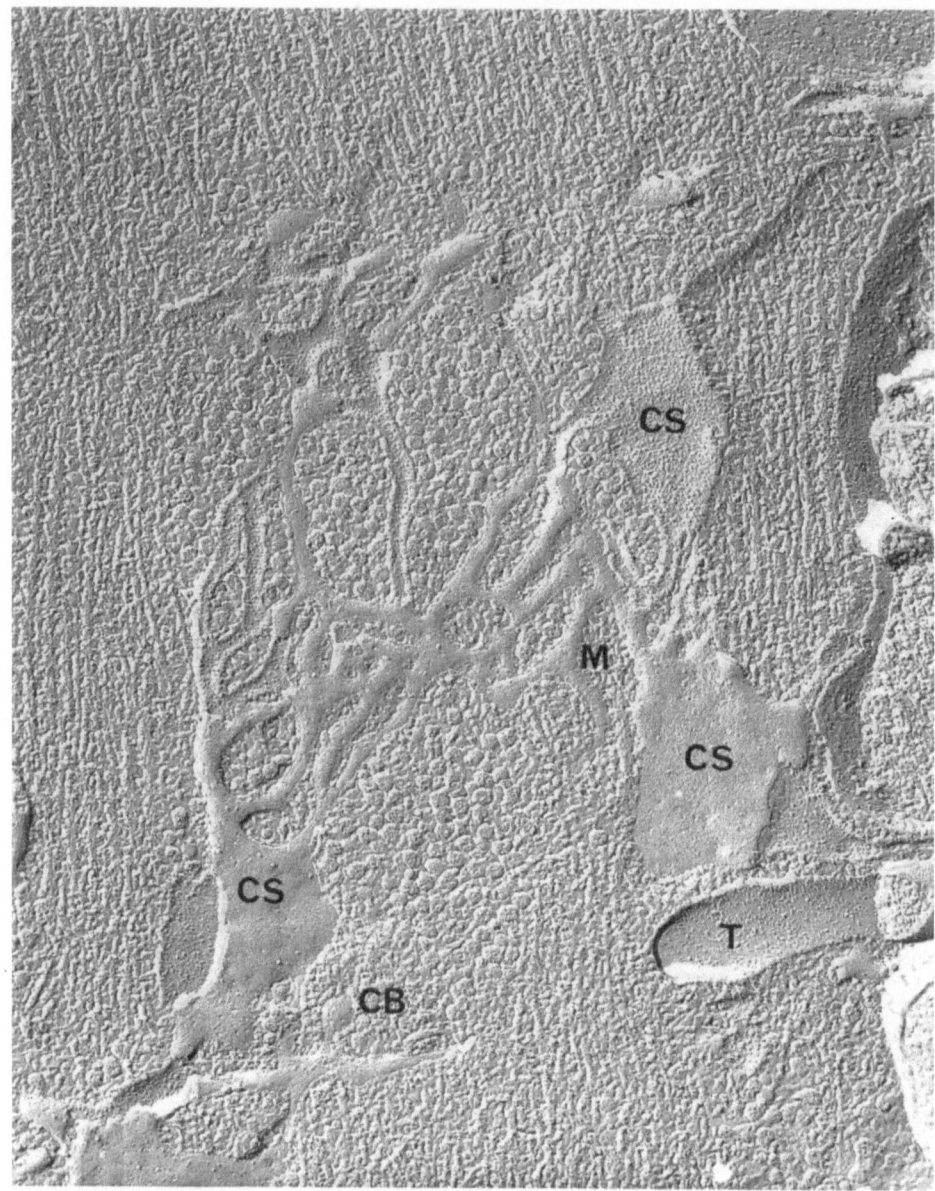

Fig. 20. Freeze fracture of dog ventricle showing three aspects of the SR: narrow tubules of the network SR, fenestrations in the M rete (*M*) and flat cisternal regions (*CS*). A T tubule is seen on the bottom right (*T*). Corbular SR (*CB*). × 75000. Micrograph courtesy of Dr. DONALD J. SCALES and THOMAS YASUMURA. (From J Mol Cell Cardiol 13:373–380, 1981)

Fig. 21 A, B. Rabbit trabecular muscle stained with Golgi black reaction demonstrating sarcoplasmic reticulum. A single sarcomere of SR is shown in a stereo pair in **A** and in an enlargement in **B.** The stereo image emphasizes the cylindrical nature of the SR. *CB* corbular SR, *F* fenestrated collar of SR in M rete. **A** × 27000; **B** × 59000. Micrographs courtesy of Dr. DONALD J. SCALES and THOMAS YASUMURA. (From J Ultrastruct Res 83:1–9, 1983)

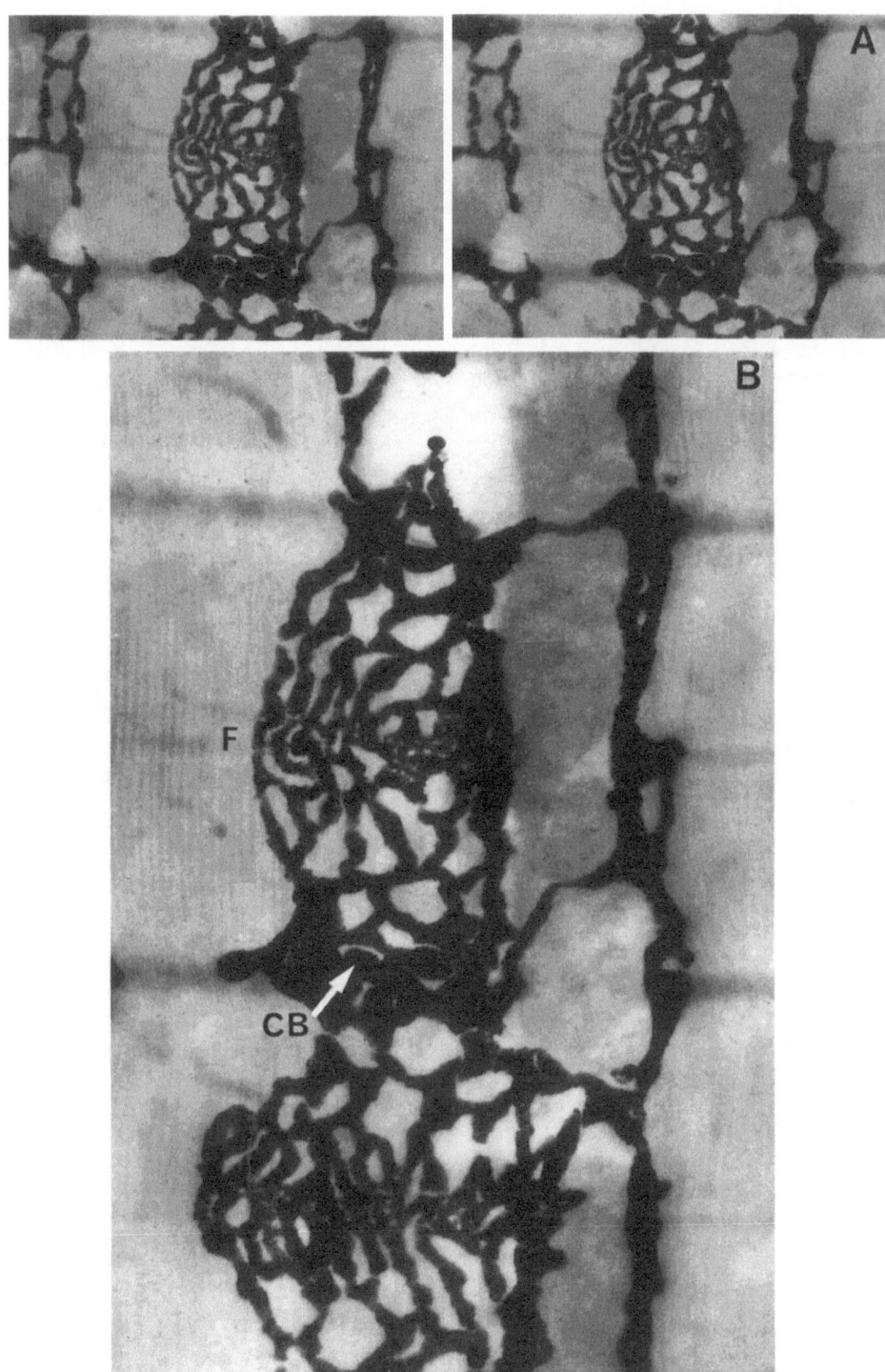

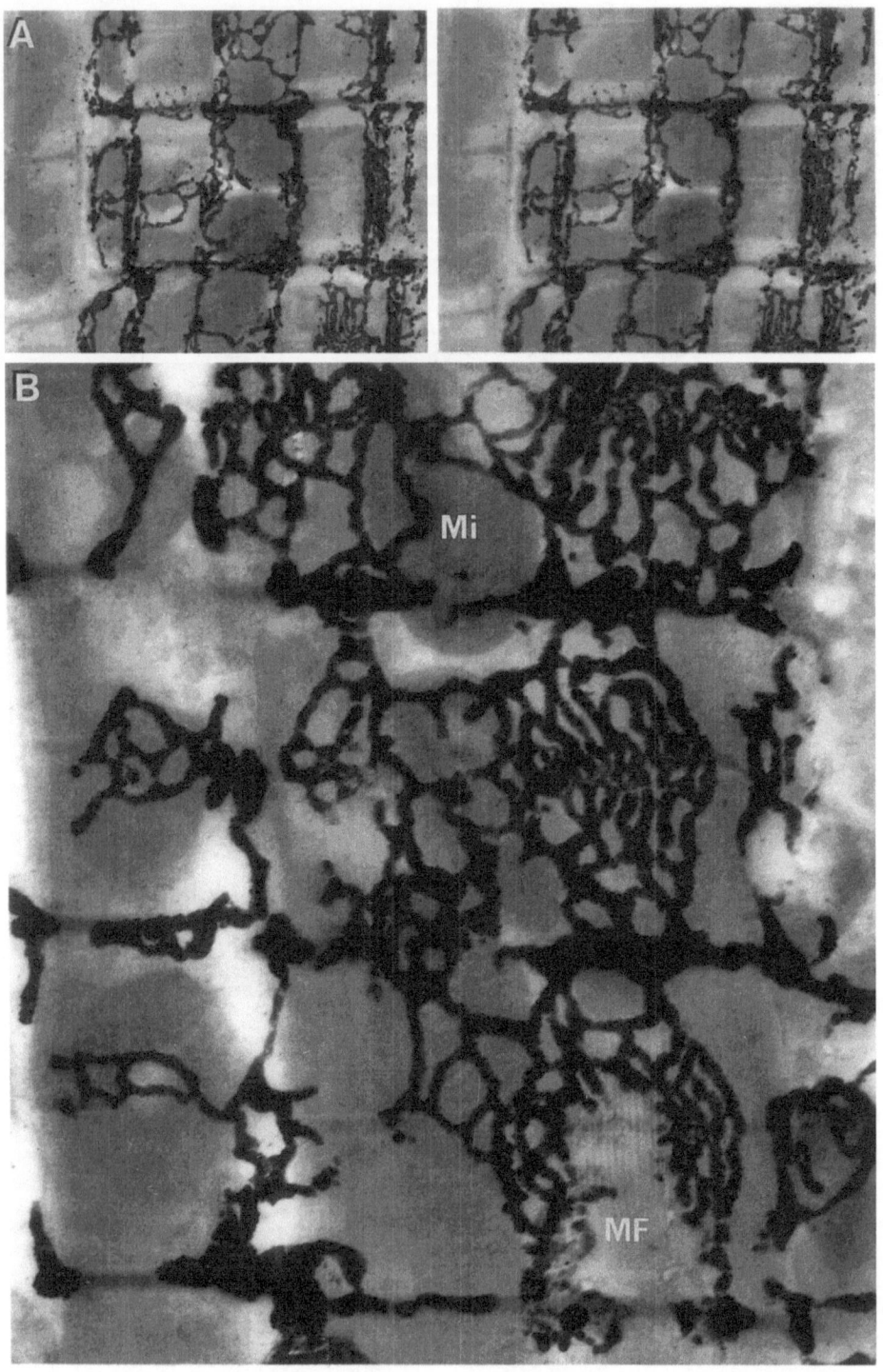

γ) *Mitochondrial Network SR*

Recent studies with the Golgi black reaction method show that SR near mitochondria always forms a simple rete (fenestrated collar) with occasional cisternae (SCALES 1983) (Figs. 21, 22). The significance of this finding remains to be determined (see FORBES and SPERELAKIS 1983 for discussion).

c) Function of SR

In skeletal muscle, release and subsequent accumulation of Ca^{++} by sarcoplasmic reticulum trigger contraction and relaxation of myofibrils respectively (MARTINOSI 1984). Evidence suggests that the calcium-binding protein calsequestrin, found largely in the SR, plays a major role in sequestering Ca^{++} during relaxation (MACLENNAN et al. 1983). Recent studies on dog heart demonstrate that calsequestrin is located in junctional SR, i.e. at peripheral and interior couplings (CAMPBELL et al. 1983). In chicken ventricle, where there are no T tubules, calsequestrin is located again in junctional SR at peripheral couplings but as well in corbular SR (JORGENSEN and CAMPBELL 1984). Calsequestrin has also been localized in the lumen of the peripheral junctional sarcoplasmic reticulum, as well as the lumen of corbular sarcoplasmic reticulum present in the I-band region of the myofibrils of sheep Purkinje fibres (JORGENSEN et al. 1984). These and other studies which clearly demonstrate Ca^{++} sequestration in myocardial SR (see FORBES and SPERELAKIS 1983), indicate that the transition from systole to diastole involves Ca^{++} uptake by the SR (see NAYLER and DRESEL 1984). However since Ca^{++} enters the cell on a beat-to-beat basis, controversy exists as to the role of Ca^{++} release by SR in the initiation of contraction (see WINEGRAD 1982; FABIATO 1983; LANGER 1984).

6. Myofibrillar System

a) Contractile Elements

It was recognized early by light microscopists that mammalian cardiac muscle, unlike fast skeletal muscle, falls into the category of "Felderstruktur" (KRUGER et al. 1933). That is, the contractile elements, which have been estimated to make up from 50% to 68% of the cell volume (PAGE and McCALLISTER 1973a; PAGE and FOZZARD 1973; FRANK et al. 1975; see Table 7, C.III.4), are arranged in a more or less continuous mass, partially broken up into bundles

Fig. 22. A Rat papillary muscle stained with the Golgi black reaction demonstrating sarcoplasmic reticulum. Stereo pair shows the three-dimensional arrangement of SR. Most of the tubules here are associated with mitochondria. At the lower right hand corner is the more complex arrangement of network SR around the myofibrils. × 13 000. **B** Rabbit trabecular muscle stained with the Golgi black reaction, demonstrating the different arrangements of SR over vertical columns of mitochondria (*Mi*) and myofibrils (*MF*). × 46 500. Micrographs courtesy of Dr. DONALD J. SCALES and THOMAS YASUMURA. (From J Ultrastruct Res 83:1–9, 1983)

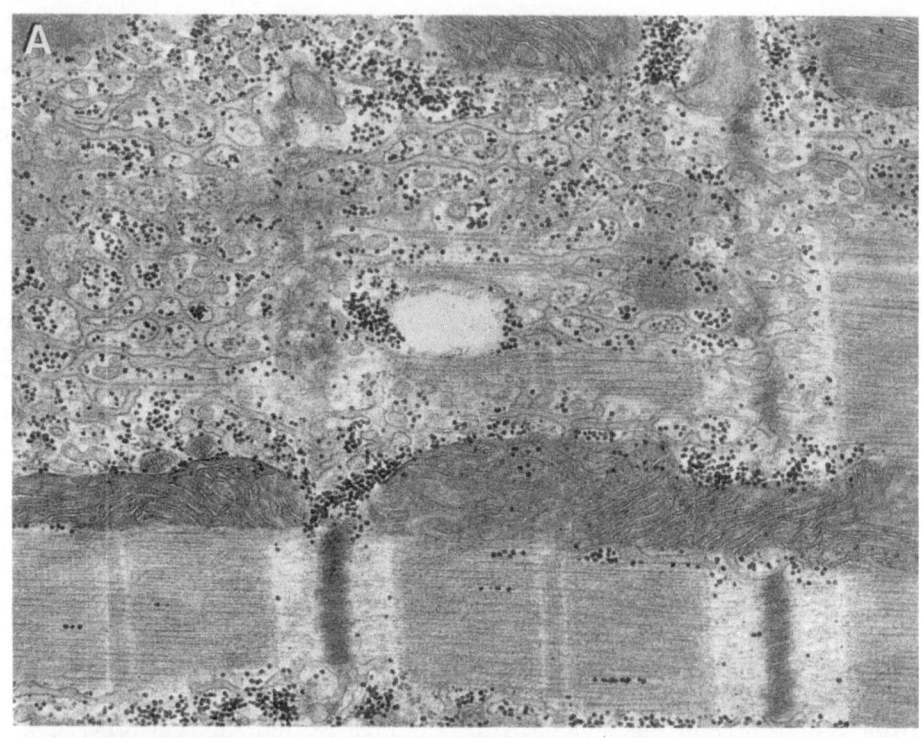

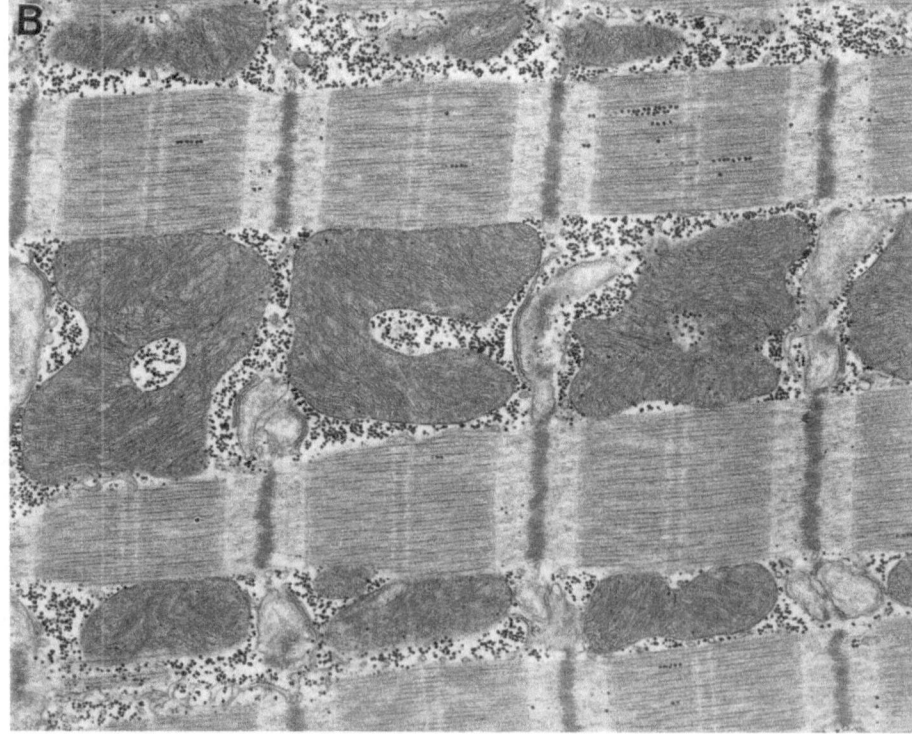

of varying size by organelles such as mitochondria, SR and T tubules (McNutt and Fawcett 1974) and sometimes glycogen (Fig. 24).

Although there are notable biochemical and immunological differences between the contractile proteins of the myofibrils of cardiac and skeletal muscle, ultrastructurally they appear similar. As well, X-ray diffraction studies show that the molecular packing of myosin and actin in the myofilaments is the same in skeletal and cardiac muscle, and changes in the diffraction patterns occurring when cardiac muscle contracts or goes into rigor are almost the same as those in skeletal muscle (Matsubara 1980). These findings suggest that a cross-bridge sliding filament mechanism for contraction operates in cardiac muscle, similar to that in skeletal muscle (Huxley and Niedergerke 1954; Huxley and Hanson 1954; Huxley 1969; see Squire 1983). The reader is therefore referred to the review on skeletal muscle by Schmalbruch in Volume II/6 of this series for a detailed description of the contractile apparatus.

b) Sarcomeric Arrangement of Myofilaments

The myofibrils are composed of myofilaments and show a banded or striated appearance in longitudinal section (Figs. 9, 12, 25). This banding consists of regularly-spaced, alternating light, and dark zones. The light or I (isotropic) band, which is bisected by a dark Z disc, consists of many thin (actin-containing) myofilaments. There are two sets of thin filaments which are attached one on each side of the Z disc. On either side of the Z disc an ill-defined N line, of obscure origin, may be evident. The dark or A band is formed principally by an orderly array of thick (myosin-containing) myofilaments. The thick filaments are cross-linked at the centre of the A band, and this gives rise to a dense M band (Fig. 25). The thick filaments also display short radial spines along most of their length which attach to inderdigitating thin filaments. These spines are absent in two narrow strips which flank the M band termed the L or pseudo-H zones. Most of the remainder of the A band consists of overlapping thin and thick filaments. In completely-contracted myofibrils, poorly defined H zones may be evident. These lie adjacent to each L zone and comprise areas containing thick filaments with radial spines but which are devoid of thin filaments. In transverse section, the A band may show thick filaments only (L and H zones), or thick and thin filaments (Fig. 26). The thick filaments are arranged in a regular hexagonal pattern. The arrangement of overlapping thick and thin filaments is also regular with each thick filament being surrounded by six thin filaments, and each thin filament located at the trigone of three thick filaments. The I band shows thin filaments only, and electron-dense material at the Z disc.

Fig. 23. A Thick glancing section of myocyte from sheep heart demonstrating how complex arrangement of network (free) SR surrounds the myofibrils. Round profiles indicate the three-dimensional nature of this network. × 18000. B Longitudinal section, sheep myocyte. Note the couplings between SR and T tubules at the level of the Z disc. Some mitochondria in this section have a shape reminiscent of a doughnut. × 13000

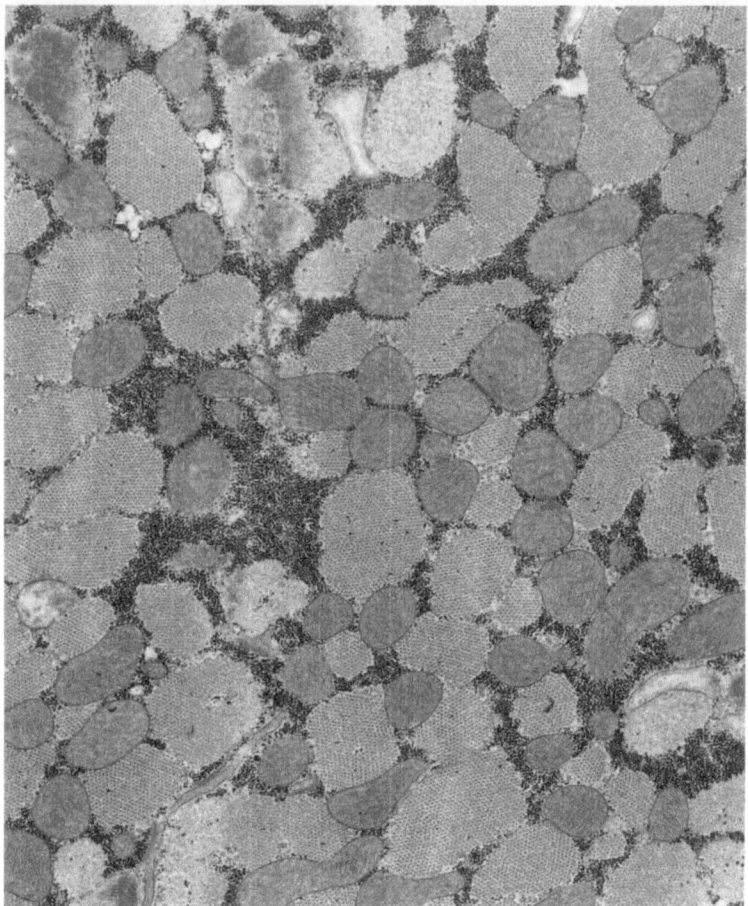

Fig. 24. Transverse section, sheep myocyte. Unlike Fig. 13 in which myofibrils are not broken up into distinct bundles by organelles (Felderstruktur), here myofibrils are separated by small amounts of SR and mitochondria and a large quantity of glycogen granules. × 9000

The functional unit of the myofibrils is the sarcomere, which is defined as the area between two adjacent Z discs (Fig. 25). The length of the sarcomere is dependent on the degree of overlap between thick and thin filaments, and is inversely related to the state of contraction of the myofibrils. The width of the A band is constant at all times, but that of the I band is greatest with relaxation, and progressively decreases with contraction.

The H zone is difficult to demonstrate in cardiac muscle (SOMMER and JOHNSON 1979). This is due to the fact that thin filaments vary in length, from 0.8 to >1.3 μm in frog atrial muscle and from 0.6 to >1.1 μm in rat atrial tissue (ROBINSON and WINEGRAD 1979), so that their composite overlap within a section appears diffuse. Thin filaments have a constant 6–8 nm diameter and profile throughout their length. On the other hand, thick filaments have a constant length of 1.5 to 1.65 μm and diameter of 12–15 nm but their profile varies

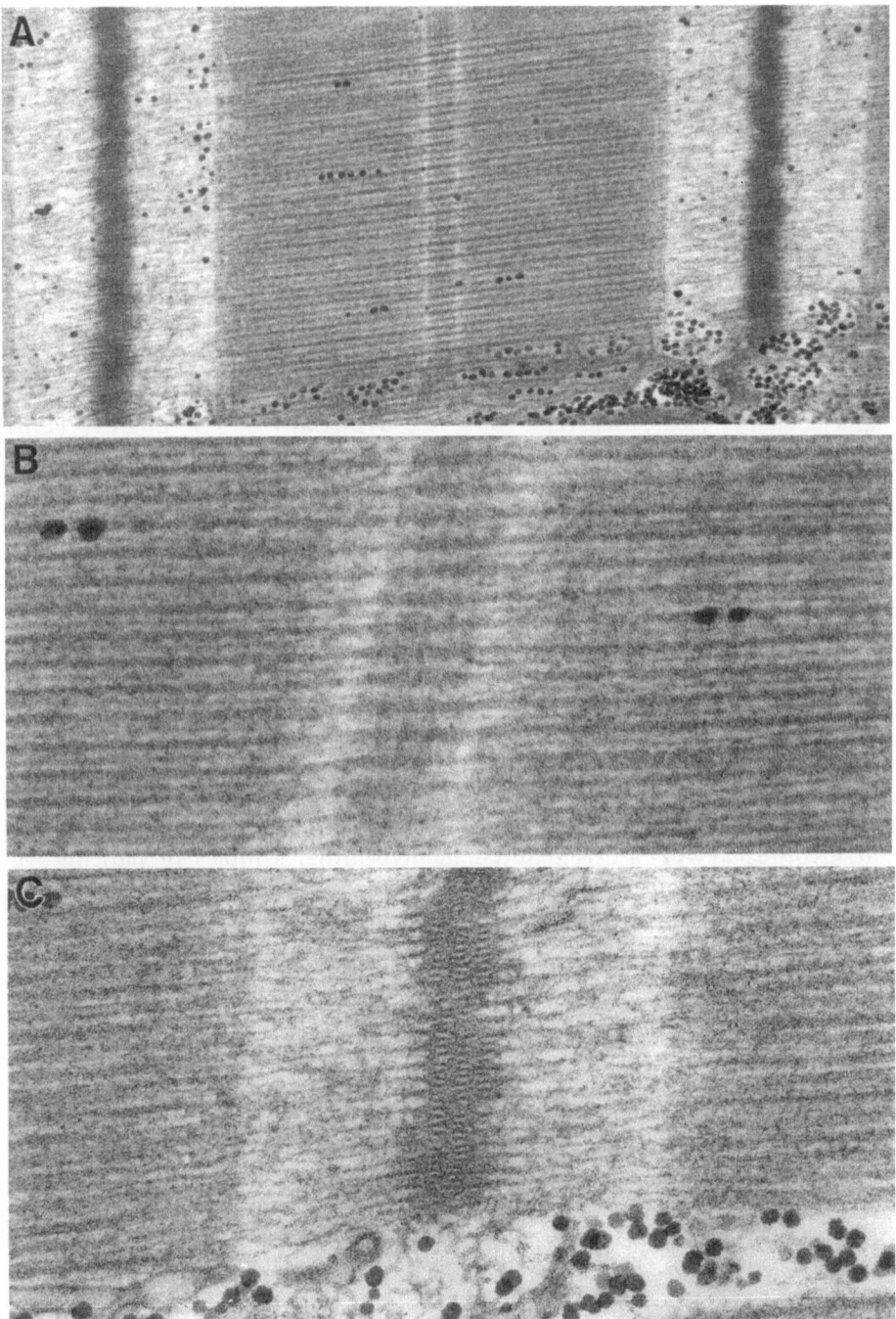

Fig. 25. A Sheep myocyte sarcomere. The I band is bisected by the Z disc. In the centre of the sarcomere is the A band representing the length of the thick myosin-containing filaments. In the centre of the A band is the M band. ×37000. **B** Sheep myocyte M band. Note the thick myosin-containing filaments pass the length of the micrograph but the thin actin-containing filaments which are between them at the edges of the micrograph terminate at the H zone. Note also the pattern of lines which make up the M band. ×83000. **C** Sheep myocyte Z disc. Note the apparent filamentous subunit structure of the Z disc, with thin actin-containing filaments attaching to it. ×81000

considerably throughout their length. They are generally round in the M band, triangular adjacent to the M band, and round again in the overlap region. At the junction of the A and I bands, the ends are tapered and triangular in shape (PEPE 1975; ROBINSON and WINEGRAD 1979).

c) Myosin ATPase Activity

Both cardiac muscle and skeletal muscle derive energy from ATP for the movement of the myosin filaments over the actin filaments, and it is the ability of myosin to hydrolyze ATP under the stimulation of actin (the actin-activated myosin ATPase system) that allows for the transformation of chemical to mechanical energy (see ADELSTEIN 1983; see WINEGART 1984). BARANY (1967) showed that there is a correlation between the shortening velocity of a skeletal muscle and its myosin Ca^{++}-activated ATPase activity (see also BARANY and BARANY 1981). These findings have now been extended to cardiac muscle (CAREY et al. 1979; SCHWARTZ et al. 1981; EBRECHT et al. 1982). For instance, atrial muscle has a faster shortening velocity (URTHALER et al. 1975, 1978) and higher myosin Ca^{++}-activated ATPase activity (LONG et al. 1977; FLINK et al. 1978; YAZAKI et al. 1979) than ventricular muscle, although some heterogeneity of fibre types exists (see THORNELL and FORSGREN 1982).

Differences between atrial and ventricular myosins have also been demonstrated biochemically and immunohistochemically (SARTORE et al. 1978, 1981; DALLA LIBERA et al. 1979; SYROVY et al. 1979; WIKMAN-COEFFELT and SRIVASTAVA 1979; SCHIAFFINO et al. 1980; THORNELL and FORSGREN 1982; BANERJEE 1983; BOUVAGNET et al. 1984). These findings, together with the observation that myofibrillar ATPase is reduced in myocardial samples of patients who died of cardiac failure (ALPERT and GORDON 1962), has stimulated a great deal of research into the notion that the efficiency with which a heart contracts may be partially determined by the type of myosin present (see SCHEUER and BHAN 1979; SAMUEL et al. 1983; see GORZA et al. 1984).

d) Myosin Subunits

Recent studies have shown that individual ventricular muscle cells can contain more than one type of myosin (SAMUEL et al. 1983). HOH et al. (1977) found in rat ventricle three isozymes of myosin that differed structurally in their heavy chains and functionally in both Ca-activated and actin-activated ATPase activity. These isozymes have been separated on nondissociating pyrophosphate gels (HOH et al. 1977) and can be distinguished by peptide mapping (HOH et al. 1979; SCHWARZ et al. 1980) or immunological techniques (SARTORE et al. 1981). They have been designated V1, V2 and V3 according to their electrophoretic mobility. Each myosin isozyme differs in the heavy chain portion of their molecules. Two distinct types of heavy chain are present, α and β. V1 and V3 are homodimers consisting of the monomers $\alpha\alpha$ and $\beta\beta$ respectively and V2 is an $\alpha\beta$ heterodimer (HOH et al. 1979; CHIZZONITE et al. 1982). V1 has the highest ATPase activity (and hence highest shortening velocity of fibres), and V 3 the lowest (HOH et al. 1977). The relative proportions of these myosin isozymes in ventricular cells varies according to the species (SYROVY et al. 1979; LOMPRÉ et al. 1981; CLARK et al. 1982; see SCHWARTZ et al. 1983), stage of development (LOMPRÉ et al. 1981; WHALEN et al. 1981; SCHIER and ADELSTEIN 1982; SCHWARTZ et al. 1982; ZAK et al. 1982), or functional and hormonal status of the animal. For instance the isozymic pattern can be changed under the influence of isoproterenol (SRETER et al. 1982). Hyperthyroidism increases the concentration of V 1 and hypothyroidism increases the concentration of V 3 (HOH et al. 1977; CHIZZONITE et al. 1982; LITTEN et al. 1982; MARTIN et al. 1982; WEISBERG et al. 1982; BANERJEE 1983; see SCHWARTZ et al. 1983; CHIZZONITE and ZAK 1984; YAZAKI et al. 1984; HOLUBARSCH et al. 1985).

Actin (see McKENNA et al. 1985), intermediate filaments (see PRICE 1984) and cardiac troponin T (RISNIK et al. 1985) also demonstrate differing isoforms. Whether these change under different conditions also has yet to be determined.

e) Z Disc (Band)

The Z disc is a filamentous structure with associated electron-dense material which varies in width from 80–160 nm. It delimits the sarcomere and, being in the middle of the I band, represents the site of interaction of thin filaments of adjacent sarcomeres (Fig. 25). Z discs of adjacent myofibrils tend to be in register and appear in many instances to be connected to each other and the sarcolemma by small bundles (from 2–50) of intermediate filaments (FERRANS and ROBERTS 1973a, b; BEHRENDT 1977). In spite of a large number of studies, the true geometry and chemical composition of the Z disc in skeletal and cardiac muscle has not been completely elucidated (see GOLDSTEIN et al. 1982). Ultra-structural studies and optical diffraction analyses show that just to one side of the Z disc the thin filaments are arranged in a large regular square array with a spacing of 24 nm. At the Z disc itself, this thin filament array plus a similar tetragonal array of thin filaments from the adjacent sarcomere interdigitate to give a centred square array (Fig. 25). These two sets of axial filaments are then held together by a regular network of cross-connecting filaments, forming the Z disc lattice (GOLDSTEIN et al. 1977, 1979).

Recent studies suggest that the filamentous structures of the Z disc are composed largely of actin and alpha-actinin. The axial filaments are mostly actin (YAMAGUCHI et al. 1983) with some alpha-actinin (ZIMMER and GOLDSTEIN, personal communication), while the cross-connecting filaments of the Z lattice are largely alpha-actinin (ZIMMER and GOLDSTEIN, personal communication). The amorphous electron-dense material is not alpha-actinin as previously proposed but another protein called amorphin (CHOWRASHI and PEPE 1982). A variety of other proteins such as tropomyosin (SCHOLLMEYER et al. 1974), eu-actinin (KURODA and MASAKI 1980), a filamin-like protein (KOTELIANSKY et al. 1982) as well as desmin and vimentin of intermediate filaments (CAMPBELL et al. 1979; GRANGER and LAZARIDES 1979; see FUSELER et al. 1981; TOKUYASU et al. 1983; TOKUYASU 1983; SASHIDA et al. 1984) have been localized at the Z disc.

Variation in the thickness of the Z disc is occasionally observed in normal myocardium, but occurs much more frequently in hypertrophied heart (MARON

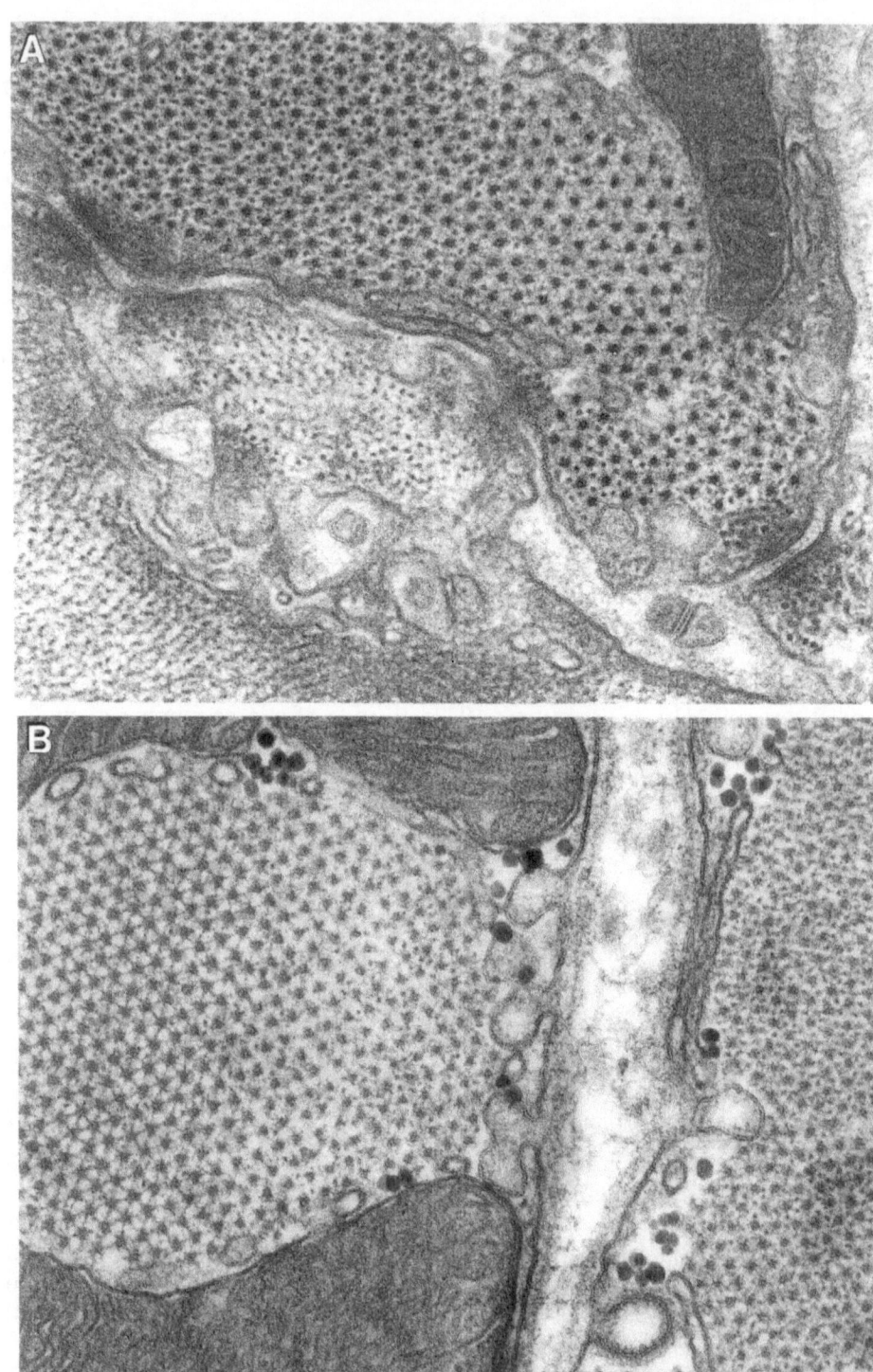

et al. 1975) and conduction system cells (see D.III.3 d). Anomalous Z-disc structures (termed Z rods or bodies) have also been observed in aging (FAWCETT 1968; MUNNELL and GETTY 1968a, b; LEKNES 1981) and hypertrophy (BISHOP and COLE 1969; COTE et al. 1970; ROY and MORIN 1971).

7. Cytoskeleton

In cardiac muscle the intracellular cytoskeleton is composed of two major elements, intermediate filaments and microtubules. Leptofibrils have been suggested to make up a third less well-understood cytoskeletal structure.

a) Intermediate Filaments

Intermediate filaments, 8-11 nm in diameter, although recognized earlier (HEUSON-STIENNON 1965), were first described in detail in developing skeletal muscle (ISHIKAWA et al. 1968). Intermediate filaments are now considered ubiquitous in animal cells and to date five major classes have been defined according to their subunit composition: (1) the major muscle type which was first isolated from smooth muscle independently by LAZARIDES and HUBBARD (1976), who called them desmin filaments, and SMALL and SOBIESZEK (1977) who called them skeletin filaments; (2) prekeratin tonofilaments found in epithelial cells (FRANKE et al. 1978); (3) vimentin filaments found in fibroblasts, developing muscle and other cells of mesenchymal origin (BENNETT et al. 1978; FRANKE et al. 1979); (4) neurofilaments of neurons (SHELANSKI et al. 1971; ERIKSSON et al. 1980); and (5) glial filaments which are present in astrocytes (UYEDA et al. 1972; SHELANSKI and LIEM 1979).

Intermediate filaments have been reported in a number of ultrastructural studies of adult cardiac muscle where they can be seen associated with the Z disc, desmosomes and the sarcolemma (LINDNER and SCHAUMBURG 1968; VIRÁGH and CHALLICE 1969; FERRANS and ROBERTS 1973a; OLIPHANT and LOEWEN 1976; JUNKER and SOMMER 1977). Extraction of cells with urea has confirmed these studies (RASH et al. 1968). Intermediate filaments are particularly prominent in hypertrophied myocardial cells (FERRANS and ROBERTS 1973a), animals treated with anabolic steroids (BEHRENDT 1977), cardiac myxomas (FERRANS and ROBERTS 1973b), and familial cardiomyopathy (PORTE et al. 1980) where their relationship with the Z disc is also maintained. Cells of the conducting system also contain numerous intermediate filaments (ERIKSSON and THOR-

Fig. 26. A Transverse section of rat ventricle myocyte. In the region of overlap in the A band, each thick filament is surrounded by a regular arrangement of six other thick filaments. These are each in turn surrounded by six thin filaments. × 90000. Micrograph courtesy of Dr. SEIJI MATSUDA and Professor YASUO UEHARA. **B** Transverse section of sheep ventricle. Note in the region of the M band each thick filament is interconnected by M bridges to its neighbours. Flask-shaped surface invaginations of the sarcolemma can be seen and in one instance there are two caveolae to a single neck. A close association of SR and the sarcolemma can also be observed. × 88000

NELL 1979; ERIKSSON et al. 1979; THORNELL and ERIKSSON 1981) (see D.III.3g), some of which are associated with glycogen (RYBICKA 1981).

Immunofluorescence studies have provided us with the best understanding of the intermediate filament cytoskeletal organization. Intermediate filaments are present at the periphery of each myofibril at the level of, and surrounding, each Z disc (GRANGER and LAZARIDES 1978; LAZARIDES 1980; LAZARIDES et al. 1982; THORNELL et al. 1984). Interconnections between adjacent myofibrils occur, such that with immunofluorescence the intermediate-filament network at the Z disc appears as a honey-comb. Alpha-actinin is localized in the openings of this honey-comb (i.e. the Z disc itself) (GRANGER and LAZARIDES 1978; LAZARIDES 1980). Connections run from the honey-comb network to the intercalated disc, sarcolemma, nucleus, and mitochondria (LAZARIDES et al. 1982; TOKUYASU et al. 1983; TOKUYASU 1983; DANTO and FISCHMAN 1984). Costameres (Latin costa, rib; Greek, meros, part), which are sites of localization of vinculin (PARDO et al. 1983), a 130000-dalton cytoskeletal protein (EVANS et al. 1984), appear to occur where intermediate filaments running from the Z disc attach to the sarcolemma (CHIESI et al. 1981; PARDO et al. 1983). These connections of Z disc and sarcolemma are probably responsible for the prominent Z grooves seen on the surface of contracted muscles (CHIESI et al. 1981; see PARDO et al. 1983). Vinculin is also present in the intercalated disc (TOKUYASU et al. 1981; THORNELL et al. 1984).

Intermediate filaments are prominent in developing cardiac muscle where they appear in two forms: (1) a complex network of cytoplasmic filaments, and (2) a dense network found in the Z disc of the sarcomere (CAMPBELL et al. 1979; FUSELER et al. 1981; DANTO and FISCHMAN 1984). This arrangement changes with development and it has been suggested that the intermediate filaments form a template allowing for the accurate assembly of the subunits of the sarcomeres in the proper register with adjacent myofibrils. The stability of this template would also permit the incompletely-developed sarcomeres to initiate and sustain contractile activities while the sarcomeres located peripherally on the same myofibril are still undergoing assembly (LAZARIDES 1980; FUSELER et al. 1981; CARLSSON et al. 1982).

b) Microtubules

Microtubules are a consistent feature of cardiac muscle although until recently they have only been mentioned in a few reports (PAGE 1967; SANDBORN et al. 1967; SIMPSON and RAYNS 1968; FERRANS and ROBERTS 1973a; RYBICKA 1978a; SAMUEL and BERTIER 1984). More detailed examination has suggested that these cytoskeletal organelles are more numerous than generally recognized (GOLDSTEIN and ENTMAN 1979; GOLDSTEIN and CARTWRIGHT 1982; CARTWRIGHT and GOLDSTEIN 1985). Ranging in size from 24–29 nm (GOLDSTEIN et al. 1979; FORBES and SPERELAKIS 1983), microtubules appear to encircle the nucleus and form helical enwrapments around the myofibrils (GOLDSTEIN et al. 1979). They are also closely associated with mitochondria, SR and T tubules near the Z disc (see GOLDSTEIN et al. 1979; FORBES and SPERELAKIS 1983).

c) Leptofibrils

Leptofibrils consist of small bundles of fine filaments (about 5 nm in diameter), with a periodic cross banding of electron-dense material similar in appearance to a Z disc, at intervals of 140 to 160 nm. The bands are thinner than Z discs, and enzyme-digestion studies (BOGUSCH 1976) also suggest a difference in composition. In cardiac muscle leptofibrils often are orientated transversely across the cell in close association with the Z disc. Leptofibrils are prominent in cells of the conducting system (CAESAR et al. 1958; VIRAGH and CHALLICE 1969, 1973a; BOGUSCH 1975; WALKER et al. 1975) although they have been reported in the working fibres of many species (THOENES and RUSKA 1960; JOHNSON and SOMMER 1967; MYKLEBUST and JENSEN 1978). They can also be found in skeletal muscle in the intrafusal fibres of muscle spindles (KARLSSON and ANDERSSON-CEDERGREN 1968), extraocular muscles (MUKUNO 1966), and in myotendinous junctions (MAIR and TOME 1972). It is thought they anchor myofibrils together and to the sarcolemma in high-stress areas (see D.III.3h).

8. Mitochondria

Mitochondria occupy from 17 to 37% of the myocyte volume (cf. 2% for frog skeletal muscle) making them second only to myofibrils (PAGE et al. 1971; see Table 7, C.III.4). They are usually located in one of three sites: in longitudinal rows between myofibrils (Figs. 9, 11, 23) or in a subsarcolemmal or perinuclear region (SHIMADA et al. 1984). In living cells the shape of mitochondria is constantly in a state of flux and in section they are elongated and pleomorphic (Fig. 23). Their length varies from 2–8 µm but often approximates that of a sarcomere; their diameter is usually 1–3 µm. As in other tissues, cardiac muscle mitochondria consist of an inner and outer membrane, with the inner membrane arranged into closely-packed transverse cristae, which enclose a matrix containing small electron-dense granules 30–40 nm in diameter (see LANGER et al. 1982). The matrix occupies 47% of the total mitochondrial volume and the cristae and inner membrane 42% (SMITH and PAGE 1976). The number of cristae in a mitochondrion of a cardiac muscle cell is three-fold greater than that found in a mitochondrion of a liver cell, presumably reflecting the greater demand for ATP in muscle. Apart from the production of ATP, much of which is hydrolyzed for contraction, cardiac muscle mitochondria can also accumulate Ca^{++}. Although this Ca probably does not play a role in normal excitation-contraction coupling, the mitochondrion plays an important role in cellular Ca buffering (LANGER et al. 1982). As mentioned before, the mitochondrion in the living state is in a state of flux and it has been clearly demonstrated that its size, configuration of cristae, density of matrix material, and numbers of electron-dense granules respond rapidly to changes in the physiological state of the cell (HACHENBROCH 1972). This accounts for the many observations of mitochondrial changes in pathological conditions such as anoxia (JENNINGS and GANOTE 1976; SCHAPER et al. 1977).

Fig. 27. Peri-nuclear region of sheep myocyte demonstrating Golgi apparatus and lipofuscin granules (*L*). ×12000

9. Lysosomes

These are commonly seen in cardiac muscle, as are residual bodies of lysosomal origin (often called lipofuscin granules). Lipofuscin granules are irregular in shape, approximately 0.05–2.0 μm in diameter, membrane-bound and contain an electron-dense matrix (MALKOFF and STREHLER 1963; see SIMPSON et al. 1973). Together with lysosomes they are usually only found at the nuclear poles (Fig. 27). Reports indicate that they increase with age to comprise up to 10% of the intracellular volume (FAWCETT and McNUTT 1969; WHEAT 1965; UEHARA et al. 1976).

10. Atrial Specific Granules

Atrial granules were named by JAMIESON and PALADE (1964) who, while observing them in a number of different mammalian species, found none in ventricular cells. Similar granules are, however, present in the ventricle of lower vertebrates (BENCOSME and BERGER 1971, 1972), chick embryos, fetal rats, and newborn dogs (see F.V.2d). The granules are round or oval with a dense homogeneous glycoprotein content (YUNGE et al. 1979), and are formed by the Golgi apparatus (THERON et al. 1978). The granules have been divided into three different types (A, B and D) according to their morphological characteristics and location (CANTIN et al. 1975, 1979) and have been the subject of a great deal of research (see FERRANS and THIEDEMANN 1983). Changes in atrial specific granules occur in experiments that modify salt and water balance (MARIE et al. 1976; DEBOLD 1979; see GRANTHAM and EDWARDS 1984), and the number of these atrial granules is related directly to salt loading and blood volume (LANG et al. 1985). Recent experiments suggest the atrial granules represent the site of storage of several biologically-active peptides that exert potent effects on kidney function (ACKERMANN and IRIZAWA 1984; CANTIN et al. 1984; FORSSMANN et al. 1984; BEASLEY and MALVIN 1985) and regional vascular resistance (GARCIA et al. 1984; OSHIMA et al. 1984; FORSSMANN et al. 1984; WINQUIST et al. 1984). This atrial natriuretic factor has now been purified and sequenced (ATLAS et al. 1984; CURRIE et al. 1984; MAKI et al. 1984; YAMANAKA et al. 1984). Synthetic peptides shown to have the same biological activity as the native peptides produce a marked natriuretic response in the kidney (BURNETT et al. 1984) and release of vasopressin from rat posterior pituitary (JANUSZEWICZ et al. 1985).

11. Other Cell Organelles and Inclusions

The reader is referred to reviews by SIMPSON et al. (1973), McNUTT and FAWCETT (1974), UEHARA et al. (1976), SOMMER and JOHNSON (1979) and FERRANS and THIEDEMANN (1983) for detailed descriptions of other cardiac muscle organelles and inclusions.

III. Histochemistry

Over the last ten or more years immunocytochemistry, cytochemistry, and histochemistry have been areas of growth and innovation. Sensitivity, specificity, applicability, tissue preparation, and presentation are continuously being improved and an even wider choice of procedures is becoming available to the investigator. In particular, advances have been made in immunocytochemistry (HEITZ 1982; BOSMAN 1983) and quantification of histochemical data, including calculation of colour differences (GLICK 1981; MARSHAL and GALBRAITH 1984), cytophotometry (CABRINI 1981; PUNKT et al. 1984) and the inevitable application of microcomputers to these quantitative methods (FUJITA 1983; FUKUDA 1983; HOSHINO 1983).

The increasing sophistication and diversity of histochemical techniques used to study cardiac muscle puts a comprehensive review beyond the scope of this chapter. Some idea of the breadth of applications can be gained from Table 2. The contribution of histochemical and cytochemical methodology to our understanding of many aspects of structure, function, and development of cardiac muscle has been treated within the relevant chapters throughout the book, and will not be repeated here.

IV. Extracellular Matrix

In rabbit ventricle, 24.6% of the total tissue volume in interstitial space is filled largely with extracellular matrix and connective tissue cells (FRANK and LANGER 1974). The connective tissue cells comprise 7% of the total interstitial volume and consist of fibroblasts, undifferentiated mesenchymal cells, Anitschkow cells, histiocytes, macrophages, and mast cells (NAG 1980; FERRANS and THIEDEMANN 1983). Of the remainder 4% is collagen, 59% vascular elements, 23% proteoglycans, with 6% classified as empty space (FRANK and LANGER 1974).

The connective tissue elements of heart and their organization have been studied by light microscopy (HOLMGREN 1907; PUFF and LANGER 1965), scanning electron microscopy (CAULFIELD and BORG, 1979; BORG and CAULFIELD 1981; BORG et al. 1981; SATO et al. 1983), and conventional and high-voltage transmission electron microscopy (BATTIG and LOW 1961; HANAK and BOCK 1971; MELAX and LEESON 1972; ROBINSON 1980; ROBINSON and WINEGRAD 1981; ROBINSON et al. 1983; SATO et al. 1983).

The endomysium consists of a network of intercellular twisted bundles of collagen (struts), 120 to 150 nm in diameter, which constitute a fibrous skeleton. They interconnect adjacent myocytes and insert at the levels of the Z disc. Between the struts enveloping the myocytes is a weave of collagen fibrils (CAULFIELD and BORG 1979). Elastic fibres interconnect cells and helically wind around cells. These two elements are embedded within an amorphous ground substance of proteoglycans which is continuous with the basal laminae of the cells. The

Table 2. Histochemistry of cardiac muscle

Parameter studied	Methodology	References
Localisation of adenylate cyclase	Cytochemistry	SLEZÁK and GELLER (1979)
Localisation of lysosomal acid proteinase	Immunocytochemistry	DECKER et al. (1980)
Metabolic changes in sympathectomised hearts	Histochemistry	JONES and CANNON (1980)
Localisation of ATPase	Cytochemistry	MALOUF and MEISSNER (1980)
Localisation of phosphofructokinase	Histochemistry	MEIJER and STEGEHUIS (1980)
Myosin Ca^{++}-ATPase accumulation as a measure of cultured heart cell maturation	Histochemistry	TAYLOR and JONES (1980)
Localisation of nickel ions in pathological conditions	Cytochemistry	BALOGH et al. (1981)
Adenylate cyclase activity following hypoxia	Cytochemistry	BALOGH et al. (1981)
Glycosomes in Purkinje cells	Cytochemistry	RYBICKA (1981)
Measurement of beta-hydroxyacyl CoA dehydrogenase activity	Quantitative histochemistry	CHAMBERS et al. (1982)
Concanavalin A and wheat germ agglutinin binding	Cytochemistry	GROS et al. (1982)
Localisation of adenylate cyclase	Cytochemistry	SCHULZE (1982)
Diagnosis from post-mortem material	Cytochemistry	SOMOGYI et al. (1982)
Localisation of calcium ions	Cytochemistry	SÓTONYI et al. (1982)
Localisation of albumin	Immunocytochemistry	YOKOTA (1982)
Differentiation of the conduction system	Immunohistochemistry	FORSGREN et al. (1983)
Localisation of guanylate cyclase	Cytochemistry	SCHULZE and KRAUSE (1983)
Demonstration of cardiac glycosides	Cytochemistry	SOTONYI and SOMOGYI (1983)
Subendocardial atrial baroreceptors	Cytochemistry	YOKOTA et al. (1983)
Localisation of calcium ions	Cytochemistry	BORGERS et al. (1984)
Localisation of atrial natriuretic factor	Immunocytochemistry	CANTIN et al. (1984)
ATPase estimation	Histophotometry	PUNKT et al. (1984)
Localisation of adenylate cyclase and guanylate cyclase	Cytochemistry	SLEZÁK and TRIBULOVÁ (1984)
Effect of chlorpromazine on ischaemic changes	Histochemistry	TRIBULOVÁ et al. (1984)

epimysium consists of large collagen fibres which form a weave pattern at slack length but are well aligned in states of stretch along the long axis of the muscle. The perimysium consists of very large bundles of collagen fibrils that connect the endomysial collagen weave around myocytes to the collagen fibres of the epimysium (ROBINSON et al. 1983).

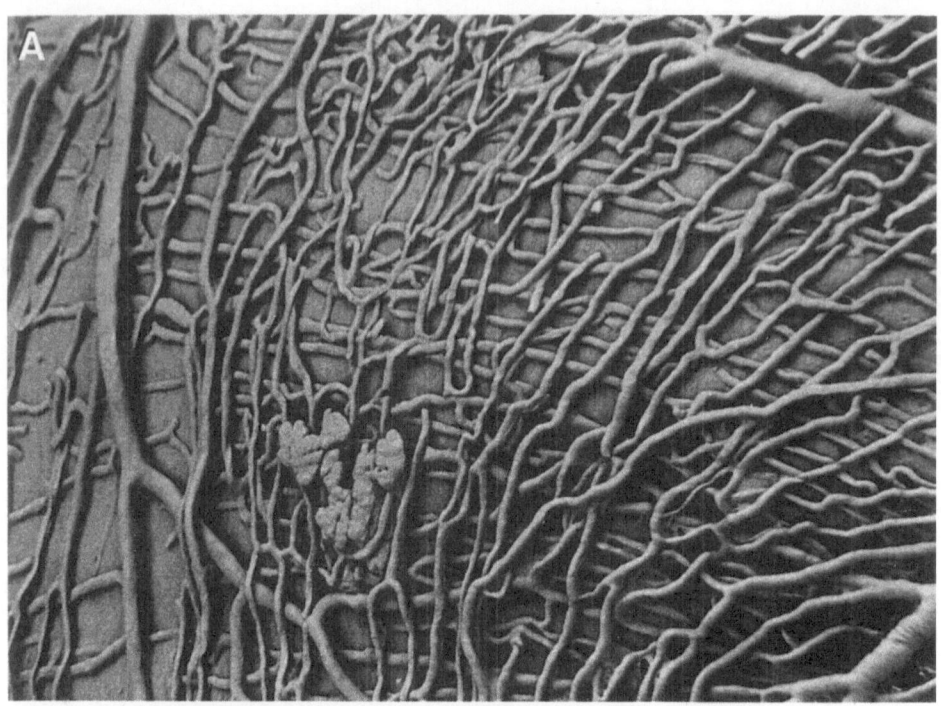

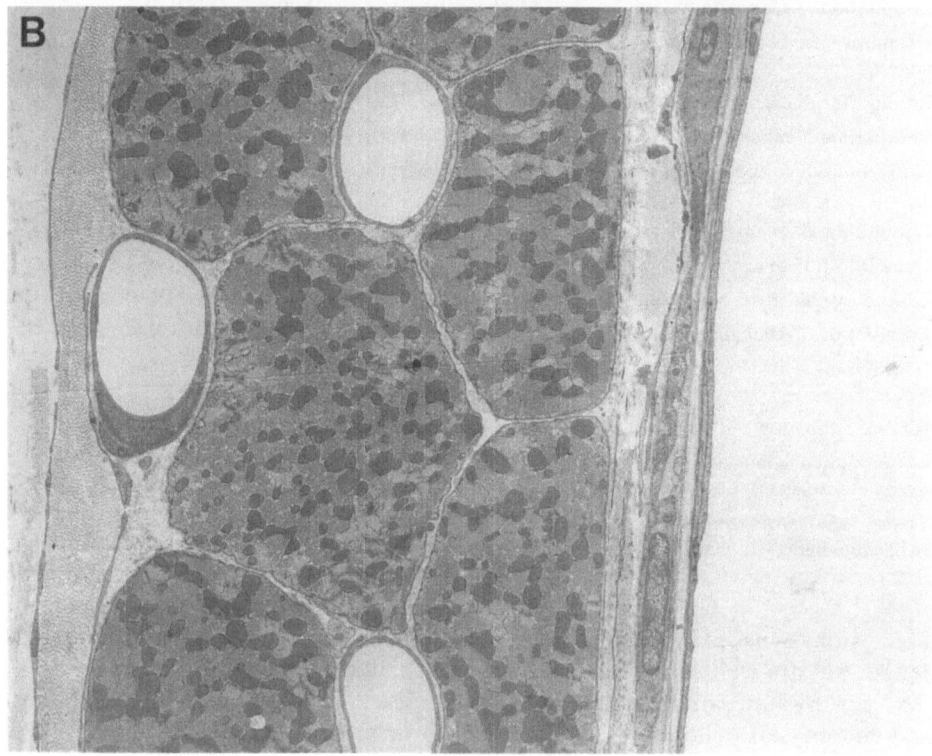

V. Venous Cardiac Muscle

In humans, sleeves of cardiac muscle continuous with the atrial myocardium extend into the superior vena cava, the four pulmonary veins, the coronary sinus, and the basal portions of the mitral valvular leaflets (see FERRANS and THIEDEMANN 1983). This extension of myocardium is a characteristic feature of many mammals (NATHAN and GLOOBE 1970), although in some species the muscle extends much further up the vessels. This is particularly so in rodents in which there are intrapulmonary extensions of the myocardium into the smallest venules (KRAMER and MARKS 1965), known as "cœur pulmonaire" (GUIEYSSE-PELLISIER 1945). Extensions of myocardium along the inferior vena cava and azygos vein also occur in these animals (Fig. 28). The myocytes in these vessels have a similar appearance to those of the atrium (KARRER 1959; LUDATSCHER 1968) (Fig. 28), are connected by intercalated discs (KARRER 1960), have similar electrophysiological characteristics (PAES DE ALMEIDA et al. 1974) and are innervated by adrenergic and cholinergic nerves (HUNG and LOOSLI 1977). In the rat azygos vein close to the inferior vena cava the media is composed almost entirely of cardiac muscle 4–6 cells wide. The muscle cells vary in orientation although there is a tendency for the inner fibres to run circumferentially and the outer ones longitudinally. The media is highly vascularized with capillaries usually running parallel to the direction of the fibres (CULLINAN et al. 1986). When corrosion casts are produced by a method similar to that of ANDERSON and ANDERSON (1980) and viewed with the scanning electron microscope, the arrangement of vessels produces a criss-cross pattern on the surface (Fig. 28).

In mouse pulmonary veins, action potentials in the cardiac muscle propagate from the left atrium towards the lung, i.e. in the opposite direction to blood flow, and it has been proposed by CHALLICE et al. (1975) that these myocytes may be important in the control of blood flow within these vessels. However it is not clear whether a wave of contraction is present.

Fig. 28. A Scanning electron micrograph of methacrylate cast of capillary network in wall of rat vena cava. Note there appears to be two layers of capillaries almost at right angles to each other. × 250. Micrograph courtesy of Dr. PETER MOSSE. B Lower power electron micrograph of transverse section through wall of perfusion-fixed rat azygos vein. A single layer of smooth muscle cells underlies the endothelium. Beneath this are two layers of well vascularized cardiac muscle cut in transverse section. × 4400. Micrograph courtesy of Ms. VIOLET CULLINAN

C. Morphometry of Cardiac Muscle

I. Methodology

Quantification or 'morphometry' of cardiac muscle can be performed using a variety of indices. In recent years much attention has focused on 'stereology', and many excellent texts and articles describing the methodology and underlying theory are now available (see AHERNE and DUNNILL 1982; LOUD and ANVERSA 1984; WEIBEL 1979; WILLIAMS 1977). This technique is applicable at the tissue, cellular and subcellular levels, and provides a wide range of quantitative parameters including volume fractions, surface area to volume ratios, and numerical densities.

The dimensions of myocytes most commonly measured are diameter and cross-sectional area. Length is more difficult to assess, particularly in adult myocytes, due to problems in obtaining true longitudinal sections, and also because of the irregular step-like character of the intercalated discs. Myocyte volume can be determined with a variety of techniques (see LOUD and ANVERSA 1984), and is the truest indicator of cell size. In growing tissue, myocyte diameter or cross-sectional area reflect the overall cell growth only if the length and shape of the cell remain constant (ANVERSA et al. 1978), which is not the case in developing myocytes. ANVERSA et al. (1981) found in newborn rats that the length to width ratio increased from 2.9 at day 1 to 4.6 at day 5 to 5.3 at day 11. There is apparently no information available about the length-to-width ratio in the fetus.

A number of points need to be made about the definitions and methodology of the various morphometric parameters.

Myocyte density is defined as the number of myocyte cross-sections in an area of 1 mm² of myocardium. This parameter may be obtained from either light (RAKUSAN and POUPA 1963; RAKUSAN et al. 1980; WACHTLOVA et al. 1965, 1967) or transmission electron micrographs (ANVERSA et al. 1978, 1979). The ratio of myocytes and capillaries can be expressed as either the number of myocytes per capillary (RAKUSAN and POUPA 1963; RAKUSAN et al. 1980; ROBERTS and WEARN 1941; WACHTLOVA et al. 1965, 1967), or conversely the number of capillaries per myocyte (ANGELAKOS et al. 1964; ANVERSA et al. 1978, 1979; OLIVETTI et al. 1980; SHIPLEY et al. 1937; SMITH and CLARK 1979; WEARN 1928). Either of these ratios by themselves do not give any information about the length of the diffusion path from capillary to myocyte.

Diameter may be estimated from myocytes cut longitudinally (HOSHINO et al. 1983), but the value obtained is an underestimate which requires a mathematical correction (see RAKUSAN et al. 1978). Diameter is usually determined from transversely-sectioned myocytes, either directly or from point counting. There is, however, no uniformity in regard to the diameter measured, which may be the minimum (GREGORY et al. 1983; HIRAKOW and GOTOH 1975; MUNNELL and GETTY 1968b), maximum (KORECKY and RAKUSAN 1978; SHERIDAN et al. 1979), median (ANGELAKOS et al. 1964), mean of the shortest, and longest diameters (ASHLEY 1945; LUND and TOMANEK 1978; LUND et al. 1979; RAKUSAN and POUPA 1963; ROBERTS and WEARN 1941; SHIPLEY et al. 1937), or mean determined from point counting (HIRAKOW and GOTOH 1980; HOERTER et al. 1981; GOTOH 1982). Measurement of cross-sectional area overcomes the difficulties associated with the irregular cross-sections of myocytes. Indeed AHERNE and DUNNILL (1982) in skeletal muscle concluded that cross-sectional area was the optimal parameter of myofibre size.

The volume fraction of myocytes within the myocardium is usually obtained by point counting of transmission electron micrographs (ANVERSA et al. 1976, 1978, 1979; FRANK and LANGER 1974; GERDES and KASTEN 1980; LAZARUS et al. 1976; MALL et al. 1980, 1982; MARINO et al. 1983), but light microscopy can also be used (LAGUENS 1971; RAKUSAN et al. 1978). The volume fractions of the various components within the myocyte are the most commonly measured stereologic parameters. However, there is a lack of uniformity in the methodology. Volume fractions are a relative measurement compared to a reference volume, which for myocytes has consisted of the whole cell (GERDES and KASTEN 1980; LAZARUS et al. 1976; MARINO et al. 1983), the sarcoplasm (ANVERSA et al. 1976, 1978, 1979; BREISCH et al. 1980; MALL et al. 1980, 1982) or the sarcoplasm excluding the perinuclear region (SINGH et al. 1981; TOMANEK et al. 1979). The values obtained using these three approaches are not exact equivalents, but in adult myocardium the nucleus constitutes only 1–5% of the cell volume (ANVERSA et al. 1976, 1978; GERDES and KASTEN 1980; LAZARUS et al. 1976; SHERIDAN et al. 1977), and the perinuclear region a much smaller proportion, so that the volume fractions derived from these different methods do not differ substantially. However, in the developing myocardium, the nucleus may account for 10–12% of the myocyte volume (ANVERSA et al. 1975a; COLGAN et al. 1978; SHERIDAN et al. 1977), so that the numerical values of the volume fractions referred to the whole myocyte and the sarcoplasm are not directly comparable. In addition, apart from myofibrils and mitochondria, there is much variation in the way that the remaining cytoplasmic components of the myocyte, variously labelled as the matrix (ANVERSA et al. 1975a, b, 1976, 1978, 1979; MALL et al. 1980) or sarcoplasm (LUND and TOMANEK 1978; DAVID et al. 1979; TOMANEK 1979; TOMANEK et al. 1979) are grouped for estimation of their volume fraction.

The mitochondrial/myofibrillar ratio is obtained by combination of the respective volume fractions and is a coarse index of energy supply or availability, in the form of adenosine triphosphate, to energy utilization within the myocyte.

The relative nature of volume fractions imposes limitations on assessment of the nature of underlying absolute changes. For example, an increased volume fraction of a component can be associated with its hyperplasia, hypertrophy, or both, or even no change if the reference volume is diminished. Consequently, to determine the precise nature of any change, absolute parameters such as size and number are also necessary (see LOUD et al. 1978, 1983).

II. Factors Influencing Morphometric Parameters

Many factors from fixation through to microscopy influence the final appearance of cardiac muscle, and therefore also affect morphometric parameters to varying degrees.

The characteristics of the fixative solution, particularly with respect to tonicity and ionic composition, may cause structural and volume changes within the myocyte affecting T tubules, sarcoplasmic reticulum and mitochondria (see LEGATO et al. 1968; PAGE and UPSHAW-EARLEY 1977; SPERELAKIS and RUBIO 1971; TOMANEK and KARLSSON 1973). Fixation of cardiac muscle is commonly performed by immersion or perfusion. Immersion fixation is dependent on the diffusion of the fixative from the outer to the inner portions of the tissue sample, and this results in a gradient in the quality of fixation (FAWCETT and McNUTT 1969). The sequelae of this gradient may include alterations in relative and absolute size of cells and organelles, as well as difficulties with identification of cellular components. MARINO et al. (1983) compared cat papillary muscle fixed by immersion and perfusion fixation. The myocytes fixed by immersion showed clear cytoplasmic areas, displacement of organelles, and poor preservation of organelles, particularly mitochondria. In the perfused tissue, the preservation of cellular components was much improved. The myocytes fixed by immersion also showed an increase in the mitochondrial and nuclear volume fractions, compared to the perfused tissue. Furthermore, the increase in the mitochondrial volume fraction resulted in an increase in the mitochondrial/myofibrillar ratio from 0.36 to 0.55.

In skeletal muscle, the cross-sectional area of the myofibres is dependent on sarcomere length (CLANCY and HERLIHY 1978). Thus, in shortened muscle with a short sarcomere length, the cross-sectional area of the myofibre is proportionately increased compared to the lengthened muscle with a long sarcomere length, but the product of sarcomere length and cross-sectional area remains the same. Although a similar experiment has apparently not been performed in cardiac muscle, the implications of the above study are clear: where a difference in myocyte cross-sectional area or diameter is anticipated, all tissues should be fixed in a defined and reproducible state to avoid errors resulting from differences in sarcomere length. The corollary of this is that fixation should be by perfusion. MARINO et al. (1983) found that myocytes fixed by immersion showed poor alignment of myofibrils with contraction bands, whilst those fixed by perfusion had sarcomeres which were evenly spaced and well aligned.

The regimen of tissue processing for transmission electron microscopy is of importance because during all steps in the procedure, from fixation through to dehydration and embedding, volume changes occur (EISENBERG and MOBLEY 1975; GERDES et al. 1982; KUSHIDA 1962) such that the final volume of the embedded tissue may be less, greater or equal to that of the tissue prior to fixation. The direction and magnitude of the change is dependent on factors such as the osmolarity of the fixative, the composition of rinsing solutions, the inclusion or exclusion of propylene oxide or en bloc staining with uranyl acetate, and the nature of the embedding medium. The change in tissue volume primarily affects absolute measurements such as myocyte diameter, cross-sectional area and density. Relative parameters such as myocyte fraction within the myocardium, and the volume fractions of the myocytes, are not affected unless there are differential volume changes in tissue or cellular components.

Cardiac myocytes prepared for light microscopy by embedding in paraffin may show marked shrinkage (ASHLEY 1945; BLACK-SCHAFFER et al. 1965; SHIP-LEY et al. 1937) which is helpful in the identification of myocyte profiles for calculation of myocyte density, but reduces absolute dimensions such as diameter and cross-sectional area. Comparing paraffin and a plastic resin, IMAMURA (1978) found that the diameters of myocytes embedded in the plastic resin averaged 20% more than those embedded in paraffin.

The degree of compression of sections during ultramicrotomy is inversely related to the thickness of the section, but volume fractions are relatively unaffected compared to the other stereologic parameters. The problem may be ignored by assuming that the compression is counterbalanced by the effects of image distortion related to spherical aberration in the lens system of the electron microscope (LOUD et al. 1965), or resolved by computation of correction factors (ANVERSA et al. 1979).

The type of microscopy can modify some morphometric parameters. LOUD et al. (1978) compared myocyte cross-sectional area obtained by light and transmission electron microscopy in the same tissue. The volume obtained with light microscopy was twice that determined with transmission electron microscopy and the difference was ascribed to difficulty in resolution of the margins of adjacent cells with light microscopy (see LOUD et al. 1984), resulting in overestimation of cell dimensions. This problem may be overcome by avoiding areas of myocardium containing tightly-packed myocytes (GERDES et al. 1979).

An accurate measurement of the magnification factor is essential in the determination of surface area to volume ratios and numerical densities, but not volume fractions. In addition, BOSSEN et al. (1978) found that the surface area to volume ratio of the plasmalemma was influenced by magnification, a lower magnification (x 7500) producing a smaller numerical value than a higher magnification (x 25000).

The technique of tissue preparation has a profound influence on estimates of the extent of binucleation in adult myocytes. In rat myocardium embedded in a plastic resin and sectioned for light microscopy, the incidence ranges from 1.9–10.6% (ANVERSA et al. 1979, 1980; LOUD et al. 1978; WEINER et al. 1979). However, in isolated myocytes prepared by enzymatic digestion, about 85% of myocytes are binucleated (BISHOP et al. 1979a, b, 1980). It should be noted that in the postnatal period, similar results are obtained with both techniques (see F.VI.4e).

III. Morphometric Parameters in Adult Myocardium

1. Myocyte Density and Ratio of Myocytes and Capillaries

Most available estimates of myocyte density have been performed in the left ventricle of the rat (Table 3). ANVERSA et al. (1978) found a significantly

Table 3. Myocyte density (number of myocyte cross-sections/1 mm^2) in adult ventricular myocardium

Species	Myocyte density	References
Rat	5470, 6520	Anversa et al. (1978)
Rat	4710	Anversa et al. (1979)
Rat	2465, 2504	Rakusan and Poupa (1964)
Rat	3105, 3284	Rakusan et al. (1980)
Rabbit	2889	Wachtlova et al. (1967)
Hare	3743	
Laboratory rat	2666	Wachtlova et al. (1965)
Wild rat	3332	

greater myocyte density in the epimyocardium, compared to the endomyocardium, but Rakusan et al. (1980) were unable to detect a difference between the mid-wall and the endomyocardium. Myocyte density is greater in species with a high level of physical activity (Wachtlova et al. 1965, 1967) but does not appear to change with ageing (Rakusan and Poupa 1964).

The ratio between the numbers of myocytes and capillaries is close to unity in a wide range of species (Angelakos et al. 1964; Anversa et al. 1978, 1979; Rakusan and Poupa 1964; Rakusan et al. 1980; Roberts and Wearn 1941; Shipley et al. 1937; Smith and Clark 1979; Wachtlova et al. 1965, 1967; Wearn 1928). Anversa et al. (1978) found a difference between the epimyocardium and endomyocardium of the rat left ventricle, but comparisons of the left and right ventricles are at variance. Angelakos et al. (1964) in their studies of rat, and Wearn (1928) in man, cat and rabbit found no difference, whilst Shipley et al. (1937) in rabbit and Smith and Clark (1979) in rat, found a greater number of capillaries per myofibre in the left ventricle.

2. Dimensions of Myocytes

The diameters of adult myocytes fall within relatively narrow limits (Table 4). Differences in heart size between species are mainly due to differences in cell number (Black-Schaffer et al. 1964). Left ventricular myocytes have a larger diameter than right ventricular myocytes in man (Ashley 1945; Gregory et al. 1983; Hoshino et al. 1983), and rabbit (Shipley et al. 1937), but reportedly not in rat (Angelakos et al. 1964). However, Bishop et al. (1979a, b) found that isolated myocytes from rat left ventricle were larger than those in the right ventricle both in length and width.

Hoshino et al. (1983) divided the left ventricular wall into endomyocardial, midwall and epimyocardial segments, and found a progressive increase in myocyte diameter from the epimyocardium to endomyocardium. In agreement with this, the cross-sectional area of left ventricular myocytes in the endomyocardium appears to be greater than in the epimyocardium (Table 5), although Tomanek

Table 4. Diameters in μm, of myocytes in left and right ventricles of normal adult species

Species	Left ventricle	Right ventricle	References
Man	19.5	16.1	ASHLEY (1945)
Man	12.6	11.1	GREGORY et al. (1983)
Man	11.2–13.0	9.9	HOSHINO et al. (1983)
Rabbit	19.0	17.2	SHIPLEY et al. (1937)
Rat	11.8	11.5	ANGELAKOS et al. (1964)

Table 5. Cross-sectional area (μm^2) of myocytes in the endomyocardium and epimyocardium of the left ventricle

Species	Endomyocardium	Epimyocardium	References
Man	617	403	STOKER et al. (1982)
Dog	303	255	GERDES and KASTEN (1980)
Rat	155	122	ANVERSA et al. (1978)
Rat	364	224	GERDES et al. (1979)
Rat	301	238	LOUD et al. (1978)
Rat	193	186	TOMANEK et al. (1979)

Table 6. Volume fraction, expressed as a percentage, of myocytes within adult myocardium

Species	Myocyte fraction	References
Cat	73.5	MARINO et al. (1983)
Dog	83.2, 83.9	GERDES and KASTEN (1980)
Hamster	76.1	LAZARUS et al. (1976)
Rat	87	LAGUENS (1971)
Rat	83.7	ANVERSA et al. (1976)
Rat	84.9, 79.4	ANVERSA et al. (1978)
Rat	80.1	ANVERSA et al. (1979)
Rat	80.9	MALL et al. (1980)
Mouse	76.4	BOSSEN et al. (1978)
Rabbit	75.4	FRANK and LANGER (1974)
Rabbit	85.9	MALL et al. (1982)

et al. (1979) found little difference in rat. The average length of adult myocytes has been reported as 62 μm in man (KAWAMURA 1982) and 71 μm in dog (LAKS et al. 1967). In rat, KORECKY and RAKUSAN (1978) found the length-to-width ratio to be 5.3 and constant over a range of cell lengths and diameters. In adult man, the length-to-width ratio has been reported to be 4.1 (KAWAMURA 1982).

3. Composition of the Myocardium

The volume fraction of myocytes within the myocardium is similar in different species (Table 6), and ranges from 73.5 to 87%. ANVERSA et al. (1978) found that in rat, the volume fraction of myocytes in the endomyocardium of the left ventricle was significantly greater than in the epimyocardium, but in dog, GERDES and KASTEN (1980), found no difference. MARINO et al. (1983) concluded that this parameter was not influenced by the mode of fixation.

4. Composition of Myocytes

a) Myofibrils

Myofibrils occupy the largest volume fraction within adult myocytes (Table 7). There does not appear to be a difference between the endomyocardium and epimyocardium of the left ventricle, but SINGH et al. (1981) found a significantly greater volume fraction of myofibrils within the right ventricle compared to the left ventricle and septum.

b) Mitochondria

After myofibrils, mitochondria comprise the next largest volume fraction within the myocyte (Table 7). Together these two elements make up 70–90% of the myocyte volume.

Mitochondria in the epimyocardium of the left ventricle have a greater cross-sectional area than those in the endomyocardium (ANVERSA et al. 1978), but their volume fraction is smaller. However, the volume fraction of mitochondria in the right ventricle is less than that in the left ventricle (LAGUENS 1971; SINGH et al. 1981), and according to LAGUENS (1971), this is associated with a smaller absolute volume per mitochondrion (0.57 vs 0.53 μm^3) as well as reduced numbers per unit volume.

c) Nucleus and Matrix

The nucleus constitutes between 1.1–2.8% of the cell volume (ANVERSA et al. 1976, 1978, 1979; GERDES and KASTEN 1980; LAGUENS 1971; LAZARUS et al. 1976; MARINO et al. 1983), but may be as high as 5.4% (SHERIDAN et al. 1977).

The matrix comprises between 5.2% and 6.8% of the nonnucleated portion of the adult myocyte (ANVERSA et al. 1976, 1978, 1979; GOTOH 1983) but a value of 20.9% has been reported (SHERIDAN et al. 1977).

d) Sarcoplasmic Reticulum and T System

The T system occupies between 0.8% and 1.6% of the myocyte volume (see BOSSEN et al. 1978; McCALLISTER et al. 1978; PAGE and McCALLISTER 1973b; SINGH et al. 1981; WEINER et al. 1979) and increases the total surface area of the myocyte by about one third (PAGE and McCALLISTER 1973b).

The reported volume fractions for sarcoplasmic reticulum range from 0.9% to 4.9% (BOSSEN et al. 1978; McCALLISTER et al. 1978; PAGE and McCALLISTER 1973b; SINGH et al. 1981; WEINER et al. 1979), but only a small proportion

Table 7. Volume fractions, expressed as a percentage, of myofibrils and mitochondria within adult myocytes

Species	Myofibrils	Mitochondria	References
Swine	65.6–67.9	21.1–22.2	SINGH et al. (1981)
Rabbit	50.1	30.9	HOERTER et al. (1981)
Rabbit	60.3	26.8	MALL et al. (1982)
Cat	47.8	22.2	SHERIDAN et al. (1977)
Cat	60.0, 58.1	28.1, 29.3	GOTOH (1983)
Cat	49.8	17.3	MARINO et al. (1983)
Dog	54.0	26.0	McCALLISTER et al. (1978)
Dog	57.4, 57.8	25.4, 24.1	GERDES and KASTEN (1980)
Ferret	62.6–66.6	21.7–22.4	BREISCH et al. (1980)
Rat	59.2–60.9	33.7–34.6	ANVERSA et al. (1971)
Rat	46.7, 43.2	36.6, 29.3	LAGUENS (1971)
Rat	47.0	35.8	PAGE and McCALLISTER (1973b)
Rat	56.0	31.4	ANVERSA et al. (1976)
Rat	53.8, 54.5	33.8, 32.9	ANVERSA et al. (1978)
Rat	62.7, 62.0	32.5, 31.3	LUND and TOMANEK (1978)
Rat	53.6	34.8	ANVERSA et al. (1979)
Rat	59.2–63.1	29.9–32.7	TOMANEK (1979)
Rat	61.2	25.3	MALL et al. (1980)
Mouse	54.3	37.5	BOSSEN et al. (1978)
Hamster	45.9	28.2	LAZARUS et al. (1976)
Frog	46.15	13.78	BOSSEN and SOMMER (1984)
Lizard	50.05	25.78	BOSSEN and SOMMER (1984)

of this is made up of junctional sarcoplasmic reticulum (BOSSEN et al. 1978; PAGE and McCALLISTER 1973b). The volume fraction and surface density of the sarcoplasmic reticulum in the atria exceeds that of the ventricle in the mouse, frog, and lizard (BOSSEN and SOMMER 1984).

The volume fractions of the T system and the sarcoplasmic reticulum show regional fluctuations but do not appear to vary between the endomyocardium or epimyocardium of the left ventricle (WEINER et al. 1979), or the left ventricle, right ventricle and septum (SINGH et al. 1981).

D. The Conduction System

I. The Sinoatrial Node

1. Location

The sinoatrial (SA) node, within which the pacemaker normally resides, contains nodal cells, transitional cells, and intercalated clear cells arranged into a well-defined mass. The location and extent of the SA node varies slightly between mammalian species. In the rabbit, cat, mouse, rat, guinea-pig and monkey, the SA node is located in the wall of the superior vena cava (SVC), above or extending down to the crista terminalis (VIRÁGH and CHALLICE 1973a; VIRÁGH and PORTE 1973a; LEV and THAEMERT 1973; TRANUM-JENSEN 1976). The SA node of the guinea-pig (ANDERSON 1972a), golden hamster (WALLS 1942), and rabbit (JAMES 1967) occupies the full thickness of the atrial wall. In most human hearts the SA node occupies a lateral position at the junction of the SVC and right atrium, but can also be draped over the junction of the SVC and the right atrial appendage (Fig. 29) (ANDERSON et al. 1979).

There is also variation in the size and configuration of the SA node between species. The SA node of the bat is a flattened, elongate structure measuring 25 μm by 160 μm (KAWAMURA et al. 1978). It is of similar shape in the rabbit, but is thicker and more prominent (histologically) in ungulates, monkey, gorilla, and man (TRUEX and SMYTHE 1965a). The node is 6 mm long and 1 to 1.5 mm wide in the monkey (VIRÁGH and PORTE 1973a) and 6 to 7 mm long, 2 to 3 mm wide, and about 200 μm thick in the rabbit (BOJSEN-MØLLER and TRANUM-JENSEN 1972; BROWN 1982). In cattle, characteristic nodal fibres extend over a relatively broader area (BORELLI 1975) and may extend 30 mm along the sulcus terminalis (HAYASHI 1962).

2. Histology

Histologically, the mammalian SA node is readily identifiable, enmeshed in a network of collagenous connective tissue distinguishing the nodal region from the surrounding atrial working myocardium (TRUEX and SMYTHE 1965a). Ganglion cells are commonly observed in the subepicardial connective tissue near the node and less frequently within the nodal area. The nodal cells stain palely and may have faint striations, or no striations. The cells are noticeably smaller than atrial working myocardial cells and are less well organised, forming

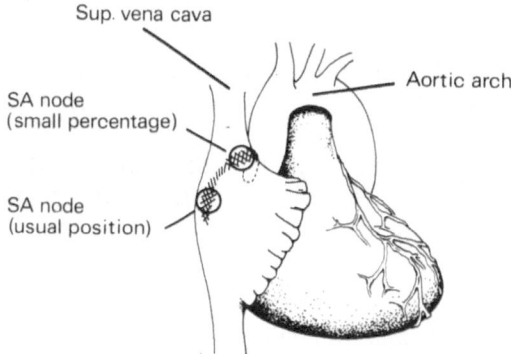

Sup. vena cava

Aortic arch

SA node
(small percentage)

SA node
(usual position)

Fig. 29. A diagrammatic representation of the variable position of the SA node in human hearts. In the majority the node is a lateral structure, but in a small percentage of individuals it is more medial overlying the crest of the atrial appendage. (Adapted from ANDERSON et al. 1979 and DAVIES et al. 1983)

small and irregularly-intertwining bundles through the connective tissue framework.

3. Ultrastructure of SA Nodal Cells

The SA nodal cells have relatively clear or slightly dense cytoplasm, a poorly-developed contractile apparatus with fibrils occupying less than half of the cell volume, sarcomeres with ill-defined Z, H and M bands, and an irregular, highly branched sarcoplasmic reticulum, not necessarily related to myofibrils (LASKOWSKI and D'AGROSA 1983).

SA nodal cells, in a wide range of species including bat (KAWAMURA et al. 1978), rabbit (TORII 1962; TRAUTWEIN and UCHIZONO 1963; CHALLICE 1966; TRANUM-JENSEN 1976; IRISAWA 1978; MASSON-PÉVET et al. 1979a, b), steer (RHODIN et al. 1961), cow (HAYASHI 1962), mouse (LEV and THAEMERT 1973), rat (CHENG 1971; MERRILLEES 1974), monkey (VIRÁGH and PORTE 1973a; COLBORN and CARSEY 1972), dog (KAWAMURA 1961b; JAMES 1962; JAMES et al. 1966), and man (JAMES et al. 1966; JAMES 1961b, 1977) have a similar fine structure.

a) SA Nodal Cell Size

In the rabbit, SA nodal cells are 3 to 9 µm wide in the nuclear region and 15 to 20 µm long (TORII 1962; TRANUM-JENSEN 1976). The corresponding cell diameter in man is 5 µm (TRUEX 1976), and 4 to 6 µm in monkey (VIRÁGH and PORTE 1973a), but less than 4 µm in the bat (KAWAMURA et al. 1978). However, measurement of cell diameter across the nuclear region may overestimate cell size (LASKOWSKI and D'AGROSA 1983).

b) Myofibrillar System

The myofibrils of the SA nodal cell frequently exhibit a number of discontinuous and thickened Z discs and indistinct or absent M bands (TRANUM-JENSEN

1976) (Fig. 30). Disorganised actin and myosin filaments not assembled into myofibrils are often seen in the cytoplasm (TRANUM-JENSEN 1976). Some disorganised thin filaments are continuous with myofibrils, and are often attached to the sarcolemma by a half zonula adherens (VIRÁGH and PORTE 1973a). The myofibrils in monkey nodal cells (VIRÁGH and PORTE 1973a) occupy only 40% of the total volume of the cells. The myofibrils in nodal cells appear to be arranged in a random manner (JAMES 1977; TRANUM-JENSEN 1976; TORII 1962). However, at least in monkey SA node (VIRÁGH and PORTE 1973a), some myofibrils are arranged spirally (Fig. 31).

c) T System and Sarcoplasmic Reticulum

There are no T tubules in the nodal cells of mammals (see SOMMER and JOHNSON 1970). The sarcoplasmic reticulum (SR) is poorly developed but has a similar arrangement around myofibrils to that observed in working myocardium. Tubules of SR are often observed within the substance of the Z disc and contribute to its discontinuity. They may form slightly-widened sacs of subsarcolemmal couplings (VIRÁGH and PORTE 1973a), separated from the sarcolemma by only 15 to 20 nm and containing particulate dense material (TRANUM-JENSEN 1976). The couplings are commonly thought to be involved in excitation-contraction coupling. Tubules of SR are often found in the cytoplasm, not associated with myofibrils. HAYASHI (1962) and TORII (1962) also observed rough-surfaced profiles of SR in nodal cells.

d) Mitochondria, Other Organelles and Glycogen

The cytoplasm of SA nodal cells also contains mitochondria, a high density of glycogen particles, free ribosomes, rough- and smooth-surfaced SR, and perinuclear Golgi complexes (TORII 1962; HAYASHI 1962). Mitochondria are identical in appearance to those in working myocardium except that they are smaller, being 0.4 to 0.6 μm in length and 0.2 to 0.4 μm in width (TORII 1962; HAYASHI 1962). The mitochondria are pleomorphic and randomly distributed, but form aggregates in some areas (VIRÁGH and PORTE 1973a).

e) Caveolae

The sarcolemma of SA nodal cells forms many ellipsoid caveolae 30 to 105 nm in diameter, all of which communicate with the extracellular space (Fig. 32) and approximately double the sarcolemmal surface area of the nodal cells (HAYASHI 1962; MASSON-PÉVET et al. 1979b). These caveolae do not have a pinocytotic function.

Fig. 30 A, B. SA nodal cells of rabbit heart, in longitudinal section, illustrating typical imperfections of the myofibrils. **A** Discontinuous Z discs (Z), indistinct H zones (H), and myosin filaments (M) which appear "free" in the cytoplasm. × 17700. **B** Other irregularities of the myofibrils include widening of the Z discs (Z). Note the large number of glycogen particles (G). × 17900. (From TRANUM-JENSEN 1976)

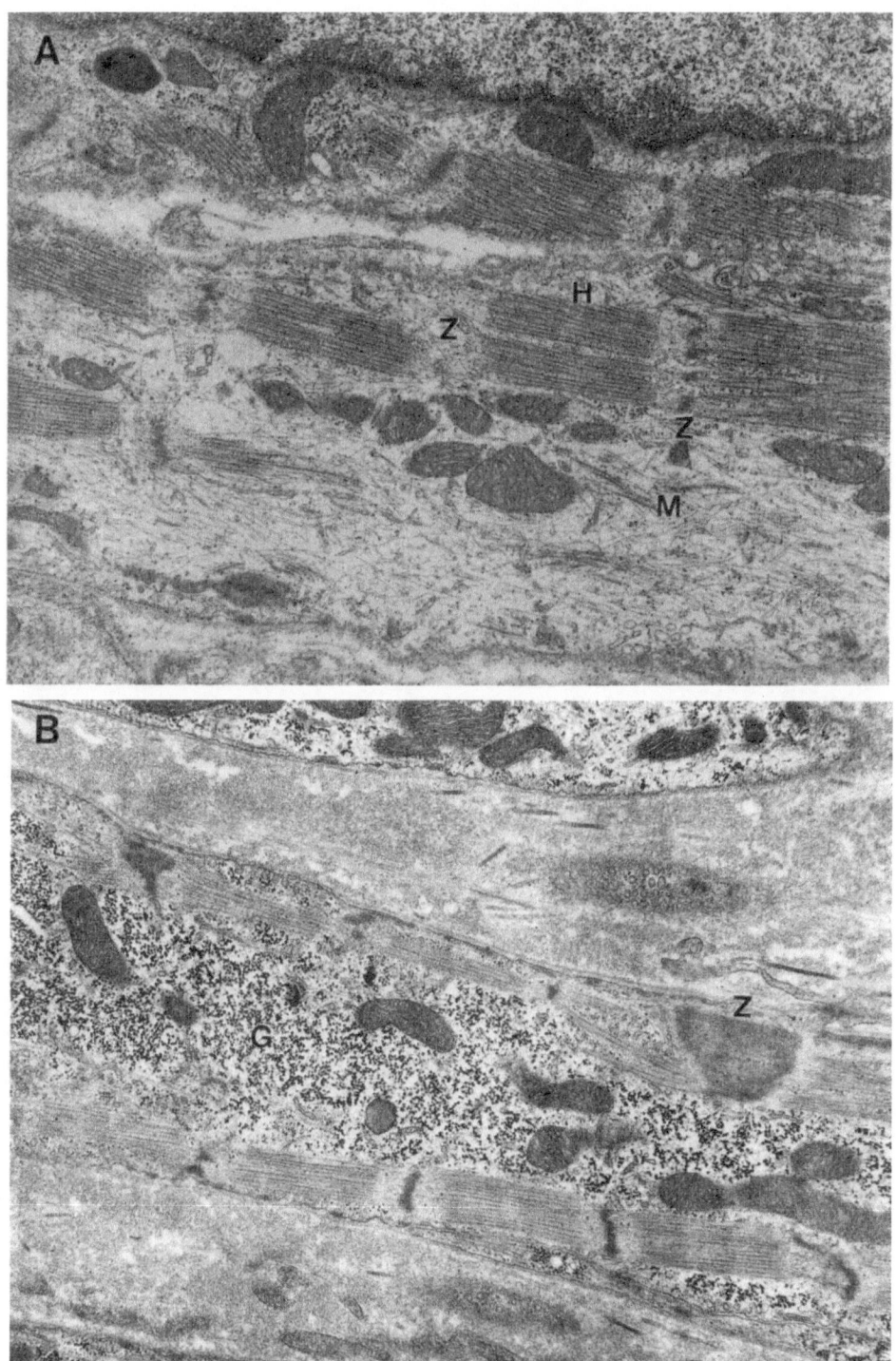

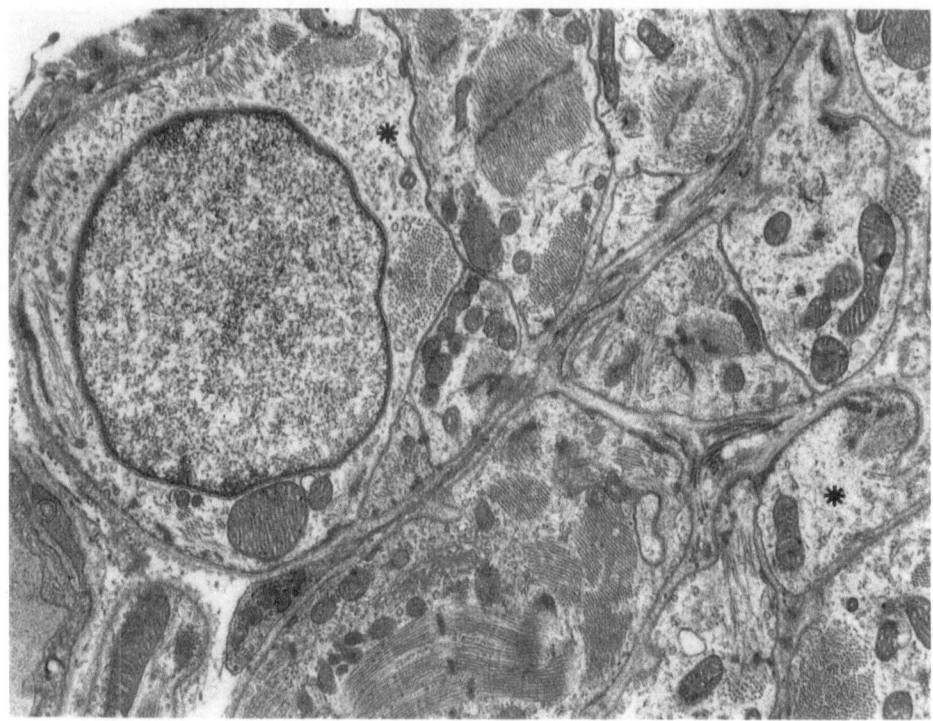

Fig. 31. SA nodal cells of monkey heart. The fusiform nodal cells twist around each other irregularly. The myofibrils lack definitive orientation and many cytoplasmic areas are lacking myofibrils (*asterisk*). Some myofibrils seem to be spirally arranged (cell at bottom-centre). × 16 500. (From VIRÁGH and PORTE 1973a)

f) Intercellular Junctions and Basal Lamina

In contrast to working myocardium, SA nodal cells form intercellular junctions between lateral membranes as frequently as between end-to-end membranes. Most specialised junctions are of the fascia adherens type with some small desmosomes, but nexuses are few and small (IRISAWA 1978). MASSON-PÉVET et al. (1979b) found only 0.2% of the membrane area to be of the nexal type of junction in SA nodal cells, which is approximately one tenth that of working myocardium. The nodal cells possess a basal lamina where the cell surface faces connective tissue. This basal lamina is not present between closely-apposed non-specialised junctional membranes separated by less than 20 nm (IRISAWA 1978; LASKOWSKI and D'AGROSA 1983).

4. Transitional Cells

Transitional cells are intermediate between SA nodal and working atrial cells, with a myofibrillar volume fraction greater than 50% and well-aligned

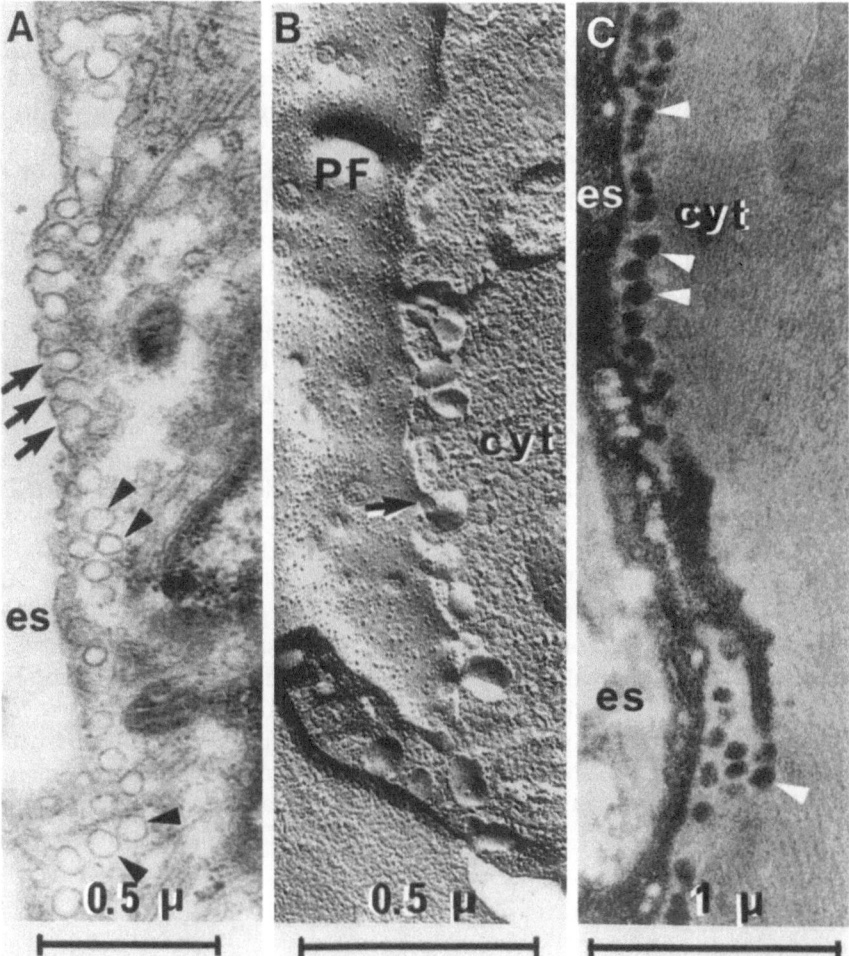

Fig. 32A–C. Caveolar invaginations in SA nodal cells. **A** Some caveolae *(arrowed)* show openings to the extracellular space *(es)*, whereas others *(arrowheads)* appear detached. **B** Freeze-fracture image reveals cytoplasm *(cyt)* on the right side in which caveolar invaginations can be observed. The opening of a caveola is indicated by an *arrow*. P face *(PF)*. **C** Following lanthanum impregnation post-fixation, all the caveolae contain electron-dense material, irrespective of whether their connection to the extracellular space is apparent. Section unstained. (From MASSON-PÉVET et al. 1979b)

myofibrils with distinct Z, H and M bands (LASKOWSKI and D'AGROSA 1983). In man their diameter is 6 to 10 µm compared to 5 µm for nodal cells (TRUEX 1976). Transitional cells have fewer myofibrils than atrial working myocardium, are joined by nexuses more frequently than the nodal cells and are distributed in the periphery of the node proper. Although there is usually a gradual morphological transformation from typical nodal to transitional to atrial working myocardium, some areas of direct communication between nodal and atrial working myocardium have been reported in the monkey heart (VIRÁGH and PORTE 1973a).

5. Intercalated Clear Cells

Intercalated clear cells have fewer myofibrils and organelles than the typical nodal cell. VIRÁGH and PORTE (1973a) stated, "In some clear cells the contractile apparatus is so rudimentary in development that the muscular nature of the cell is scarcely detectable". These cells have been described as Purkinje or Purkinje-like cells or P-cells (JAMES 1977). However, the term "intercalated clear cell" is more accurate (VIRÁGH and CHALLICE 1973a), leaving the term "Purkinje fibre" or "Purkinje cell" for the constituents of the specialised conducting network of the ventricles. Intercalated clear cells can be distinguished by light and electron microscopy occurring singly or sometimes in small groups, but never forming fibres (Fig. 33). In monkey heart their diameter varies from 6 μm to 10–12 μm (VIRÁGH and PORTE 1973a). In man, the diameter of clear cells shows a marginal increase from 13.2 μm in infants to 14.2 μm in adults (TRUEX 1976). Similar cells have been described in the SA node of the hamster (WALLS 1942) and dog (KAWAMURA 1961b; JAMES 1962). Although TRUEX (1976) has reported intercalated clear cells in the rabbit SA node, they have not been observed by others (TORII 1962; TRAUTWEIN and UCHIZONO 1963; TRANUM-JENSEN 1976; IRISAWA 1978; BLEEKER et al. 1980). There is also doubt about their existence in the rat, bat, and young dogs and piglets (MERRILLEES 1974; KAWAMURA et al. 1978; COPENHAVER 1981; LASKOWSKI and D'AGROSA 1983).

The number of intercalated clear cells identified in a given species is influenced by the quality of fixation (BOJSEN-MØLLER and TRANUM-JENSEN 1972), and cells subject to fixation artefacts may be mistakenly identified as intercalated clear cells. However, most of the ultrastructural features attributed to true intercalated clear cells, i.e. a high density of glycogen particles, intact membranes and sparsity of myofibrils (VIRÁGH and PORTE 1973a) are inconsistent with poor fixation.

6. Correlations Between Electrophysiological Pacemakers and the Morphological SA Node

A number of studies have successfully combined morphological and electrophysiological techniques to test the premise that pacemaker current originates in cells that are morphologically classified as typical nodal cells. The techniques and associated problems have been reviewed by JANSE et al. (1978).

The leading pacemaker group can be localised to a small area of the midportion of the SA node (TRAUTWEIN and UCHIZONO 1963; WOODS et al. 1976; MASSON-PÉVET et al. 1978; BLEEKER et al. 1980). For instance, in rabbit a pacemaker site has been localised to a group of about 5000 typical SA nodal cells, each less than 8 μm wide with few myofibrils and large amounts of glycogen, covering an area of approximately 0.1 mm² (BLEEKER et al. 1980).

The cytoplasmic composition of the cells electrophysiologically localised as leading pacemakers has been quantified (MASSON-PÉVET et al. 1979a) (Table 8). Organised structures such as the nucleus, mitochondria, myofilaments, and sar-

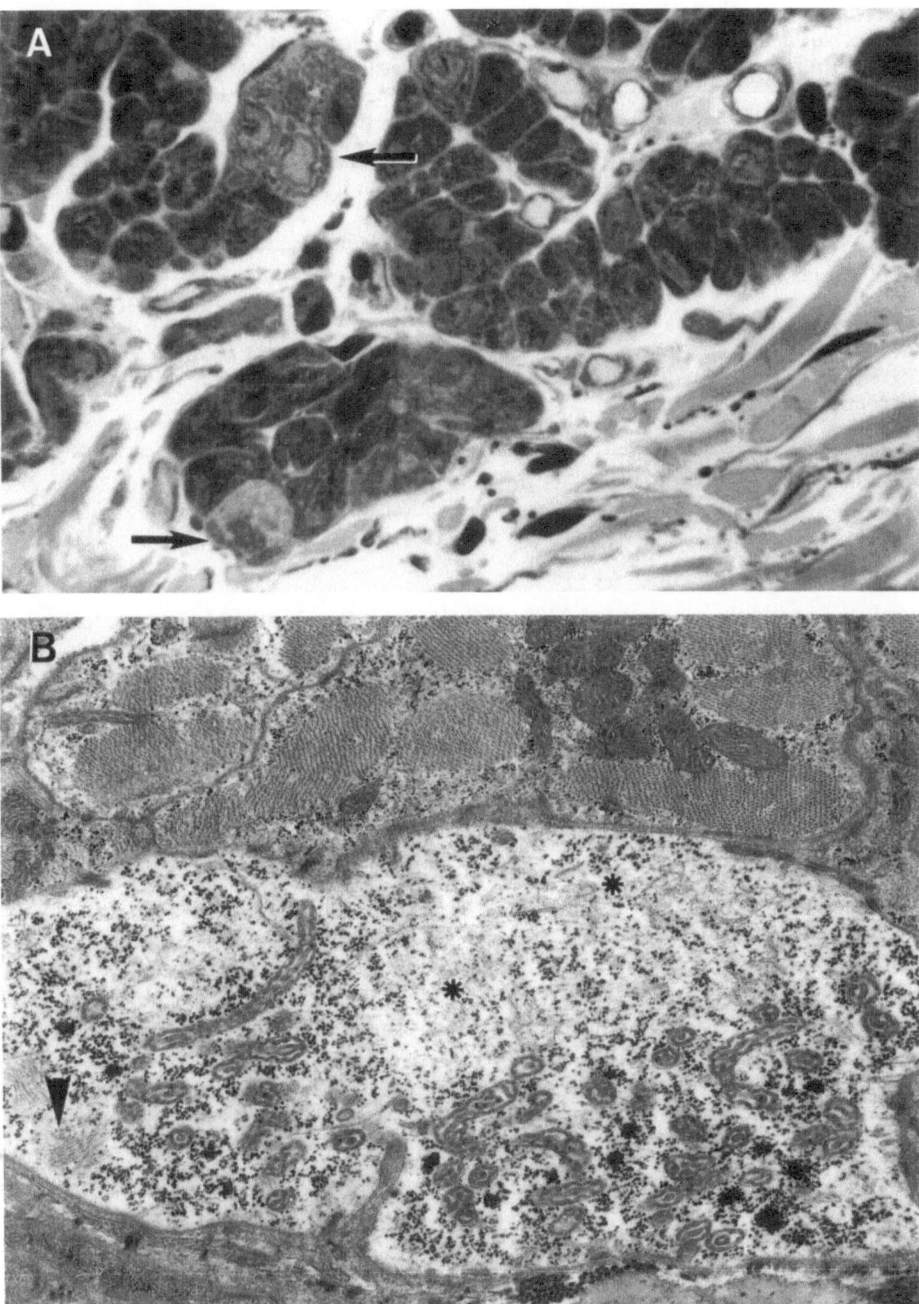

Fig. 33A, B. The SA node of monkey heart. **A** Intercalated clear cells (*arrowed*) are present amongst the nodal fibres. Connective tissue separates cords of nodal fibres. × 1000. Araldite-embedded, toluidine blue. **B** An intercalated clear cell in contact with several nodal cells. Few myofibrils (*arrowhead*), fine filamentous material (*asterisk*), small elongated mitochondria, and glycogen particles can be observed in the cytoplasm. × 15000. (From VIRÁGH and PORTE 1973a)

Table 8. Mean relative volume densities in percents of myofilaments, mitochondria, nucleus, sarco-plasmic reticulum (SR) tubules, smooth-surfaced (SS) vesicles, SR total (SR tubules + SS vesicles) and cytoplasm in leading pacemaker cells and atrial cells. (From MASSON-PÉVET et al. 1979a)

Structures quantified	Percentage of cell volume occupied by the different structures in:			
	Leading pacemaker cells mean value ± S.E.M.	Atrial cells mean value ± S.E.M.	t^*	Significance P (two tailed) < 0.05
Myofilaments	14.7 ± 3.1	53.6 ± 1.5	14.0	+
Mitochondria	14.7 ± 0.9	23.0 ± 2.5	3.4	−
Nucleus	13.1 ± 1.3	4.2 ± 0.6	6.3	+
SR tubules	0.7 ± 0.4	2.6 ± 0.3	15.5	+
SS vesicles	0.3 ± 0.1	0.2 ± 0.0	0.7	−
SR total	1.0 ± 0.3	2.8 ± 0.3	30.0	+
Cytoplasm	56.5 ± 3.7	16.4 ± 1.6	9.8	+

t^* is the test value of the Student t-test for paired observations, in each case with 2 degrees of freedom. The chosen significance level is 0.05

coplasmic reticulum comprise 43% of the total cell volume in leading pacemaker cells in contrast to 84% in atrial working myocardial cells.

These studies support the concept that pacemaker cells are morphologically specialised. However, there are far more typical nodal cells in the SA node than can be accounted for in electrophysiologically-determined pacemaking areas (MASSON-PÉVET et al. 1979a).

7. Shifting Locations of the Leading Pacemaker

The pacemaker site is not fixed. As early as 1914 MEEK and EYSTER found changes in pacemaker site under the effects of vagal stimulation and cooling of the SA node in dog heart. It was suggested a hierarchy of pacemakers operated within the node in such a way that the leading group changed under various physiological conditions. The normal location of the pacemaker is in the rostral region or "head" of the node but this shifts caudally with vagal stimulation. Similar shifts within the SA node, and from the SA node to extrano-dal myocardium and AV bundle, accompany noradrenaline administration, low external calcium concentrations, as well as vagal and sympathetic stimulation (BOUMAN et al. 1968, 1978; BROOKS and LU 1972; SANO and IIDA 1968; GOLD-BERG et al. 1973; GOLDBERG 1975; STEINBECK et al. 1978). Furthermore cells have been identified in the SA ring bundle (see E.III.1b,c) which have some pacemaker-like properties (PAES DE CARVALHO et al. 1959; DE MELLO and HOFF-MAN 1960; TAKAYASU et al. 1969).

In dog atrium BOINEAU et al. (1980) found a number of anatomically different points of origin (O-points) of the pacemaker, each corresponding to a particular range of heart rate. Three O-points designated A, B and C were found in consistent anatomic positions and another three (D, E and F) were found

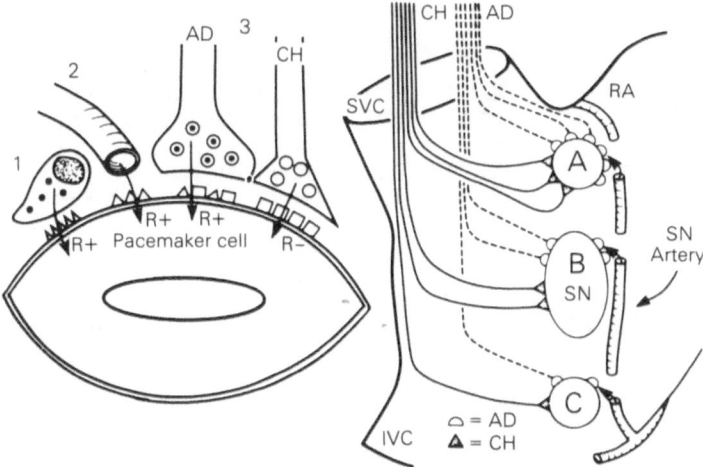

Fig. 34. Basis for multicentric impulse origin. Mechanism to explain multicentric initiation, link between 0-point and rate, and effects of autonomic intervention is proposed. Left panel illustrates section of membrane of cell at one of pacemaker centres or 0-points. Receptors are indicated by *R*, and *arrows* indicate excitatory (+) and inhibitory (−) effects of autonomic mediators. Basis for differentiated response of separate centres may be a difference in number or density of receptors at each site (right), with area A containing the most and area C the least. Slope of phase 4 of action potential is affected by spatial and temporal summation of both excitatory and inhibitory receptor inputs. Inputs to receptors at the pacemaker centre are from (*1*) atrial stores, (*2*) circulating, and (*3*) neurally released mediators that provide background and slow and fast rate response systems. *AD* adrenergic receptors; *CH* cholinergic receptors; *ESP* excitatory synaptic potential; *ISP* inhibitory synaptic potential; *RA* right atrium; *IVC* inferior vena cava; *SVC* superior vena cava; *SN Artery* sinus node artery. (From BOINEAU et al. 1980)

less consistently. Of these, only B was located within the SA node. The 0-points were 3–4 mm apart and had a maximum diameter of 5 mm. On the basis of these results it was proposed that the SA node did not always contain the dominant pacemaker but was one of several possible leaders in a pacemaker complex and that different densities of adrenergic and cholinergic receptors on the cells in the 0-points was the underlying mechanism of changing dominance (Fig. 34). Furthermore, most of the 0-points did not maintain a pacemaker function after surgical isolation. Pacemaker function may be altered by neural sensory feedback (MORAVEC and MORAVEC 1982, 1984), or different membrane properties among nodal cells (BOUMAN et al. 1978).

8. Internodal Myocardium

An ongoing controversy, well reviewed by JANSE and ANDERSON (1974), is whether tracts of specialised conducting cells exist between the SA and AV nodes. Whilst there is agreement that electrophysiologically-demonstrable preferential pathways of conduction exist between the SA node and the AV node along well-developed muscle bundles (e.g. crista terminalis) (SANO and YAMAG-

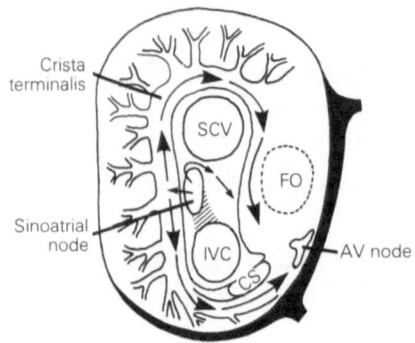

Fig. 35. The spread of excitation from the SA node to the AV node is illustrated. *Large arrows* show relatively fast conduction within the crista terminalis. *Small arrows* indicate areas of slow conduction. The *hatched area* represents a region of extremely slow conduction. Superior vena cava (*SCV*), inferior vena cava (*IVC*), ostium of coronary sinus (*CS*) and fossa ovalis (*FO*). (Adapted from SANO and YAMAGISHI 1965)

ISHI 1965; OISHI 1967; HOLSINGER et al. 1968; SPACH et al. 1969; 1971; GOODMAN et al. 1971) the debate centres on whether specialised conducting fibres exist within these preferential conduction pathways.

No internodal tract has been found that fulfils the criteria of ASCHOFF (1910) and MONKEBERG (1910), namely that fibres within the tract be identical and that a connective tissue sheath should enclose the system (cited in JANSE and ANDERSON 1974). However, the criteria for specialised tracts in the ventricles cannot necessarily be applied to the atria (SHERF and JAMES 1979; HOFFMAN 1979). Support for the existence of specialised internodal tracts has been based on morphological evidence from light and electron microscopy (JAMES 1963; SHERF and JAMES 1979; PAVLOVICH and CHEROVA 1981; LICHNOVKSY et al. 1982a,b). This, however, has been disputed (TRUEX and SMYTHE 1965; VIRÁGH and CHALLICE 1973a; VIRÁGH and PORTE 1973a; JANSE and ANDERSON 1974; TRUEX 1976; COPENHAVER 1981; ANDERSON et al. 1981). SHERF and JAMES (1979) described six specialised cell types in the human atria, but TRUEX (1976), VIRÁGH and PORTE (1973a) and ANDERSON et al. (1981) observed only one, corresponding to the intercalated clear cell. These cells are found not only in the presumed internodal tracts but in all parts of the atrial musculature (VIRÁGH and PORTE 1973a; JANSE and ANDERSON 1974; ANDERSON et al. 1981), and there is no direct evidence that they are physiologically specialised for conduction. In fact, conduction velocity in the crista terminalis is only slightly faster than other working myocardium (SANO and YAMAGISHI 1965). An area of conduction block, flanking the SA node on one side allows impulses travelling within the preferential pathways to reach the AV node first, even though the pathway through the area of block is much shorter (Fig. 35) (SANO and YAMAGISHI 1965; BLEEKER et al. 1982).

On available evidence it appears therefore that preferential pathways of conduction without specialised tracts exist between the SA and AV nodes. There is little doubt that the anatomic variation and arrangement of atrial working myocardium alone is capable of producing different conduction velocities, and even conduction block in the atria (SCHER and SPACH 1979; SPACH et al. 1982a, b; SPACH 1982; SPACH and KOOTSEY 1983).

9. The Nodal Artery

The SA node of most animals has a prominent artery which serves as a landmark and supplies blood to much of the right atrial musculature as well as to the SA node (HARDIE et al. 1981). The size of the SA nodal artery is variable, and in small mammals it is only an arteriole. It has one layer of medial smooth muscle cells in the bat (KAWAMURA et al. 1978) and 3 to 4 layers in the rat (MERRILLEES 1974).

The source of the nodal artery varies. JAMES (1978) found that some humans have their nodal artery arising from the right coronary artery while in others it arises from the left coronary artery. A small group has a dual supply of nodal arteries from both left and right coronary arteries (VIEWIG et al. 1975), whilst others are without a prominant SA nodal artery (ANDERSON et al. 1979).

The SA node may receive additional vessels from an extracoronary source in a variety of mammals. Sometimes the nodal artery itself may have an extracoronary origin (DOMENECH-MATEU and ORTZ-LLORCA 1976). Ligation of the nodal artery does not produce ischaemia in the SA node because of the multiple vascular supply and anastomosis within the node (ESPERANCA et al. 1975; ELISKA 1983).

II. The Atrioventricular Junctional Tissues

In normal hearts the atrioventricular (AV) junctional tissues are the only functional myocardial connection between the atria and ventricles. The AV junctional tissues have been defined as comprising the nodal approaches, the AV node, plus the penetrating and non-branching part of the AV bundle (HECHT et al. 1973). The AV node, which is under continuous neural and humoral influence, delays ventricular activation allowing time for atrial contraction and ventricular filling.

1. Location

The AV node is located at the base of the interatrial septum, anterior to the opening of the coronary sinus, and subjacent to the fibrous annulus of the septal leaflet of the tricuspid valve, in a region rich in connective tissue (TRUEX and SMYTHE 1965a). It is slightly superior to the annulus of the tricuspid valve but lateral or inferior to the mitral annulus (Fig. 36), lying behind the non-coronary aortic valve leaflet. Because it is a completely internal structure, the AV node cannot be positively identified without histological sections or electrophysiological measurements. However, the general position of the AV node can be deduced from surface features of the right atrium forming the triangle of KOCH (after DAVIES et al. 1983) (Fig. 36).

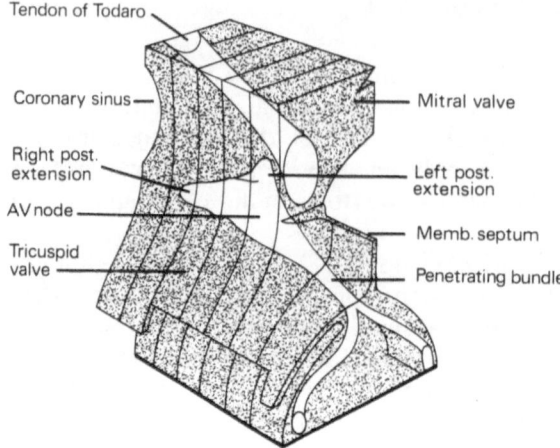

Tendon of Todaro

Coronary sinus

Right post. extension

AV node

Tricuspid valve

Mitral valve

Left post. extension

Memb. septum

Penetrating bundle

Fig. 36. The location of the AV junctional specialised tissues is in the apex of the triangle of Koch, which is formed from the continuation of the Tendon of Todaro behind the membranous part of the septum to the base of the tricuspid valve. The coronary sinus forms the base of the triangle. (Adapted from JAMES 1961 a)

2. Definition of the AV Nodal-Bundle Junction

Distally, the AV node is continuous with the AV bundle and left and right bundle branches. In humans, because there is a very gradual histological transition from nodal to AV bundle cells, the identification of the proximal AV bundle depends on the appearance of a complete connective tissue investment by the central fibrous body and loss of atrial contacts (BECKER and ANDERSON 1976; LEV 1968). There is no central fibrous body in mouse hearts but a connective tissue plane which separates atrial myocardium from the conduction cells serves to mark the beginning of the AV bundle (LEV and THAEMERT 1973). In monkey hearts there is wide individual variation in morphology at the junction of the AV node and bundle (VIRÁGH and PORTE 1973b). Some exhibit abrupt histological changes but others undergo a gradual transition to typical AV bundle cells. In animals such as the cow, the transition is marked by an abrupt change in the histology of the conducting cells (HAYASHI 1962). Here, thin nodal fibres 7–12 μm in diameter join directly to Purkinje cells of the AV bundle which are over 25 μm in diameter. Usually 1 to 3 nodal cells are joined to a single Purkinje cell. In the ferret heart, TRUEX et al. (1974) and MARINO (1979) defined the beginning of the AV bundle as occurring proximal to the right fibrous trigone (central fibrous body) on the basis of histological changes in the constituent cells.

3. Non-Junctional Atrial Myocardium Near the AV Junction

A projection of atrial myocardium has been described between the right atrial endocardium and the AV node (Fig. 37) (JAMES 1961a, 1983; BECKER and ANDERSON 1976; TRUEX and SMYTHE 1965a). The functional significance of this extension of atrial myocardium is not clear, in part because of the variability of its connections. Most recent studies do not find any connections between this extension and conduction tissue (VIRÁGH and PORTE 1973b; BECKER and ANDERSON 1976). Even so, it is possible that if such fibre bundles form true "bypass tracts" they may assume critical importance in pathological conditions by participating in AV junctional arrhythmias (JAMES 1961a, 1983).

4. Converging Preferential Pathways

Posteriorly, connections or inputs of atrial myocardium converge on the AV junctional region (JAMES 1963; VIRÁGH and CHALLICE 1973a). These extensions of atrial myocardium, which pass from the posterior wall of the atrium and from the interatrial septum are the terminations of the anterior, middle and posterior preferential pathways of conduction in the right atrium (see D.I.8). The anterior and middle pathways converge on the interatrial septum, superior to the AV node (JAMES 1963). The posterior pathway travels within the Eustachian ridge to enter the node from behind (Fig. 37). In the monkey, the subendocardial portions of the anterior, middle, and posterior nodal inputs by-pass the AV node to form the subendocardial layer of atrial myocardium which overlays the node and terminates in the base of the tricuspid valve (VIRÁGH and PORTE 1973b).

5. Transitional Cell Zone

Where the preferential pathways converge onto the AV junction there is a gradual transition from working atrial muscle to nodal cells. The predominant myocardial cell type in this region is the transitional cell which forms a large part of the AV node. These have also been called atrionodal cells (VIRÁGH and CHALLICE 1973a; VIRÁGH and PORTE 1973b). A second part of the AV node, which is surrounded by the loosely-organised transitional cells is commonly called the compact part of the node (TRUEX and SMYTHE 1965a, b; VIRÁGH and CHALLICE 1973a; VIRÁGH and PORTE 1973b; BECKER and ANDERSON 1976; MARINO 1979). Three bands of transitional cells predominate, converging on the compact node (BECKER and ANDERSON 1976). A superficial band of transitional cells passes on the endocardial side of the Tendon of Todaro to join the compact part of the node and extends some fibres into the base of the tricuspid valve. On the other side of the Tendon of Todaro a deeply-positioned group of transitional fibres connects the left side of the interatrial septum to the compact node. The third band is a posterior accumulation of transitional fibres which connect the atrial myocardium above and below the coronary sinus to the compact node (Fig. 37).

6. The Compact Node

There are two subdivisions of the compact node: a superficial and deep layer (TRUEX and SMYTHE 1965a, b; VIRÁGH and CHALLICE 1973a; VIRÁGH

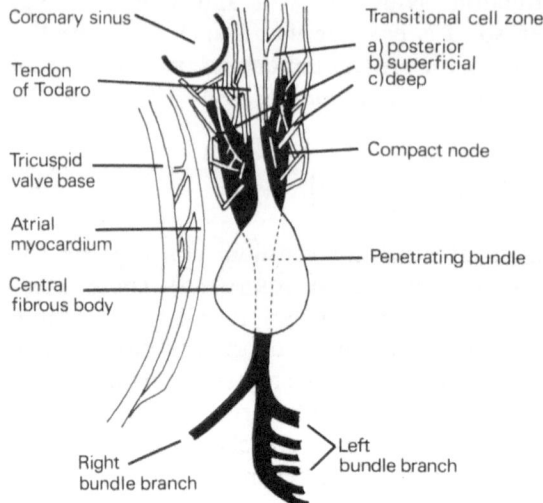

Coronary sinus

Tendon
of Todaro

Tricuspid
valve base

Atrial
myocardium

Central
fibrous body

Right
bundle branch

Transitional cell zone
a) posterior
b) superficial
c) deep

Compact node

Penetrating bundle

Left
bundle branch

Fig. 37. The AV junctional tissues of human heart and the branching part of the AV bundle are shown. The AV node has a compact part with left and right posterior extensions. The (a) posterior, (b) superficial and (c) deep transitional cell zones are terminations of the preferential atrial pathways and form connections with the compact node. The penetrating bundle is distal to the node and receives no atrial connections, being surrounded by connective tissue of the central fibrous body. Distally the bundle bifurcates into a single right bundle branch and a number of left bundle branches. Nonjunctional atrial myocardium terminates in connective tissue at the base of the tricuspid valve

and PORTE 1973b; ANDERSON et al. 1975a; BECKER and ANDERSON 1976; MARINO 1979; MARINO et al. 1981). By using reconstructions of serial histological sections the precise orientation of the subdivisions of the compact part of the human AV node and their relationship to the central fibrous body have been elucidated (TRUEX and SMYTHE 1967). Only the distal part of the compact node can be subdivided into superficial and deep layers (BECKER and ANDERSON 1976). The nodal cells of the superficial part of the AV node are generally arranged in a cranio-caudal direction and the cells forming the deep portion have an oblique or horizontal orientation, and run in a similar direction to the annulus fibrosus (TRUEX and SMYTHE 1967). The AV bundle is only continuous with the deep portion of the compact part of the node. There are atrionodal connections with the superficial portion of the AV node and an atrionodal connection to the superior surface of the deep portion of the node from the interatrial septum. Posteriorly the compact AV node usually bifurcates (Figs. 36, 37). Strands of nodal cells from the compact node can be followed towards the tricuspid and mitral aspects of the central fibrous body (ANDERSON et al. 1975a; BECKER and ANDERSON 1976).

7. Individual Variation

The organisation of parts of the AV node is variable, but this seems to be of little functional consequence (TRUEX and SMYTHE 1967; ANDERSON et al. 1975a; BECKER and ANDERSON 1976). The right posterior extension of the compact node is usually better developed, and the left sometimes receives no connections from the transitional cells but becomes embedded in the annulus fibrosus (ANDERSON et al. 1975a). The group of transitional cells deep to the Tendon of Todaro does not form a consistent input to the compact node. In some hearts their connection to the compact node is well-formed, while in others they appear to connect with working atrial myocardium (ANDERSON et al. 1975a).

8. Species Differences

a) Topography

The AV node has a variable shape, depending on the species. TRUEX and SMYTHE (1965a) wrote that "It may have a compact ovoid shape (platypus, sheep, beef), be a thin compact sheet closely applied to the annulus fibrosus (monkey, man); or be an elongated slender nodal mass often separated from the annulus by several layers of adipose tissue (rat, rabbit, dog)."

The size of the AV node is quite small even in the hearts of large mammals. In man the AV nodal length is 2–4 mm at birth, 3–5 mm by the age of 1–15 years and 5–7 mm by the age of 15–40 years (WIDRAN and LEV 1951; JAMES 1961a; TRUEX and SMYTHE 1965b).

b) Connective Tissue

A consistent feature of the AV node is the abundant connective tissue surrounding the strands of transitional cells and nodal cells. However, the amount of connective tissue present is variable between species. For example the nodal tissue of cow heart is more open and looser than that of dog or rabbit (HAYASHI 1962). The nodal region of the mouse has less connective tissue than that of man. The amount of connective tissue seems to reflect heart size and the degree to which the heart's fibrous skeleton is developed (LEV and THAEMERT 1973).

c) Subdivisions of the AV Junction

The organization of AV node of rat and mouse conform to the general plan of the rabbit, while the AV node of the ferret, monkey, and dog fit the general plan of the human (VIRÁGH and PORTE 1973b; DAVIES et al. 1983; MARINO 1979) (see D.II.5,6).

The conduction system of the ferret heart is particularly well differentiated despite the small size of the heart (TRUEX et al. 1974). The AV junctional tissues include a transitional cell zone and a compact part of the node. In contrast to humans and monkeys there are three parts of the compact node (MARINO 1979). Two portions correspond to the superficial and deep layers described in human and monkey heart (VIRÁGH and PORTE 1973b; TRUEX and SMYTHE 1967; BECKER and ANDERSON 1976). The third area consists of a group of large cells associated with the coronary sinus and positioned posterior to the superficial part of the compact node. This area of the AV node is the only region where superficial nodal cells do not connect with the transitional cell zone. In the anterior part of the node some transitional cell groups connect

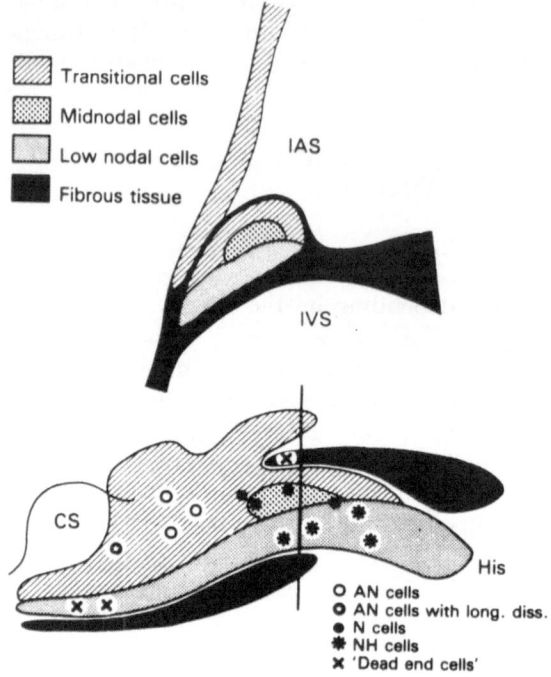

Transitional cells
Midnodal cells
Low nodal cells
Fibrous tissue

IAS

IVS

CS

His

O AN cells
O AN cells with long. diss.
● N cells
✳ NH cells
✗ 'Dead end cells'

Fig. 38. The results of microelectrode studies of the rabbit atrioventricular junction correlated to ultrastructural cell types. (From JANSE et al. 1976)

directly with the deep portion of the AV node similar to human and monkey heart (BECKER and ANDERSON 1976; VIRÁGH and CHALLICE 1973b).

The rabbit AV node (Fig. 38) represents the greatest departure from the organisation of the human AV node. In addition to components such as the transitional cell zone found in other species, a portion of the rabbit AV node, designated the "enclosed" part is surrounded by a fibrous collar which extends from the central fibrous body (TRANUM-JENSEN 1976). According to the criteria of ANDERSON et al. (1975a) and BECKER and ANDERSON (1976) the region of the node surrounded by the connective tissue collar would be regarded as the penetrating AV bundle since it is at this point that the node loses all atrial connections. However, the tissue surrounded by the fibrous collar is clearly nodal (ANDERSON et al. 1974; TRANUM-JENSEN 1976). The more proximal part of the transitional cell zone is called the "open node" (TRANUM-JENSEN 1976). The transitional cells within the enclosed node converge on a knot of nodal cells which have been called the midnodal cell zone or knot (ANDERSON et al. 1974; TRANUM-JENSEN 1976). Below the midnodal cell knot there lies a zone of lower nodal cells which is directly continuous with the penetrating AV bundle. Posteriorly it can be traced into the lower regions of the transitional cell zone of the open node where it thins out. Communication between the transitional cells and lower nodal cells occurs within the enclosed node but not in the open node (TRANUM-JENSEN 1976).

9. Ultrastructure of the AV Nodal Tissues

a) Transitional Cells

The transitional cell zone is predominantly oriented towards the compact node and forms an interconnected network of cells. Histologically, the transitional cells are smaller than the atrial myocardium, are paler in their staining and their nuclei have a wider heterochromatic periphery (MARINO 1979). Similar to the transitional cell zone of the SA node (D.I.4), there is a gradual morphological transition from the atrial muscle to the compact node (human, monkey, dog, ferret) or enclosed node (rabbit, rat, mouse). T tubules are rarely observed in transitional cells (MARINO 1979).

Although bundles of transitional cells appear more widely separated by connective tissue than those of atrial myocardium (BECKER and ANDERSON 1976), the adjacent cells are more tightly packed, with close associations over much of their lateral membranes (VIRÁGH and PORTE 1973b; TRANUM-JENSEN 1976; MARINO 1979). In spite of this large area of non-specialised junctional membrane, only 0.7% of the membrane area consists of gap junctions compared to 2.6% in atrial myocardium (MARINO 1979).

b) AV Nodal Cells

AV nodal cells have few myofibrils which are poorly aligned and exhibit a variety of imperfections, including the appearance of incomplete or thickened Z discs and free myofilaments within the cytoplasm. The AV nodal cells are smaller than atrial muscle cells and form irregular complex junctional areas of closely-apposed membranes. They contain abundant glycogen but very little sarcoplasmic reticulum (SR); however, around myofibrils the SR forms a tubular network similar to that in other myocardial cells. The AV nodal cells do not possess a T system or a basal lamina between closely apposed cells. The intercellular junctions, which include few gap junctions, are randomly distributed and poorly organised (HAYASHI 1962; TORII 1962; LEV and THAEMERT 1973; TRANUM-JENSEN 1976; MARINO 1979; VIRÁGH and CHALLICE 1973a; BHATNAGAR and SPOONAMORE 1979).

AV nodal cells may have fewer myofibrils and less connective tissue surrounding them than cells of the SA node (VIRÁGH and PORTE 1973b; KAWAMURA et al. 1978; JAMES 1983).

There are quantitative differences between the superficial and deep nodal cells of the compact node in ferret heart (MARINO 1979). While superficial nodal cells are similar in size to transitional cells, deep nodal cells are smaller. Myofibrils comprise 37% of the sarcoplasmic volume in superficial nodal cells and 32% in deep nodal cells. However, the major difference is that the proportion of membrane forming gap junctions in deep nodal cells (0.5%) is 2.5 fold greater than in the superficial nodal cells (MARINO 1979).

Although the general plan of the AV junctional tissues in rabbit heart is different from that of man, the midnodal cells of the rabbit heart may be morphologically similar to the deep nodal cells of the compact part of the human node. The nodal cells are irregular in shape, with a diameter of less than 5 μm (MOCHET et al. 1975). There is no region of the human AV node which corresponds to the lower nodal cell zone in rabbit heart. The lower nodal cells are larger and stain more intensely, having better organised myofibrils than the midnodal or transitional cells. Distally they are virtually indistinguishable from the AV bundle cells (TRANUM-JENSEN 1976). Whether there are functional differences between human and rabbit AV node as a result of their different anatomic organization is unknown.

10. Correlations Between Ultrastructurally- and Electrophysiologically-defined Zones

There is some uncertainty about how well anatomically-defined regions correlate with the electrophysiologically-defined Atrio-Nodal (AN), Nodal (N) and Nodal-His (NH) regions (DE FELICE and CHALLICE 1969; WOODS et al. 1982).

Probably the most successful correlation of morphological and electrophysiological findings in the AV junctional tissues has been that of ANDERSON et al. (1974) using rabbit heart. In agreement with DE FELICE and CHALLICE (1969) and WOODS et al. (1982), the cells which produced an AN form of action potential corresponded to transitional cells. Those with N-type action potentials were localised to the enclosed portion of the rabbit AV node. The NH action potentials were localised inferior to the midnodal cell knot, in the zone of lower nodal cells which became continuous with the AV bundle (Fig. 38).

ANDERSON et al. (1974) were not able to distinguish between the morphologically distinct midnodal cells and transitional cells of the enclosed node by electrophysiological measurements of action potentials, and it was therefore suggested that morphology is only one of many factors determining action potential configuration, with nodal architecture having a greater influence.

11. Anomalous AV Conduction Pathways

In some hearts there exists a group of variably-positioned myocardial fibres which appear to short-circuit the normal conduction route. These fibres may form connections to the ventricular myocardium from various points along the AV junctional tissue, and sometimes directly from the atrial myocardium. ANDERSON et al. (1975b) devised a nomenclature for the possible accessory and bypass tracts which they proposed was more useful than the commonly used eponymous terms MAHAIM fibres and bundles of KENT.

The best-established anomalous AV conduction pathway is the accessory atrioventricular muscle bundle (KENT's bundle) which provides the anatomic basis for the Wolff-Parkinson-White syndrome (ROSEN et al. 1980; RIGBY and GRABOYS 1981). This muscular bundle provides direct communication between the atrium and ventricle, by-passing the entire specialised AV junctional region, and therefore avoids the normal AV conduction delay. A second form of accessory muscle bundle which results in abnormal conduction is the nodoventricular muscle bundle. These fibres form a communication between the distal AV node and the ventricular myocardium and they have often been discussed in relation to supraventricular tachyarrhythmias (GALLAGHER et al. 1976; MOTTE et al. 1978; TOUBOUL et al. 1977, 1978; WARD et al. 1979; SUNG and STYPEREK 1979). GALLAGHER et al. (1981) found that although nodoventricular fibres were capable of supporting a sustained re-entrant tachycardia, fasciculoventricular fibres did not have a direct role in arrhythmias.

Fasciculoventricular fibres (and possibly nodoventricular fibres) are common in infants but are lost with subsequent development as part of a normal process of post-natal remodelling of the conduction tissues (JAMES 1970; MARINO et al. 1981). Although the accessory fibres were once implicated in sudden infant death syndrome (JAMES 1968) it now appears that they may be a normal but highly variable feature of the AV junctional region with little functional consequence in normal hearts. In particular, CASTA et al. (1980, 1983) observed dual and multiple AV pathways in arrhythmia-free children in an electrophysiological study. Unfortunately, it is not known whether the electrophysiologically-demonstrable AV pathways are anatomically separate. In contrast to the lack of correlation between arrhythmias and multiple AV pathways in children, there is some association between the two in adults (CASTA et al. 1980, 1983; DOPIRAK et al. 1980; SWIRYN et al. 1982). As well, nodoventricular and fasciculoventricular bundles have been implicated as possible causes of sudden unexpected death in adults (JAMES and MARSHALL 1976; JAMES 1983). JAMES (1983) proposed that the accessory bundles remain as a consequence of failure of the normal post-natal remodelling of the conduction system which would otherwise have led to the loss of nodoventricular and fasciculoventricular bundles. Some accessory pathways are acquired as a consequence of hypertrophy of the AV node in response to ischaemia and influence from the autonomic nervous system (OKADA 1984). Whether accessory connections are directly related to sudden death or are an incidental finding is uncertain. Many people with accessory muscle bundles remain asymptomatic (ROSEN et al. 1980).

III. The Ventricular Purkinje System

1. Introduction

The Purkinje fibres are the final components of the heart's conduction system, bringing the impulse of depolarisation which started in the SA node to the ventricular working myocardium. They provide a co-ordinated activation of the ventricular myocardium from apex to base (ESMOND et al. 1963; MYERBURG et al. 1972; LAZZARA et al. 1974). Consistent with their distribution, activation by the Purkinje fibres is most important in the apex and the ventricular

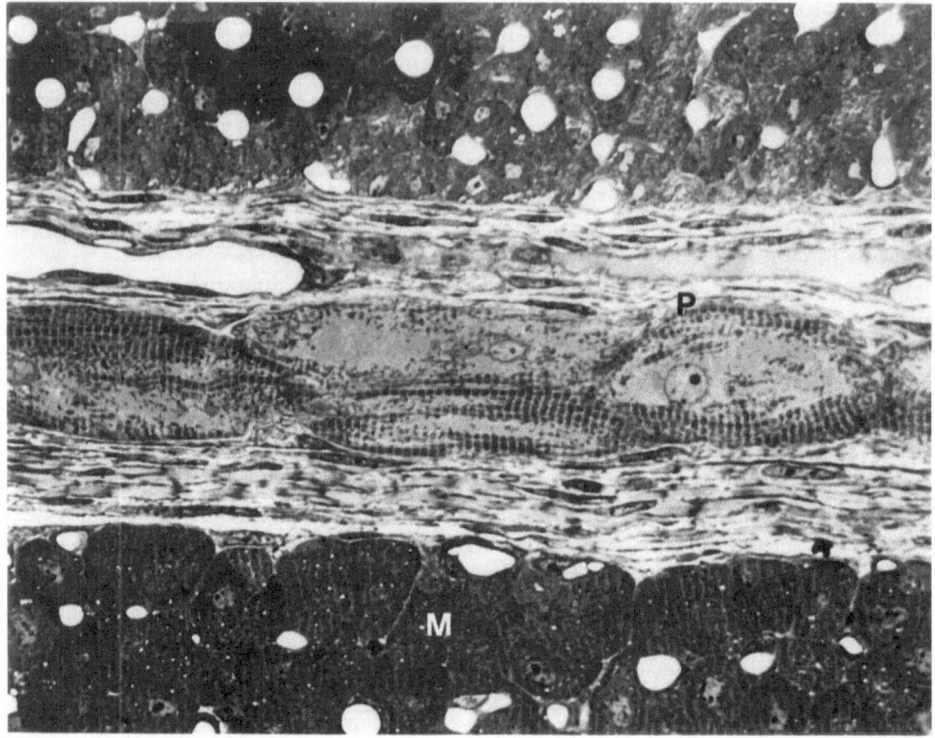

Fig. 39. An intramural Purkinje fibre (*P*) in sheep heart with a dense connective tissue sheath. Distended, empty capillaries are visible in the working myocardium (*M*) due to the perfusion-fixation procedure. Note the large size of the Purkinje cells and the scarcity of myofibrils compared to working myocardium. × 312. Plastic embedded, methylene blue

free walls (MYERBURG et al. 1972; NAGAO et al. 1981) and less significant in the upper two thirds of the septum (NAGAO et al. 1981).

a) Distribution

From the bifurcation of the penetrating AV bundle, left and right bundle branches enter the left and right ventricles respectively. The left bundle branch fans out into many fibres near the bifurcation but the right bundle branch extends through the septal myocardium to the subendocardium of the right ventricle before branching. Purkinje fibres are found within both ventricular walls (Fig. 39) but are most common in false tendons, the apical subendocardium and the base of papillary muscles. False tendons are composed of a central core of Purkinje fibres and sometimes a strand of working myocardium (BENCOSME et al. 1969) surrounded by a sheath of dense connective tissue which is in turn covered by endocardium. These structures are called false tendons because of certain similarities with chordae tendineae, such as paleness due to a high content of connective tissue, and because they traverse the ventricular cavity. Unlike chordae tendineae which join to valvular cusps, false tendons form connections between the lower septum and various parts of the apex, or from papillary muscle base to the ventricular free wall. In adult sheep heart, false tendons of varying thickness and up to 2 to 4 cm long are not uncommon (CANALE, personal observations). In the right ventricle a major part of the right bundle branch is carried onto the lower right ventricular free wall by the septomarginal band (sometimes called the "moderator band"). This band has been reported as being absent in dog hearts (ESMOND et al. 1963) but is present in most human and sheep hearts. The band differs from false tendons because it is thicker, consists mainly of working myocardium, and has a very consistent anatomical position. The septomarginal band also carries nerves supplying the ventricular wall (BOJSEN-MØLLER and TRANUM-JENSEN 1971).

b) Conduction Velocity

The conduction velocity of Purkinje fibres is 1.7 to 1.8 m/s in peripheral fibres (LAZZARA et al. 1974; NAGAO et al. 1981) and 2.1 to 2.5 m/s in false tendons (DRAPER and MYA-TU 1959; PRESSLER et al. 1982), compared to 0.4 to 0.9 m/sec (DRAPER and MYA-TU 1959) in working myocardium. Conduction velocity is proportional to fibre diameter (KATZ 1948; DRAPER and MYA-TU 1959; TOYOSHIMA et al. 1982) and this relationship seems particularly significant in the case of Purkinje fibres which are larger than working myocardial cells in many mammals including the human, dog, and rabbit (TRUEX 1961; DOWD 1969).

Purkinje fibres are separated from surrounding myocardium by connective tissue (Fig. 39) which functions as an insulator (LATHROP and BAILEY 1977) and probably has a protective role as a mechanical support.

c) Purkinje Network

Purkinje strands within false tendons are highly interconnected (CANALE et al. 1983b). In sheep and cows where Purkinje fibres have a well-developed connective tissue sheath, injection of colloidal dye into this sheath enables demonstration of the extensiveness of the subendocardial network of this system (CARDWELL and ABRAMSON 1931; HAYASHI 1962). Interconnections of Purkinje fibres have also been demonstrated by scanning electron microscopy (CANALE et al. 1983b; SHIMADA et al. 1983) (Fig. 40).

2. Species Differences

Unlike nodal tissue, the morphology of Purkinje fibres shows great variation between species. Species have been grouped according to the ease of differentiation from working myocardium, cell size and the arrangement of Purkinje fibres in false tendons (TRUEX and SMYTHE 1965a; SOMMER and JOHNSON 1970). There is a morphological gradation from highly-differentiated Purkinje cells forming strands of two or more cells wide (group I) to poorly-differentiated Purkinje

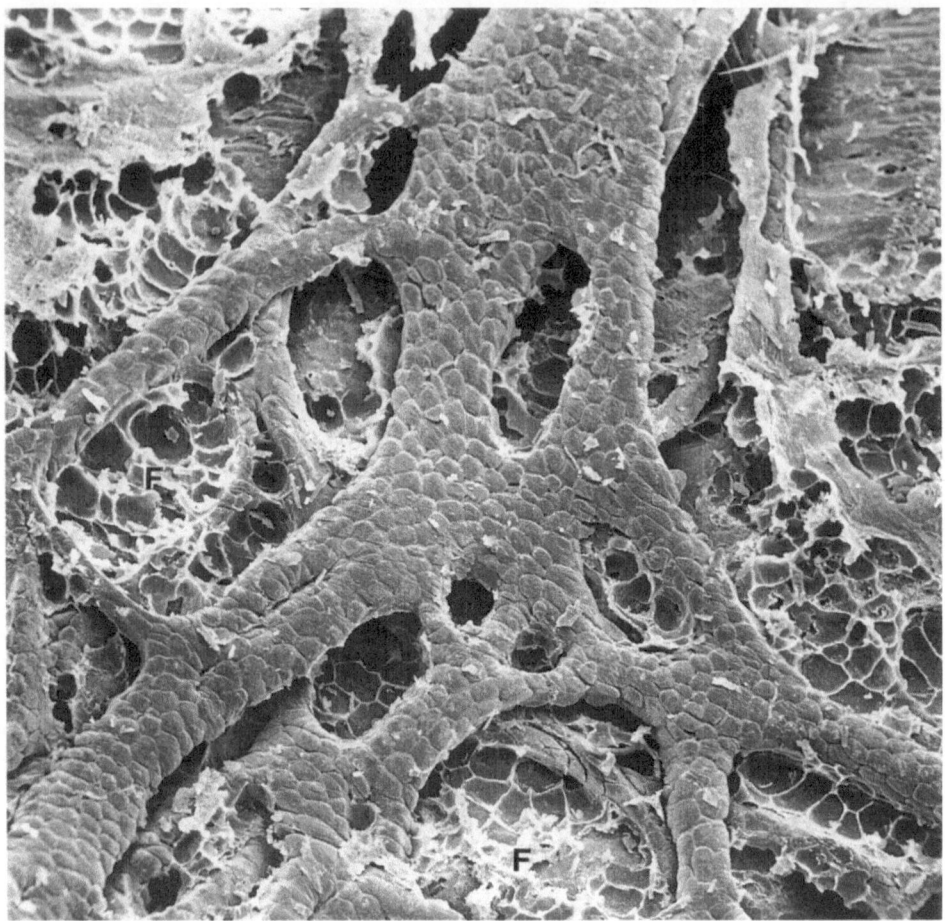

Fig. 40. The subendocardial network of Purkinje fibres, viewed after hydrolytic removal of connective tissue. The fibres are contracted and clefts on their surface do not represent cell boundaries. Note the extensive interconnections. The remnants of fat-cells (*F*) can be seen between the branching fibres. × 96

cells forming strands one cell wide (group III). Birds form a fourth group which does not fit the pattern observed in mammals. Their Purkinje fibres are well differentiated from working myocardium but form strands only one cell thick (SOMMER and JOHNSON 1970).

In the first group which includes the whale and species of the class Artiodactyla (sheep and pigs), Purkinje fibres are clearly differentiated from the working myocardium (MUIR 1957b; CAESAR et al. 1958; RHODIN et al. 1961; HAYASHI 1962; TRUEX and SMYTHE 1965a; HIRAKOW 1966; SOMMER and JOHNSON 1968a, 1970; VIRÁGH and CHALLICE 1973a; NÚÑEZ-DÚRAN 1981). Purkinje fibres in this group are much larger than the working myocardial fibres, and possess a few peripherally-placed myofibrils, large central areas of glycogen and one or two round nuclei (Figs. 39, 41).

In the second group which includes man, monkey, gorilla, dog, and cat, the Purkinje cells are less well differentiated from the working myocardium, but are still distinguishable histologically (TRUEX and SMYTHE 1965a). Purkinje cells are larger than the working myocardial cells, contain more glycogen (DRAPER and MYA-TU 1959) and fewer myofibrils, but the differences are less striking than in the first group. Purkinje strands in false tendons of these species may be one or more cells wide (SOMMER and JOHNSON 1970).

In the third group comprising the mouse, rat, and rabbit, Purkinje cells are very similar in appearance to large cells of the working myocardium and their identification is unreliable by routine light microscopy (SOMMER and JOHNSON 1968a; KIM and BABA 1971). However, mouse Purkinje fibres can be distinguished after PAS staining (VIRÁGH and CHALLICE 1982). In these animals the Purkinje strand is characteristically one cell thick (SOMMER and JOHNSON 1970).

3. Ultrastructure of Purkinje Fibres

a) Purkinje Type I, II and III

Myofibrils are generally sparse in Pukinje cells compared to working myocardial cells, but with increasing distance from the AV bundle there is a gradual increase in the density of myofibrils and a change in cell size. Three forms of Purkinje cells, type I, II, and III, have been described based on these gradual morphological changes (VIRÁGH and CHALLICE 1973a). Purkinje type I cells are found in the region of the bifurcation of the AV bundle and in the proximal parts of the bundle branches. They are smaller and possess a smaller proportion of myofibrils than the Purkinje type II cells with which they merge. Purkinje type II cells are the most abundant type of Purkinje cell in the ventricles. These form the extensive subendocardial network and false tendons, and because of their accessibility are most commonly used for study. They are the largest cells in the myocardium (TRUEX 1961), but steadily decrease in size and increase in myofibrillar content close to the Purkinje-working myocardial cell junction (VIRÁGH and CHALLICE 1973a). At this junction, a transitional form of Purkinje cell is observed (MARTINEZ-PALOMO et al. 1970), named the Purkinje type III cell (Fig. 42), which is morphologically similar to working myocardial cells, but is slightly larger, has fewer mitochondria and lacks T tubules (VIRÁGH and CHALLICE 1973a; MARTINEZ-PALOMO et al. 1970).

b) T System

The absence of a T system in conduction cells of a variety of animals has been demonstrated using extracellular markers (SOMMER and JOHNSON 1968a, b, 1979, 1980) and with routine electron microscopy (KAWAMURA 1961b; ARMIGER et al. 1979; BENCOSME et al. 1969; KIM and BABA 1971; VIRÁGH and PORTE 1973a; ARLUK and RHODIN 1974; MOCHET et al. 1975). However, it is not certain that the T system is completely absent in all Purkinje cells. AYETTEY and NAVARATNAM (1978) and PAGE (1967) observed an attenuated T system

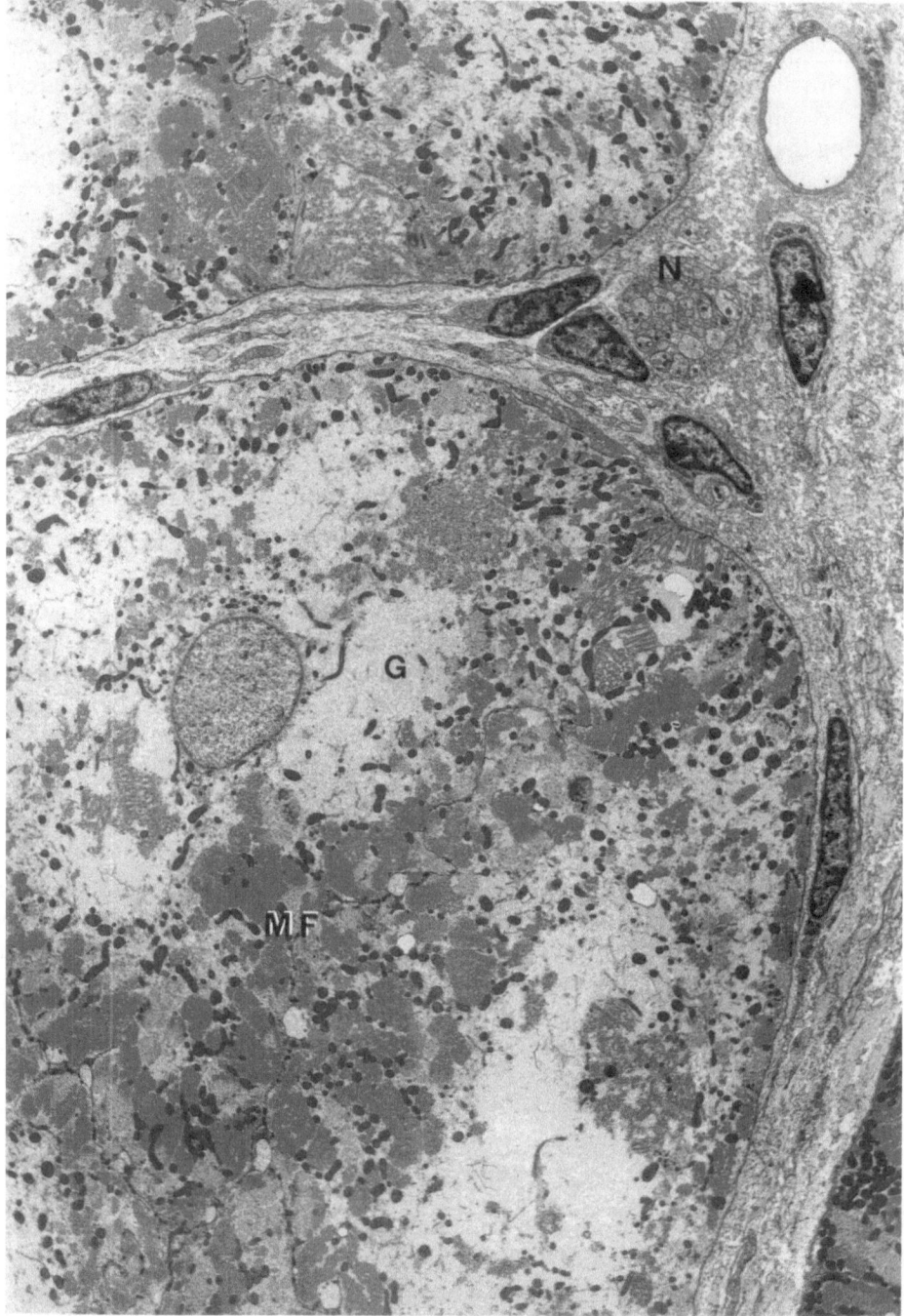

Fig. 41. Cross-section of Purkinje fibres showing central glycogen-filled region (*G*), peripheral myofibrils (*MF*), nerve bundles (*N*). × 3250

in rat and cat Purkinje cells which were morphologically very similar to large working myocardial cells. The presence of a T system may be related to an increasing structural similarity to working myocardium. In an extensive comparative study of myocardial cell membranes of vertebrates, HOWSE et al. (1970) observed that a T system developed in working myocardial cells where myofibrils were centrally located or distributed evenly within the cell, but not in cells containing myofibrils located primarily in the periphery.

c) Purkinje Myofibrillar Arrangement

In those Purkinje fibres which contain few myofibrils, such as in whale, sheep, and cattle, there is an apparent disorganisation of much of this system (MUIR 1957b; CAESAR et al. 1958; ARLUCK and RHODIN 1974; RHODIN et al. 1961; VIRÁGH and CHALLICE 1969; THORNELL 1973b; THORNELL et al. 1976). Similar disorganisation, but of a lesser degree occurs in dog (KAWAMURA 1961a; MAEKAWA et al. 1967; BENCOSME et al. 1969; MARTINEZ-PALOMO et al. 1970; HAYASHI 1971; ARMIGER et al. 1979), monkey (VIRÁGH and CHALLICE 1969), rabbit (MAEKAWA et al. 1967; THORNELL 1973a), cat (BENCOSME et al. 1969; THORNELL et al. 1976), and guinea-pig (KIM and BABA 1971), where the Purkinje cells contain a relatively greater amount of contractile material. This disorganisation may be more apparent than real. Myofibrils with a spiral course are present in avian Purkinje cells (BOGUSCH 1974). When stretched, Purkinje myofibrils appear aligned in the direction of component tension forces which form many intersecting lines at branch points (THORNELL et al. 1976). In view of the highly-branched nature of both false tendons and subendocardial Purkinje fibres, the three dimensional arrangement of myofibrils along intersecting lines of stress may not be clearly appreciated in thin sections used for transmission electron microscopy.

It has been proposed that the major function of Purkinje myofibrils is to act as passive cytoskeletal components (THORNELL et al. 1976; THORNELL and ERIKSSON 1981). However, contraction of Purkinje cells has been demonstrated conclusively in culture (CANALE et al. 1983a) and in cells fixed under conditions of high extracellular calcium (CANALE et al. 1983b). An interesting feature of false tendons is the formation of arched configurations by the myofibrils of Purkinje cells when contracted (Fig. 43). Contraction and formation of myofibrillar arches in Purkinje fibres may be involved in withstanding the sudden increase in intraventricular and subendocardial pressures during systole.

d) Alterations of the Myofibrillar System

The myofibrillar system in Purkinje fibres shows some morphologically distinguishing features (Figs. 44, 45). These include myofilament-polyribosome complexes (THORNELL 1972, 1973a) and aberrant Z discs (VIRÁGH and CHALLICE 1969; THORNELL 1973b; OLIPHANT and LOEWEN 1976), which may be unusually shaped (VIRÁGH and CHALLICE 1969), abnormally wide (MARTINEZ-PALOMO et al. 1970; THORNELL 1973b), discontinuous or completely missing (THORNELL 1973a, b; ARMIGER et al. 1979). Wide Z discs, although present in working

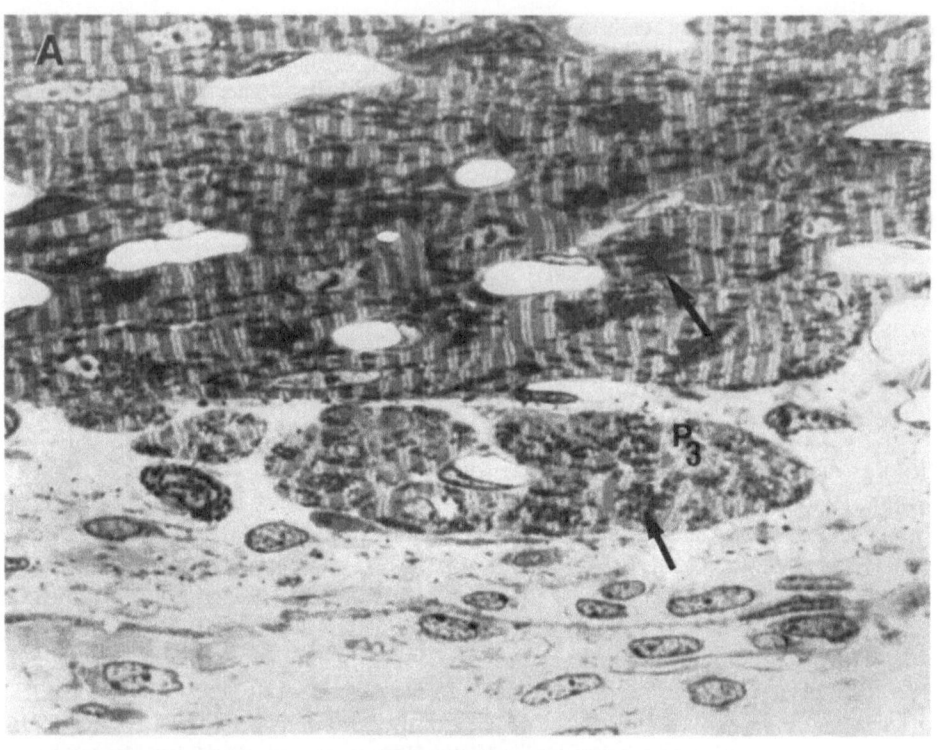

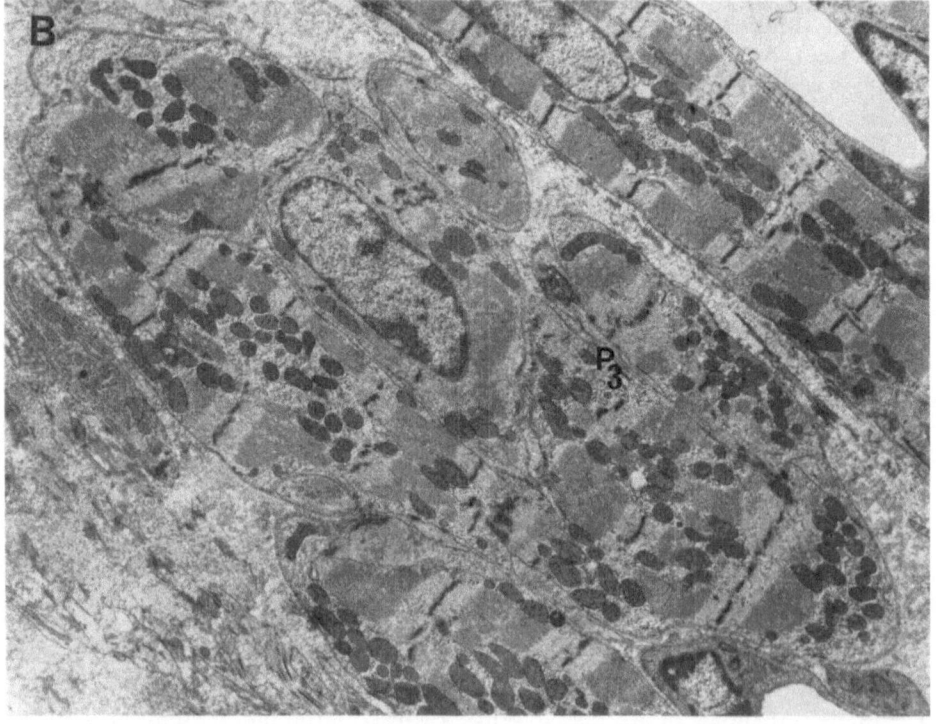

myocardium are much less common than in Purkinje cells (MUNNELL and GETTY 1968a; FAWCETT 1968).

Actin and myosin filaments are sometimes observed grouped together but not into sarcomeres. These groups may range in organisation from tightly packed bunches of parallel thick myosin filaments of normal length to a haphazardly arranged mass of randomly oriented filaments (Fig. 45a) (VIRÁGH and CHALLICE 1969; THORNELL 1972; OLIPHANT and LOEWEN 1976). Myofilaments and polyribosomes form complexes (THORNELL 1972) which are unique to Purkinje fibres and are thought to be an indication of ongoing synthesis and degradation of the myofibrillar proteins (THORNELL 1972, 1973a, b; THORNELL and ERIKSSON 1981). The suggestion that these complexes indicate arrest of myofibrillogenesis at an embryonal stage (OLIPHANT and LOEWEN 1976) has not been supported (THORNELL and ERIKSSON 1981). Furthermore, studies of the development of Purkinje fibres suggest early maturity of the Purkinje myofibrillar system (FORSGREN and THORNELL 1981; FORSGREN 1985).

e) Myosin Subunit Composition

Myosin in adult ventricular working myocardium contains two light chains with molecular weights of approximately 25000 and 18000, while Purkinje cells have an additional type of intermediate molecular weight (THORNELL et al. 1978; SAITO et al. 1981; THORNELL and ERIKSSON 1981; THORNELL and FORSGREN 1982). This extra light chain is indistinguishable from a light chain in fetal myocardium and adult atrial myocardium (WHALEN et al. 1982). The myosin heavy chain also exists in multiple isoforms in adult heart and antibodies to the heavy chain of slow tonic muscle myosin have recently been found to react specifically with muscle cells of the adult chicken heart conduction system (GONZALEZ-SANCHEZ and BADER 1985). For further information on the heterogeneity of myosin subunit types in the developing and adult heart, the reader is directed to SYROVÝ (1979), CUMMINS et al. (1980), PRICE et al. (1980), SARTORE et al. (1981), WHALEN et al. (1982), CHIZZONITE and ZAK (1984) and SWEENEY et al. (1984) (see B.II.6d).

f) Myosin ATPase Activity

Differences between working myocardium and Purkinje cells in myosin light-chain composition have been correlated with differences in their ATPase activity (THORNELL et al. 1978; THORNELL and ERIKSSON 1981). A correlation between myosin ATPase activity and shortening velocity of contraction (CAREY et al. 1979; SCHUER and BHAN 1979) suggests that the contractile properties of Purkinje fibres differ from that of working myocardium. Furthermore, although

Fig. 42. A Purkinje type III cells (P_3) adjacent to working myocardium. The mitochondria (*arrowed*) appear more densely crowded in the working myocardium. × 1000. Plastic embedded, methylene blue. B Transmission electron micrograph of part of previous figure, less than 2 μm deeper in section. The Purkinje III cells (P_3) are in different orientation to the working myocardium and more irregular in shape, with interdigitations. × 5600

Fig. 43. Purkinje cell from sheep false tendon, fixed in a contracted state to demonstrate myofibrillar arches (*arrowed*). Ridges are formed on the surface due to fibre shortening (*arrowhead*). Other myofibrils (*asterisks*) appear to form wavy patterns in a different orientation. × 3640

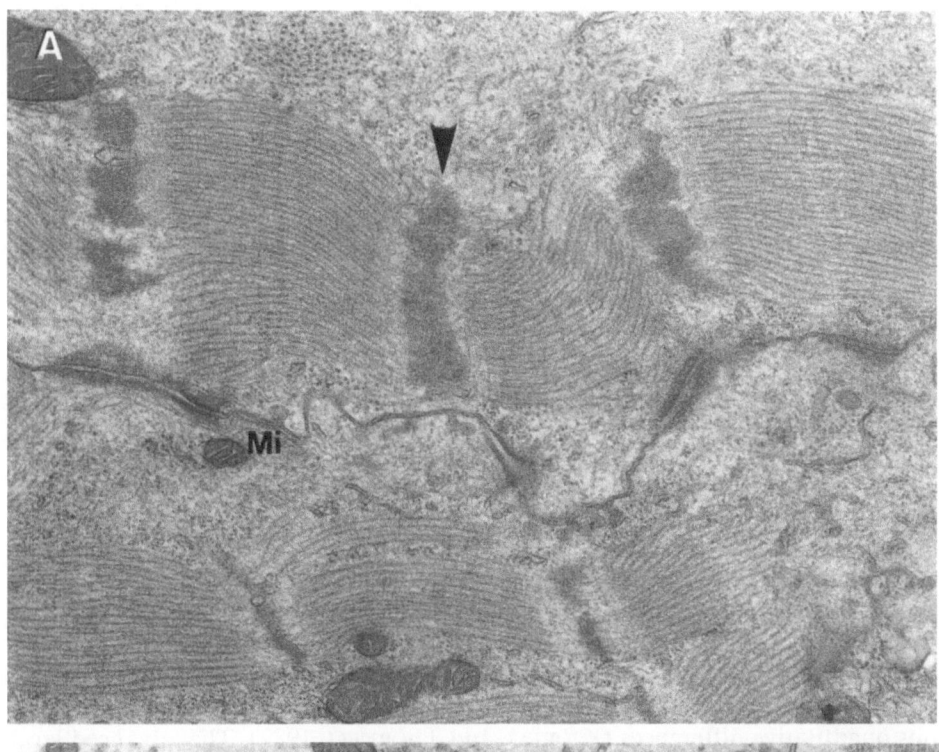

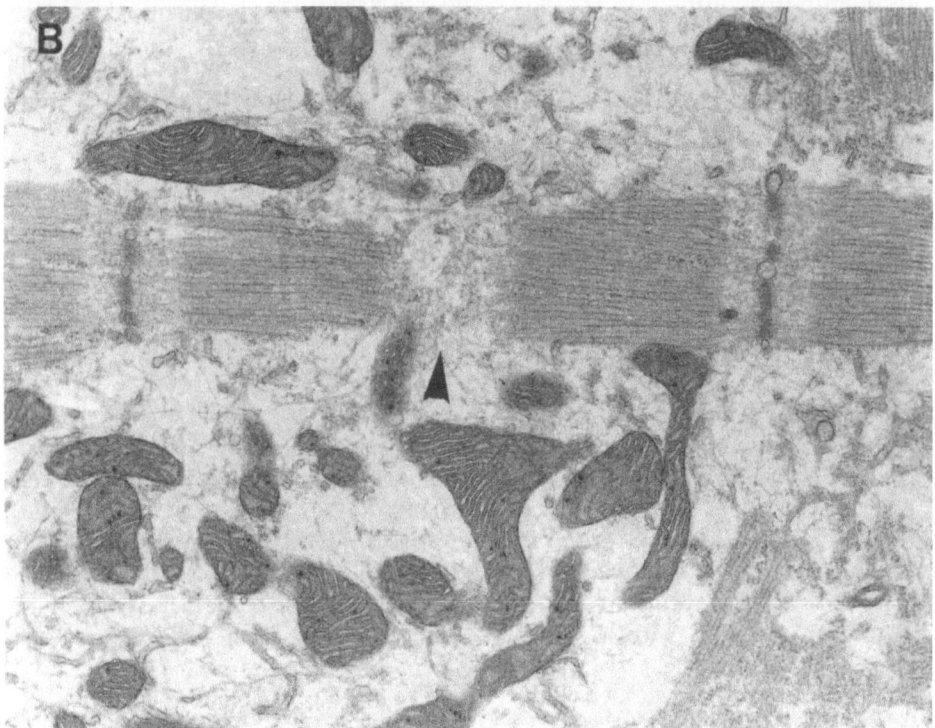

Fig. 44. A Widened Z discs (*arrowhead*) are a frequent finding in Purkinje cells. Small mitochondria (*Mi*) with only one or two cristae are common. ×24000. **B** Myofibrillar imperfections, such as missing Z disc material (*arrowhead*), are also common in Purkinje cells. ×21000

SAITO et al. (1981) found similar ATPase activities related to different myosin subunits in Purkinje cells, histochemical evidence suggests that there is a heterogeneity in myosin ATPase activity within the conduction system and within the working myocardium (THORNELL and FORSGREN 1982).

g) Intermediate Filaments

Intermediate filaments have a cytoskeletal function in all cardiac muscle cells (see B.II.7a), but this function is particularly well developed in Purkinje fibres (THORNELL and ERIKSSON 1981) where their predominance (THORNELL et al. 1978) (Fig. 45b) has made possible their purification (ERIKSSON et al. 1978; ERIKSSON and THORNELL 1979; ERIKSSON et al. 1979, 1980).

The number of intermediate filaments varies according to the species. They are most abundant in Purkinje fibres of Artiodactyla (Fig. 46) and birds, less prevalent in man and cat, and relatively few in rat and guinea-pig (ERIKSSON et al. 1979).

A number of charge-modified forms of the Purkinje intermediate filament protein, skeletin (Purkinje fibre desmin), have been reported (THORNELL and ERIKSSON 1981; KJÖRELL and THORNELL 1982). This is in contrast to skeletin in working myocardium which appears in two forms only, and suggests functional differences in the Purkinje intermediate filament system besides the obvious quantitative differences (KJÖRELL and THORNELL 1982). The advanced level of development of the intermediate filament cytoskeleton in Purkinje fibres may be related to a number of factors including the scarcity of myofibrils, the large size of the cells and the mechanical stresses on the cells (THORNELL and ERIKSSON 1981).

Similar to working myocardium (see B.II.7a), Purkinje fibre desmin is associated with Z discs. Furthermore, structural proteins such as desmoplakins, spectrin, and vinculin also interact with Purkinje fibre desmin (THORNELL et al. 1985). Vinculin is mainly located at the Purkinje sarcolemma facing connective tissue, but lighter staining is also present at junctional membranes, the difference reflecting the presence or absence of a basal lamina (THORNELL et al. 1985).

h) Leptofibrils

Leptofibrils consist of well-aligned thin filaments and show a short periodicity of Z disc-like material (Fig. 47). They have been observed in Purkinje cells of sheep (CAESAR et al. 1958), cow (THORNELL 1973b), dog (KAWAMURA 1961a), birds (HIRAKOW 1966; SCOTT 1971; BOGUSCH 1975), rabbits and monkeys (VIRÁGH and CHALLICE 1969), and cats (THORNELL et al. 1976). They have also been reported in working myocardium and skeletal muscle (see B.II.7c). They are often continuous with myofibrils coming together from various directions

Fig. 45. A Myofilament-polyribosome complexes in Purkinje cells. The ribosomes are electron-dense granules associated with myosin filaments lacking a sarcomeric arrangement. × 21 000. B A network of intermediate-sized filaments is associated with the glycogen-filled region. Ribosomes appear more electron dense than glycogen particles. × 35 000

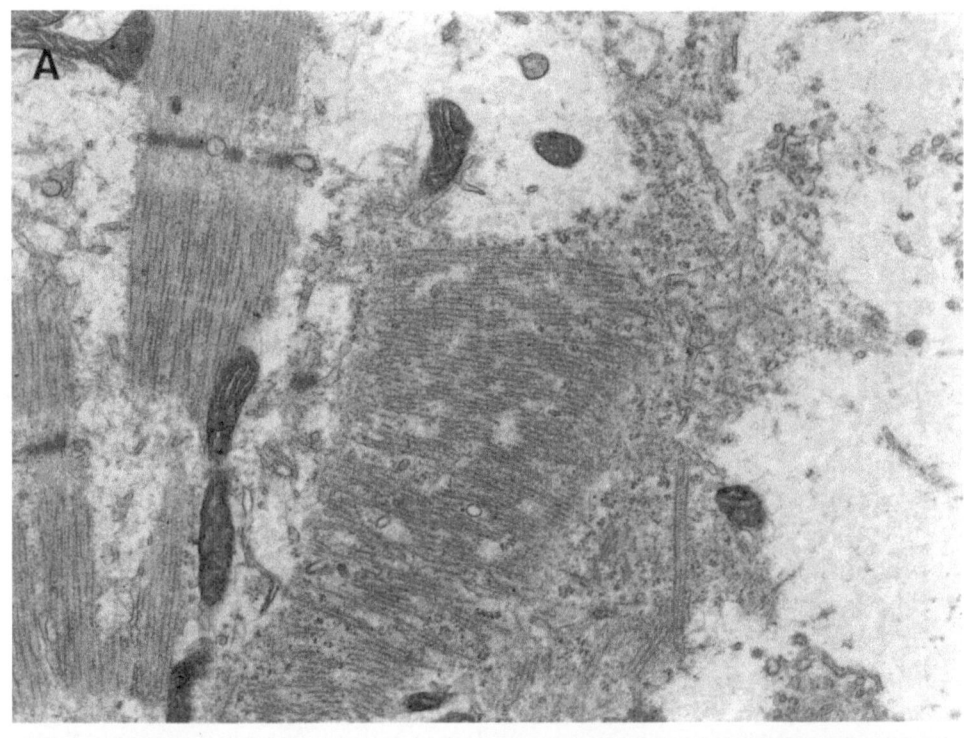

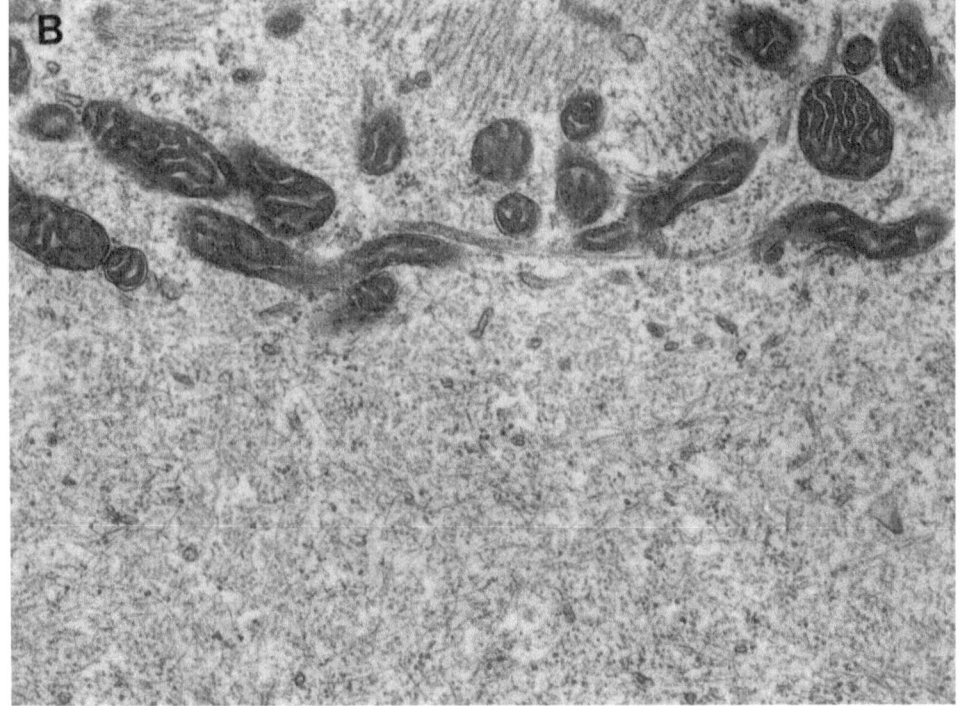

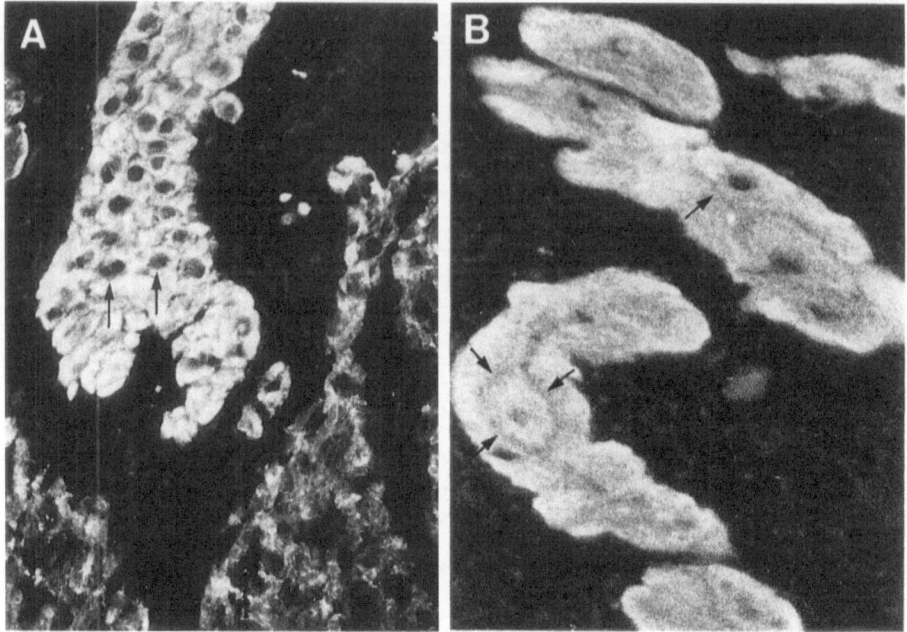

Fig. 46 A, B. Localisation of skeletin (intermediate) filaments in bovine fetal myocardium by the indirect immunofluorescence technique. **A** Fetal calf (16 cm CRL). Only Purkinje cells show an intense fluorescence while the working myocardium (*bottom*) shows a much weaker fluorescence. Nuclei are non-fluorescent (*arrows*). × 250. **B** Fetal calf (80 cm CRL). Where the bundle cells face each other the fluorescence is weak (*arrows*) compared to the central parts of the cytoplasm. × 400. (From FORSGREN et al. 1980)

or interposed between the sarcolemma and myofibrils (THORNELL et al. 1976). A detailed understanding of their function is lacking, but they probably have a passive mechanical role supplying rigidity and anchoring myofibrils to each other and to the sarcolemma (THORNELL et al. 1976).

i) Intercellular Junctions

Purkinje cells are joined together by gap junctions, desmosomes and fasciae adherentes with the same ultrastructure as in working myocardium (see B.II.4). In most mammalian species, the end-to-end junctions of Purkinje fibres are similar to intercalated discs of working myocardium but are less interdigitated and have longer gap junctions (VIRÁGH and CHALLICE 1973a; KIM and BABA 1971; KAWAMURA and JAMES 1971; HAYASHI 1971). The end-to-end junctions of Purkinje cells of species of Artiodactyla differ markedly from the intercalated discs of working myocardium (Fig. 48). This is apparently due to the large size of the Purkinje cells and the sparsity of myofibrils compared to other species (KAWAMURA and JAMES 1971). Nonetheless, they have still been classified as intercalated discs (CAESAR et al. 1958; RHODIN et al. 1961; HAYASHI 1962;

MAEKAWA et al. 1967; NÚÑEZ-DÚRAN 1981). Purkinje cells also have many lateral connections (Fig. 48) many of which are gap junctions (VIRÁGH and CHALLICE 1973a; KAWAMURA and JAMES 1971; NÚÑEZ-DÚRAN 1981), but with an area only half that of end-to-end junctions (NÚÑEZ-DÚRAN 1981). Gap junctions make up more of the intercellular surface area than either desmosomes or fasciae adherentes (MOBLEY and PAGE 1972).

IV. Histochemistry of Mammalian Conduction Systems

a) Glycogen

The conduction system has a high content of glycogen compared to the working myocardium. This has been established biochemically (YAMAZAKI 1929; JEDEIKIN 1964), by enzyme histochemistry (SCHIEBLER 1955; PANNESE 1955; HARA 1967; OTSUKA et al. 1967) and ultrastructurally (MUIR 1957b; CAESAR et al. 1958; RHODIN et al. 1961; KIM and BABA 1971; ARLUK and RHODIN 1974; THORNELL 1974a, b, c). In routine electron microscopy three kinds of particles can be distinguished, each about 20 nm in diameter, but cytochemical studies suggest that the most electron-dense type is ribosomal and the other two are glycogen granules (THORNELL 1974a, b), composed of subunits approximately 3 nm in diameter (THORNELL 1974c; RYBICKA 1981).

The glycogen granules are in the form of a protein-glycogen complex known as a glycosome, the electron density of which is dependent on the preservation of the protein component since glycogen itself has no affinity for osmium, uranium or lead. However, if the acid-sensitive protein component is removed, glycogen can still be demonstrated cytochemically (RYBICKA 1979, 1981). The adverse effects on "glycogen staining" of en bloc staining with uranyl acetate (VYE and FISCHMAN 1970) is due to the acidic action of uranyl acetate on the protein component of the glycosomes (RYBICKA 1979). Glycosomes may be associated with cellular structures, particularly intermediate filaments (THORNELL 1974a), in which case they are more stable to the action of acids (RYBICKA 1981).

The glycosomes of Purkinje cells are less soluble in water than that of working myocardium (HARA 1967) and more stable to acids (SCHIEBLER 1953). The significance or function of large quantities of glycogen in conduction cells remains obscure, but there is a high activity of glycolytic and gluconeogenetic enzymes (see Section (c) Metabolic Enzymes).

b) Lipids

Less lipid is normally present in the conduction system than in working myocardium (ROBB 1965; SNIJDER and MEIJER 1970). Because connective tissue lipid also contributes to the difference, an exact comparision is difficult (MALLOV et al. 1953).

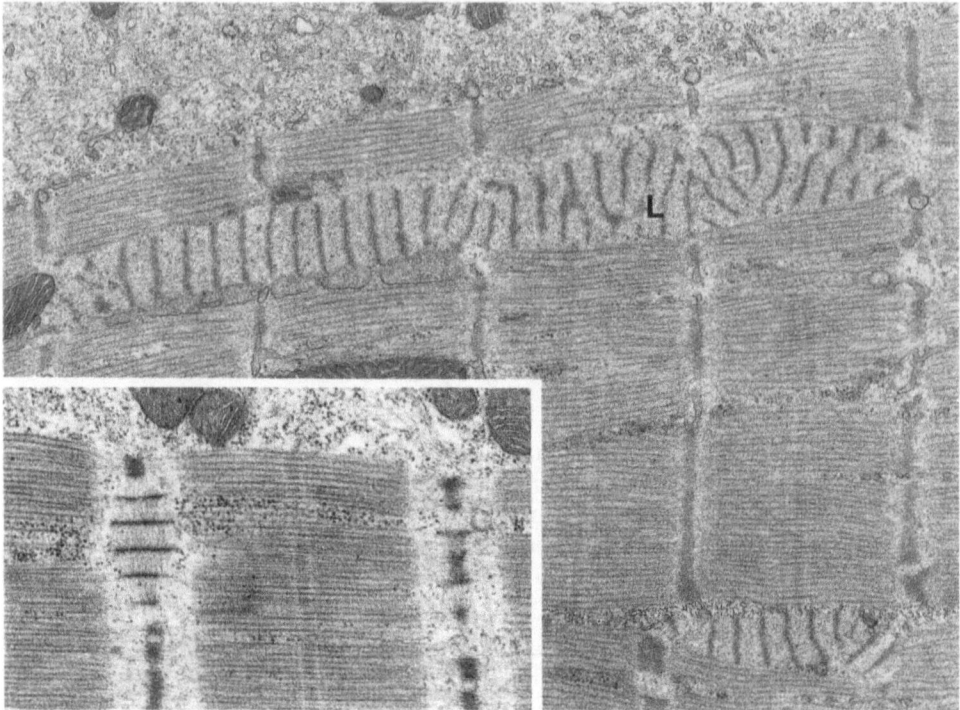

Fig. 47. Leptofibrils (*L*), in continuity with myofibrils (× 20000), and in Z discs (insert × 22600)

c) Metabolic Enzymes

Compared to working myocardium, Purkinje cells have similar or higher activities of glycolytic and gluconeogenetic enzymes and a lower activity of oxidative enzymes, suggesting a better capacity for anaerobic metabolism (ROBB 1965; OPIE 1969; SNIJDER and MEIJER 1970; MEIJER and DE VRIES 1978; KAWAMURA and JAMES 1978; MUKUDAI et al. 1978; OTSUKA et al. 1979; HIGUCHI et al. 1979; ELIAS et al. 1980).

In contrast to working myocardium the enzyme pattern within conduction cells shows both species and regional variation. Compared with the working myocardium, the activity of the glycolytic enzymes hexokinase and phosphofructokinase is much higher in the conduction system of cattle and pigs (MEIJER and DE VRIES 1978) but similar or slightly higher in that of dogs (HENRY and

Fig. 48. A End-to-end junctions in sheep Purkinje cells are less tortuous and simpler than intercalated discs of working myocardium. Desmosomes (*D*), fasciae adherentes (*F*) and nexuses (not illustrated) are common. Nearly all the nonspecialized junctional membranes are intimately apposed. Subsarcolemmal coupling (*arrowhead*). × 20000. **B** Junctions between laterally apposed membranes are comprised mostly of desmosomes (*D*) and nexuses (*N*) but also include the fascia adherens type of junction (*F*). × 20000

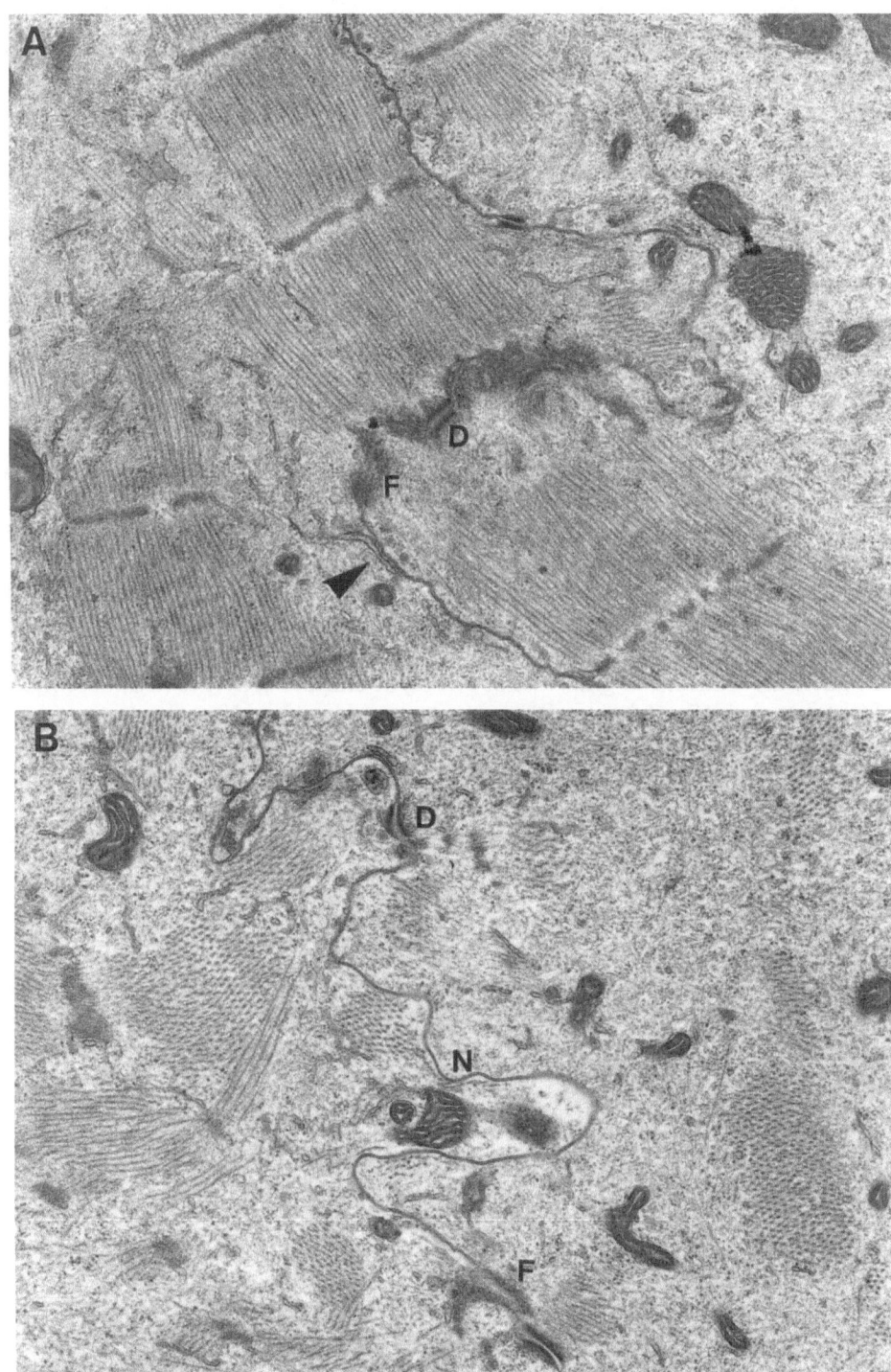

LOWRY 1983). Lactate dehydrogenase, is 40% lower in dog conduction fibres compared with working myocardium (HENRY and LOWRY 1983), similar in man (ELIAS et al. 1980), and much higher in cattle (MEIJER and DE VRIES 1978). Acid phosphatase activity also varies among species, being high in the dog and monkey, less in rabbit, guinea-pig, rat, and mouse, and low in cow and pig, while it is uniformly low in the working myocardium of all these species (MUKUDAI 1980). Subendocardial Purkinje cells show greater glycolytic enzyme activity than their intramural counterparts (HENRY and LOWRY 1983), but no histochemical differences have been observed between the AV node, AV bundle, left or right bundle branches (OHYUMI 1975; ELIAS et al. 1980).

Although conduction cells are less active in oxidative metabolism than working myocardium (YAMAZAKI 1930 a, b; MURRAY 1954), they compare favourably with other metabolically active tissues such as brain and slow (red) skeletal muscle, but the significance of this is unknown (TSUYUGUCHI et al. 1978; HENRY and LOWRY 1983).

The high activities of glycolytic enzymes in conduction cells may be important during periods of ischaemia (OPIE 1969; SNIJDER and MEIJER 1970; HENRY and LOWRY 1983). The conduction system appears to be more resistant to the effects of prolonged ischaemia than the working myocardium (COFFMAN et al. 1960; FRIEDMAN et al. 1973). However, hearts from patients who die from ventricular fibrillation almost always display histochemical changes restricted to the conduction system (ELIAS et al. 1982).

There are transmural gradients of glycolytic enzyme activities (LUNDSGAARD-HANSEN et al. 1967) and glycogen concentration (JEDEIKIN 1964) in the ventricular walls and septum, which may be related more to the transmural distribution of the Purkinje system than to histochemical differences within the working myocardium.

Conduction cells can be distinguished from working myocardium on the basis of distinct patterns of histochemical reactions in many species (MEIJER and DE VRIES 1978; SNIJDER and MEIJER 1970; KAWAMURA and JAMES 1978; ELIAS et al. 1980) even when morphologically similar to working myocardium (OTSUKA et al. 1979). Phosphorylase activity is higher in the conduction system (OHYUMI 1975) and has been suggested as a general histochemical marker for conduction cells (PATHAK and GOYAL 1978). Conduction cells can also be distinguished from working myocardium by the activity of cholinesterases. The reader is referred to sections E.III.1 and E.III.3 where this is discussed in relation to innervation.

V. Non-Mammalian Conduction Systems

1. Anatomy of the Non-Mammalian Vertebrate Heart

In fish the heart is a simple S-shaped tube divided into four chambers. A well-developed sinus venosus receives venous blood. This empties into a single atrium, from which blood enters the ventricle. The ventricle pumps the blood into the conus arteriosus from where it enters the branchial arteries. In cyclostome fish, secondary heart-like organs, the caudal heart (GREENE 1900; VOGEL 1985), and portal vein heart (AUGUSTINSSON et al. 1956), are present. For a thorough review of the fine structure of the hearts of fish the reader is referred to YAMAUCHI (1980).

In amphibians a complete interatrial septum is present resulting in left and right atria. The right atrium receives systemic venous blood from the sinus venosus and the left atrium oxygenated blood from the pulmonary veins. Both atria empty into a single ventricle.

In reptiles there is an additional partial septation of the ventricle into left and right chambers. Only in crocodiles is the interventricular septation complete.

2. Lower Vertebrates

a) Physiology

Early physiological studies of the heart of lower vertebrates established the existence of a conduction delay between the sinus, atrium (or atria), and ventricle and the need for myocardial continuity for successful conduction from one chamber of the heart to the next (GASKELL 1882, 1883). It was later shown that conduction delays occur within the junctional areas between the different heart chambers (sinus venosus, atria, ventricle, and conus) (GILSON 1942; INOUE 1959; LIBERTHSON et al. 1975; SINEVA et al. 1976). Electrocardiographic recordings from lower vertebrate hearts show the general pattern observed in mammals, having a P wave, QRS complex and T wave. In addition some species exhibit additional waves associated with contraction of the sinus venosus or conus (OETS 1950; TEBECIS 1967; cited in RANDALL 1970). Preferential pathways of conduction within the ventricle of the fish heart elicit contraction in the apex prior to other parts of the ventricle or conus (RANDALL 1970).

Studies using transmembrane microelectrode techniques demonstrate characteristic pacemaker potentials in cells of the SA junction (SAITO 1973; HUANG 1973), AV junction, and conoventricular junction (KANNO 1963; ALANIS et al. 1973). When these junctional cells are separated from the influence of atrial and ventricular myocardium, they undergo spontaneous depolarisations with a slow rate of rise (ALANIS et al. 1973). Furthermore, many sites capable of pacemaker function can be found in fish heart (McWILLIAM 1885a; KISCH 1948, cited in ROBB 1965 and RANDALL 1970). SA pacemaker cells, atrial cells, AV pacemaker cells, and ventricular cells differ in their responses to acetylcholine and adrenaline (KANNO 1963; SAITO 1973; HUANG 1973; ALANIS et al. 1973; MARTINEZ-PALOMO and ALANIS 1980). Electrophysiological evidence of an organised conduction system has been found in the African lungfish but histological examination has failed to reveal a morphologically specialised system (ARBEL et al. 1977).

On the basis of the above, there is ample evidence for the existence of preferential pathways of conduction and heterogeneity of cell membrane properties including pacemaker function in the myocardium of lower vertebrates (RAN-

DALL 1970; KANNO 1963). These are sufficiently well organized to be considered parts of a conduction system functionally analogous to the mammalian system. An important question is whether the conduction system of lower vertebrates bears any phylogenetic relationship to that of birds and mammals. The similarity of the action potentials of lower vertebrate and mammalian conduction cells as well as the similarity of electrocardiographs suggest that this may be the case (RANDALL 1970; ALANIS et al. 1973; HUANG 1973).

b) Histology

The distribution of presumed specialised conduction tissue is similar in fish, amphibian and reptilian hearts. Rings of cells have been described in the SA junction, AV junction, and sometimes in the bulboventricular junction in hearts of many lower vertebrates and are considered to constitute part of a primitive conduction system (ROBB 1965). These cells are histologically distinct from atrial and ventricular myocytes because of their small size, pale cytoplasm, and bulging round nucleus (GASKELL 1882, 1883; KEITH and FLACK 1907; MACKENZIE 1910, 1913; ROBB 1953, 1965; MORI 1955; PRAKASH 1960; SAHAI and CHAWLA 1967; NAIR 1970; KUMAR 1973, 1974; MOHSEN et al. 1976; THOMAS 1976).

The distal part of the AV junction is a short tube-like opening which leads to the ventricular chamber. This has been described as the AV funnel, canal or channel. The dorsal part of the AV canal forms a thicker connection to the ventricle than other parts in reptilian and amphibian hearts (KEITH and FLACK 1907). The myocytes making up this part of the AV canal in lower vertebrates exhibit some histological differences from the atrial and ventricular myocardium (McWILLIAM 1885b; MACKENZIE 1910; ROBB 1953, 1965; PRAKASH 1954, 1960). Typically these cells are larger than the ventricular myocytes, closely assembled, less densely staining and with fewer striations. They have been described as specialised conducting cells and are thought to form the pathway for conduction to the ventricle (KEITH and FLACK 1907; MACKENZIE 1910; ROBB 1953, 1965; PRAKASH 1954, 1960).

c) Ultrastructure

Ultrastructural investigations are few and do not support the existence of complete rings of specialised tissue. They indicate instead that only parts of the rings in the SA and AV junctions of lower vertebrates contain morphologically specialised tissue, and therefore describe the tissue as "nodal" although it is more extensive than in mammals (PRAKASH 1960; RUSKA 1965; YAMAUCHI and BURNSTOCK 1968b; YAMAUCHI 1969; YAMAUCHI et al. 1973; ABRAHAM 1982). The SA and AV junctional regions of fish heart are well-innervated. Although SA nodal tissue has been identified in trout and loach heart (YAMAUCHI and BURNSTOCK 1968b; YAMAUCHI 1969; YAMAUCHI et al. 1973), no morphological distinction between the myocytes in the SA junction and the myocytes of the sinus venosus or atrium, or muscular connection between the atrium and ventricle has been found in the heart of the plaice (SANTER and COBB 1972). Whether this is due to species differences has not been explored.

The SA nodal cells of trout and loach heart are well delineated from atrial myocytes by loose connective tissue and form an incomplete ring at the base of the venous valve. Each nodal cell forms an intimate contact with at least one axon, possesses relatively few myofibrils and has a smaller cell diameter than the average atrial myocyte (YAMAUCHI and BURNSTOCK 1968a, b; YAMAU-CHI et al. 1973). A group of cells in the SA ring of amphibian heart (RUSKA 1965) and reptilian heart (ABRAHAM 1982) has been similarly described. In addition the SA nodal cells of trout and loach heart are devoid of membrane-bound atrial-specific granules (150–300 nm diameter) which are otherwise common in atrial myocardium (YAMAUCHI and BURNSTOCK 1968a, b; YAMAUCHI et al. 1973).

Within the AV ring of the axolotl and turtle heart there are myocytes of similar morphology to the nodal cells of mammalian hearts. Compared with atrial and ventricular myocytes they are small, contain few myofibrils and have a round or oval nucleus instead of a cigar-shaped one (ALANIS et al. 1973). They also have few junctional contacts, few small mitochondria, and variable amounts of glycogen (MARTINEZ-PALOMO and ALANIS 1980).

In the AV canal of bullfrog heart myocardial cell bundles run within the main neural trunk from BIDDER's ganglion (see E.V.2b, Figs. 64, 65). These myocardial cells appear larger than those of ventricular working myocardium (BALUK and FUJIWARA, personal communication) and may be part of the AV conduction pathway described by KANNO (1963). In toad heart, conduction across the AV junction remains normal as long as two small localised areas of the AV canal remain intact. Both regions, one dorsal and one ventral, are each located at the base of an AV valve near BIDDER's ganglia. The dorsal pathway is more important and damage to it usually results in AV block (KANNO 1963).

At present, the true level of development of the conduction system and its distribution in the hearts of non-mammalian vertebrates is largely uncertain. The phylogenetic relationships are barely better understood than eighty years ago. This situation is largely due to the lack of application of ultrastructural techniques to the study of non-mammalian heart conduction systems. It is also surprising, in view of the dissimilarity in metabolic enzyme histochemistry between mammalian working myocardium and conduction tissue, that little literature pertaining to the distinction of conduction tissue in lower vertebrates by histochemistry is available. – Regarding the older literature on the conduction system in lower vertebrates the contribution by BENNINGHOFF (1930) in this handbook series should be consulted.

3. Birds

a) Sinoatrial Node

An SA node similar to that in mammalian hearts has been described in birds (DAVIES 1930a, b; PRAKASH 1956; MURAKAMI et al. 1981) although it is not as distinct and its borders not as clearly delineated (TRUEX 1961; TRUEX and SMYTHE 1965a). This may be due to comparatively less connective tissue in the avian conduction system.

The fine structure of the SA node of birds has apparently not been investigated, however a detailed histological study has been made (DAVIES 1930a). The SA node is located in the base of the right venous valve and extends from just above the orifice of the inferior vena cava, cranially to just below the opening of the superior vena cava. The thickest part of the node is midway between these points (DAVIES 1930a). This is in contrast to the mammalian heart where the SA node is typically near the junction of the superior vena cava and the right atrium.

The myocytes of the SA node are pale-staining with clear central regions containing one or two round nuclei (DAVIES 1930a). They form interlacing fibres, separated by connective tissue containing nerve fibres. No nerve cell bodies are present within the SA node but they may be observed in the connective tissue surrounding it (DAVIES 1930a). Most of the SA nodal cells are larger than the atrial working myocardial cells. Myocytes of the SA node are in continuity with working atrial myocytes and atrial Purkinje-like cells (DAVIES 1930a; MURAKAMI et al. 1981).

In contrast to mammals, the atria of avian hearts possess fibres similar in morphology and size to ventricular Purkinje fibres (HOLMES 1923; DAVIES 1930a; SCOTT 1971; BOGUSCH 1974). The atrial Purkinje-like and ventricular Purkinje cells are structurally similar, and both form a periarterial and a subendocardial network (DAVIES 1930a; SCOTT 1971; BOGUSCH 1974).

b) Atrioventricular Node

Many ultrastructural studies of the avian conduction system make no mention of an AV node (see TRUEX 1961; TRUEX and SMYTHE 1965a; HIRAKOW 1966; MIZUHIRA et al. 1967; GOSSRAU 1968; SCOTT 1971; AKESTER 1981). However, an AV node has been described in a number of birds (DAVIES 1930a, b) as an ovoid structure embedded in connective tissue in the lower posterior part of the interatrial septum but not in direct communication with the Purkinje-like cells of the atria (DAVIES 1930a; MURAKAMI et al. 1981). The cells in the upper part of the AV node are small and form strands separated by connective tissue. In the lower part the myocytes are more compactly arranged and continuous with the AV bundle (DAVIES 1930a).

c) Ventricular Purkinje System

The Purkinje system of avian hearts is more extensive and complex than that of mammals. In addition to a right and left bundle branch giving rise to a subendocardial network, the avian cardiac Purkinje system includes periarterial fibres, a middle bundle branch connected to a right AV ring of Purkinje cells and a branch to the base of the right muscular AV valve (Fig. 49) (DAVIES 1930; GOSSRAU 1968).

The AV bundle runs forward and deep into the interventricular septum reaching about one quarter the way down the septum before branching into

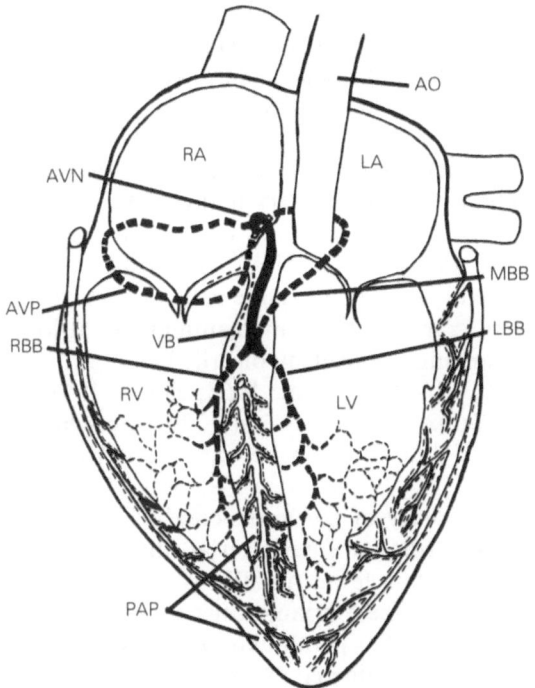

Fig. 49. A diagrammatic representation of the avian ventricular Purkinje system and AV node. As well as left (*LBB*) and right (*RBB*) bundle branches there is a middle bundle branch (*MBB*) which becomes continuous with the atrioventricular Purkinje ring (*AVP*), after circling the root of the aorta (*AO*). The AVP branches from the AV node (*AVN*). A valve branch (*VB*) passes into the right AV valve, which is a thick muscular valve in birds. The ventricular Purkinje fibres ramify extensively in the ventricles (*LV, RV*), and are continuous with the periarterial Purkinje fibres (*PAP*). The similar periarterial Purkinje fibres of the atria (*LA, RA*) have been omitted for clarity. (Adapted from DAVIES 1930a and GOSSRAU 1968)

three bundles. These are called the left, right, and recurrent (DAVIES 1930a) or middle branch (GOSSRAU 1968).

The first Purkinje fibre branch comes from the lower part of the AV node and proceeds dorsally and to the right, forming a loop around the right AV orifice (DAVIES 1930a; GOSSRAU 1968). This ring widens into a sheet, overlaying the ventricular muscle component of the large, muscular AV valve. Two types of fibres have been described making up the AV ring bundle: the ring-Purkinje cells, and perianular cells which surround them (GOSSRAU 1968). As the Purkinje bundle continues to the left and away from the AV region, it thins out and joins the distal part of the middle branch of the AV bundle. At the AV bundle this branch runs upwards and to the left, looping around the base of the aorta from the left side. It continues behind the aorta to the right where it merges with the continuation of the right AV ring bundle (DAVIES 1930a; GOSSRAU 1968) (Fig. 49).

The right bundle branch bifurcates near the endocardium on the right side of the septum. One Purkinje branch passes upwards to the base of the AV valve. The other, together with the left bundle branch, gives rise to a subendocardial network, similar to that in mammalian ventricles. However, instead of merging directly with the working myocardium most fibres become continuous with the periarterial Purkinje system, and it is from this system that branches ending in the working myocardium arise (DAVIES 1930a).

The avian Purkinje cells are morphologically similar to their mammalian counterparts in that they are much larger than the working myocytes and form connecting bundles. In chickens these cells are up to 150 μm long (MIZUHIRA et al. 1967), and average 16 μm in width (SCOTT 1971). They have very few myofibrils, infrequent small mitochondria but many intermediate filaments (TRUEX and SMYTHE 1965a; HIRAKOW 1966; MIZUHIRA et al. 1967; GOSSRAU 1968; SCOTT 1971; AKESTER and AKESTER 1971; BOGUSCH 1974, 1975; AKESTER 1981). Unlike mammalian cells, avian Purkinje fibres have very little glycogen, and myofibrils are distributed throughout the cytoplasm (HIRAKOW 1966; MIZU-HIRA et al. 1967; GOSSRAU 1968; SCOTT 1971). Nexuses have been reported as being uncommon (MIZUHIRA et al. 1967) or numerous and distributed at irregular intervals along the lateral membranes of Purkinje cells (SCOTT 1971). Regardless, they are smaller and less common than in mammalian Purkinje fibres (AKESTER 1981). Although the arrangement of myofibrils has often been described as random (HIRAKOW 1966; MIZUHIRA et al. 1967; SCOTT 1971), it appears they actually follow a spiral course (BOGUSCH 1974). The sarcoplasmic reticulum is generally poorly developed but this can be variable (MIZUHIRA et al. 1967).

Both the ultrastructure and histochemistry of the bird's conduction system display regional variation. The subendocardial and periarterial fibres are mor-phologically identical (SCOTT 1971) but contain fewer myofibrils and mitochon-dria, and more intermediate filaments than the Purkinje cells of the AV ring bundle (GOSSRAU 1968). In the AV ring bundle, perianular cells have a single nucleus, contain more mitochondria and better aligned myofibrils than the ring-Purkinje cells which they surround (GOSSRAU 1968). The latter are often binuc-leate and contain greater amounts of glycogen, Golgi zones, and leptofibrils than the perianular cells. Phosphorylase activity is high in the AV bundle, its branches, the subendocardial and periarterial networks but low in the AV ring Purkinje cells and very weak in the perianular cells (GOSSRAU 1968). In pigeon heart, acetylcholinesterase (AChE) activity is weak in the AV bundle, its branches and the subendocardial and periarterial networks, stronger in the AV ring Purkinje cells and very strong in the perianular cells (GOSSRAU 1968). How-ever, this difference is species-dependent, the perianular cells of adult chickens and young chicks having similar AChE activity to the AV bundle and its branches. In addition, perianular cells show no AChE activity in the canary, finch, and budgerigar (GOSSRAU 1968).

The ventricular Purkinje cells also show morphological differences between species. In the hearts of large birds such as the ostrich, Purkinje cells are general-ly very distinct because of their size and pale staining but less so in small birds like duck and fowl (HOLMES 1923; DAVIES 1930a). However, their identifi-cation, even in embryonic chick heart, is made easier by electron microscopy (BOGUSCH 1979).

4. Phylogenetic Correlations

The existence of a conduction system in lower vertebrates, and in particular the specialised nature of the junctional rings, has been controversial (SWETT 1923; DAVIES and FRANCIS 1940, 1952; DAVIES et al. 1952; MATHUR and HUR-

KAT 1973; THOMAS 1976). Furthermore, it has been suggested that the cardiac conduction system of birds and mammals is phylogenetically neomorphic, having no counterparts in lower vertebrates (DAVIES and FRANCIS 1940, 1946), although others disagree (MACKENZIE 1910; PRAKASH 1960; ROBB 1965; TRUEX and SMYTHE 1965a; BURNSTOCK 1969; ANDERSON et al. 1976). The controversy may be related to the poor histological differentiation of the rings and AV canal cells in comparison to the avian and mammalian conduction systems. It is interesting that in the heart of the African lungfish, ARBLE et al. (1977) could define a conducting system electrophysiologically but not histologically. Additionally, only parts of the SA and AV junctional rings in lower vertebrates are comprised of specialised cells (PRAKASH 1960; RUSKA 1965; YAMAUCHI and BURNSTOCK 1968b; YAMAUCHI 1969; YAMAUCHI et al. 1973; ABRAHAM 1982).

The development of an AV conduction system in mammals and birds may be regarded as a refinement of the system already present in the lower vertebrates. The AV pathway and the SA node become relatively smaller and more clearly defined as the cells involved become better specialized for impulse conduction. That this process is repeated in ontogeny is central to the specialised ring tissue theory of ANDERSON and colleagues (ANDERSON et al. 1974, 1976). A ring of specialised tissue around the right AV orifice has been described in the human fetal heart (ANDERSON et al. 1974). Similar rings of specialised tissue are known to persist in adult rats and guinea-pigs (ANDERSON 1972a, b, c). This suggests that at least part of the AV node is a retention and further development of a portion of the AV ring (ANDERSON et al. 1976). Furthermore, the distribution of the rings of specialised tissue in rats and guinea-pigs is similar to that of the AV ring bundle of Purkinje cells in bird hearts, suggesting there may be a common phylogenetic origin.

The broad plane of cells forming the preferential pathway of conduction in the AV canal may be considered the forerunner of the AV bundle and its branches. Evidence that ontogenic development of the AV bundle follows phylogeny can be found in the similarity between the earliest AV conduction pathway in mouse embryos (VIRÁGH and CHALLICE 1977a; G.I.2) and the preferential pathway of conduction in the AV canal of adult lower vertebrates.

There does not appear to be a similarly modified population of cells within the ventricle of the lower vertebrate heart which could be considered as the forerunner of the mammalian cardiac Purkinje system. However, the freely anastomosing arrangement of the spongy myocardium is architecturally similar to the trabecular myocardium of the mammalian fetus, which provides a template for the ontogenetic development of the Purkinje network (G.I.3, G.IV.2).

E. Innervation of the Heart

The neural control of the heart, sympathetic and parasympathetic interactions, baroreceptor reflex regulation, and the role of the central nervous system as an integrator of neural influences, have been extensively reviewed by RANDALL (1977), LEVY and MARTIN (1979), DOWNING (1979) and KORNER (1979). This chapter will discuss the distribution and morphology of neural elements in the heart and their relationships to myocardial tissue as revealed by chemical, histochemical, and ultrastructural techniques.

I. Major Nerves Supplying the Heart

There is a general intermingling of vagal and sympathetic components in the major cardiac nerves of the dog heart (RANDALL and ARMOUR 1977). The major vagal nerves contribute to the cardiac nerves at a number of levels. Two major branches of the right thoracic vagus, the craniovagal and caudovagal nerves, supply the right atrium. Most sympathetic nerve components to the dog heart arise from the middle cervical and stellate ganglia. The middle cervical ganglia send fibres which connect with their respective stellate ganglia. The right stellate cardiac nerve arises from the right stellate ganglion and supplies primarily the SA nodal region and the right atrium. The right middle cervical ganglion contributes to the right recurrent cardiac nerve which also takes components from the recurrent laryngeal nerve and the vagus. The right recurrent cardiac nerve also has many afferent fibres which arise from all four chambers of the heart. On the left side, the middle cervical ganglion contributes to most of the innominate nerve which also contains vagal and afferent fibres. The left middle cervical ganglion also contributes to the ventromedial, the ventrolateral, and the dorsal cardiac nerves. The major part of the ventromedial nerve arises from the left vagus. Most of the fibres of the dorsal cardiac nerve are distributed to the aortic arch and descending aorta and many afferents from these regions run in this nerve. The ventrolateral cardiac nerve has few afferents or vagal fibres and mainly innervates the left atrium and a large portion of the ventricles and AV nodal region. The left stellate cardiac nerve, which contains mostly afferent fibres from the left atrium, arises from the left stellate ganglion (RANDALL and ARMOUR 1977).

Cardiac nerves of primarily afferent function contain many myelinated fibres whereas predominantly efferent cardiac nerves are made up of mostly nonmyelinated fibres and thin myelinated nerve fibres. No functional efferent cardiac nerves arising from ganglia below the level of the stellate ganglia have been found (RANDALL and ARMOUR 1977).

As well as those arising from the extrinsic ganglia mentioned above, many nerves within the heart arise from intrinsic ganglia. These ganglia and their nerve fibres survive surgical denervation and are thought to be predominantly parasympathetic (PRIOLA 1980). They are generally located epicardially or near nodal tissue.

The baboon cardiac innervation pattern differs from the dog (RANDALL and ARMOUR 1977). Although present on the left, the stellate cardiac nerve is absent on the right side. As well, the middle cervical ganglia do not connect to the thoracic vagi in the baboon. A few major interconnecting branches between vagal and sympathetic components on the left side also occur. The innervation of the human heart is anatomically more similar to the dog than to the baboon (RANDALL and ARMOUR 1977), the main difference between dog and man being that the vagal and sympathetic cardiac nerves, except for a few interconnections, remain separate in man.

II. Sensory End Formations

1. Myelinated Fibres

As well as inhibitory vagal and excitatory sympathetic efferent fibres, many afferent fibres are present in cardiac nerves (RANDALL and ARMOUR 1977). All nerve fibres supplying sensory end organs in the body are myelinated and hence the observation of myelinated nerve fibres in the heart or within a cardiac nerve has generally been taken as an indication that the nerve fibres are functionally afferent (HIRSCH and BORCHARD-ERDLE 1960). Whether there is any contribution to sensory innervation from nonmyelinated fibres is unknown (FLOYD 1979).

There is convincing evidence that not all the myelinated fibres of the vagal and sympathetic cardiac nerves are actually afferents. In a quantitative analysis of degenerating myelinated nerve fibres in dog heart, 11% were efferent sympathetic preganglionic fibres which degenerated following ventral root sections at levels thoracic 1 to 8, and 76% of the myelinated fibres were in the vagal branches. Following vagotomy proximal to the nodose ganglion, 55% degenerated leaving 21% representing sensory fibres with cell bodies in the nodose ganglion. Only 13% of the myelinated fibres were from dorsal root ganglia (FUKUYAMA 1970).

2. Morphological Types of Sensory End Formations

Terminations of sensory fibres in most tissues including the mammalian heart have been classed morphologically into three categories: (a) free fibre endings which may show some branching (end-nets), or remain unbranched and are formed by small myelinated or unmyelinated fibres; (b) complex unencapsulated endings which are formed by repetitive branching of the nerve ending and appear as a discrete termination (end-organ); and (c) encapsulated endings which are also discrete terminations but are surrounded by a capsule of Schwann cells or connective tissue, e.g. Pacinian corpuscles (MILLER and KASAHARA 1964).

a) End-Nets

Two types of end-nets have been described: (1) a type of free fibre ending, or (2) a sensory formation peculiar to vascular tissue (MILLER and KASAHARA 1964). End-nets are thought to be a plexus of branched dendrites from myelinated fibres. TRANUM-JENSEN (1979) has suggested that the branches of the end-net plexus are actually composed of many nerve fibres of which a large proportion are adrenergic efferents. Many of the apparent branchings of fibres within the plexus are thought to be due to a divergence of axons and dendrites from a common pathway. End-nets are a common finding in the atrial subendocardium just above the myocardium and are supplied by nerves from the epicardium and myocardium (FLOYD 1979). Sensory endings of the simple free fibre type are also observed in the subendocardium (MILLER and KASAHARA 1964; FLOYD 1979).

Although sensory end-nets may be found anywhere in the atria they are more common near the junction of the inferior vena cava and the right atrium, the opening of the pulmonary veins into the left atrium, and near the base of the interatrial septum (MILLER and KASAHARA 1964). End-nets forming from myelinated fibres are also observed in ventricular subendocardium. No particular association between end-nets and Purkinje fibres was noted by MILLER and KASAHARA (1964). End-net formations are also observed in relation to coronary vessels and the adventitia of the pulmonary artery and aortic arch, however, this plexus cannot be reliably distinguished from the autonomic motor plexus of the blood vessels without special techniques (MILLER and KASAHARA 1964). End-nets disappear permanently after bilateral section of the vagal nerves distal to the nodose ganglia (FLOYD 1979). However, unilateral vagotomy does not lead to noticeable degeneration (HOLMES 1957).

b) Unencapsulated Endings

Unencapsulated endings are often described as compact or diffuse, however, they demonstrate a range of intermediate morphologies making this distinction of limited use (MILLER and KASAHARA 1964; FLOYD 1979). Whether functional differences exist between the "diffuse type" or the "compact type" of complex unencapsulated endings is unknown. Unencapsulated endings have been more recently called end-organs (TRANUM-JENSEN 1975, 1979) (Fig. 50), and there is good evidence that these sensory endings are the baroreceptors described physiologically (COLERIDGE et al. 1957, 1964; KAPPAGODA et al. 1972).

Unencapsulated endings have a similar distribution to the end-nets except they are not found in the ventricular subendocardium (MILLER and KASAHARA

1964) but are most frequently encountered near the junctions of major vessels with the atria (NETTLESHIP 1936; NONIDEZ 1937; HOLMES 1957; MILLER and KASAHARA 1964). These veins have working myocardium within their walls instead of smooth muscle (see NONIDEZ 1937, 1941). Unencapsulated endings on both sides of the heart range in number from 100–300 (FLOYD 1979), and together with end-nets are of vagal origin (NETTLESHIP 1936; HOLMES 1957). The contribution of spinal nerves and nonmyelinated fibres is not completely understood although there is some involvement (FLOYD 1979; EMERY et al. 1976).

c) Encapsulated Endings

These endings can be found in the walls of large veins but are relatively rare in mammalian hearts (MILLER and KASAHARA 1964), except possibly the horse (MICHAILOW 1908).

3. Ultrastructure of Sensory End Formations

To date, only a few reports on the ultrastructure of sensory end-organs in the heart are available (CHIBA and YAMAUCHI 1970; TRANUM-JENSEN 1975, 1979; YAMAUCHI 1979; YOKOTA et al. 1983). Myelination of the afferent nerve fibre ceases at a variable distance from the end-organ and the myelinated part of the fibre is 4 to 9 µm thick. Some nerve fibres branch, giving rise to a number of end-organs which are 100 to 400 µm in length and 10 to 20 µm in thickness. Running with the large afferent fibre are small-diameter fibres, some of which enter the end-organ directly from the periphery (TRANUM-JENSEN 1979). As well, sensory varicosities constitute approximately 2% of all nerve profiles outside end-organs in the atrial appendage (CHIBA and YANAUCHI 1970).

a) Afferent and Efferent Fibres Supplying End Formations

Thin nerve fibres with vesicle-filled varicosities are found at the periphery of the sensory end-organs. Based on studies of end-organs in other tissues (SANTINI 1969; SCHOULTZ and SWETT 1972), TRANUM-JENSEN (1979) proposed that these thin fibres have an efferent function. However, this suggestion remains controversial since a direct influence of catecholamines on the responses of the baroreceptors in the atria has not been observed, and the decreased discharge of afferent fibres during sympathetic stimulation is a result of changes in atrial pressure rather than any direct effect on the sensory end-organ (ZUCKER and GILMORE (1974). Similarly, WAHAB et al. (1975) concluded that the observed effect of vagal stimulation on atrial baroreceptor discharge is a result of atrial pressure changes and not mediated by direct vagal effects on the baroreceptors.

b) Cells and Nerve Endings Within the End Formation

The end-organ commences with an extensive arborization of the large nonmyelinated nerve fibre. Collagen fibre bundles are prominent and densely packed

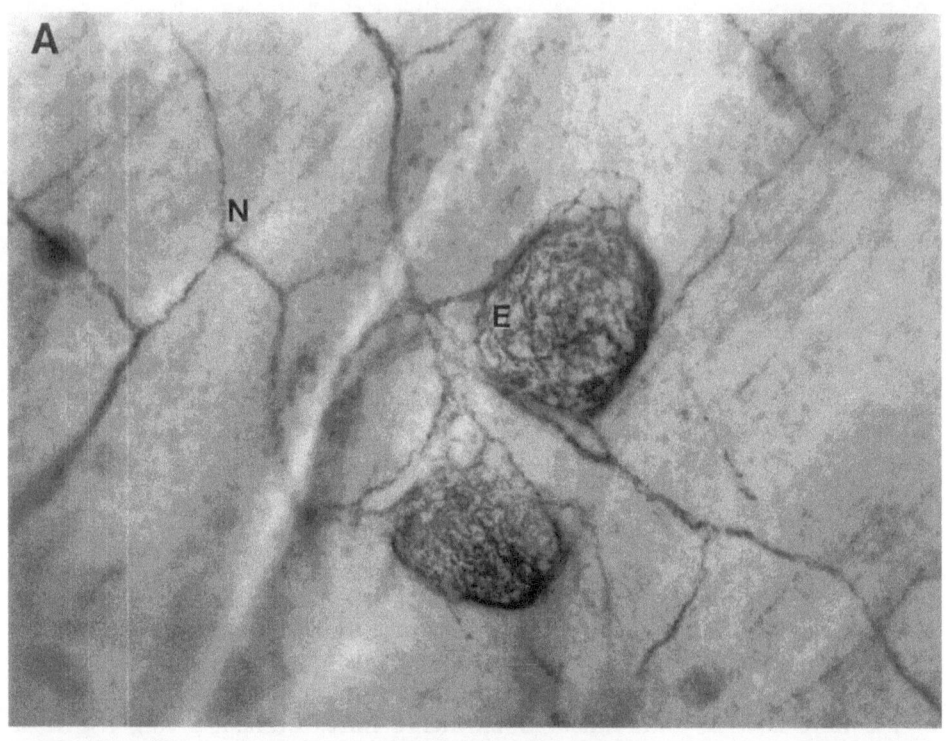

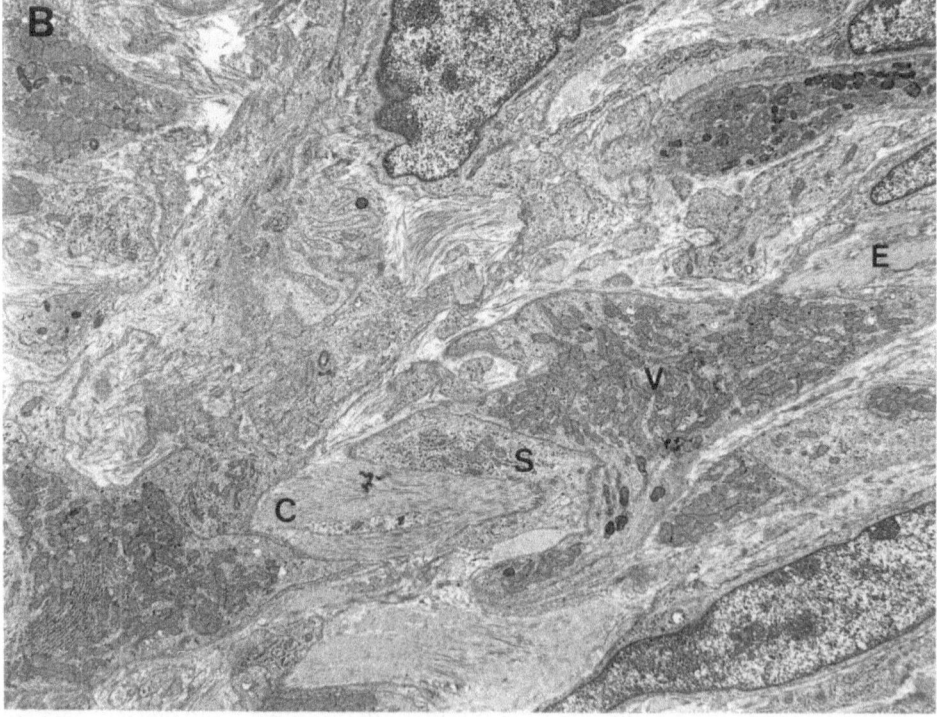

in most end-organs although elastic fibres which are abundant in the subendo-cardium generally do not extend within (Fig. 50 B). End-organs are avascular and do not have a capsule (TRANUM-JENSEN 1979).

The end-organ cells have been described as modified Schwann cells and do not appear to have a receptor function (TRANUM-JENSEN 1975, 1979). Schwann cells cover the nerve branches but many varicosities are ensheathed only with basal lamina and are in close contact with the collagen bundles. Non-specific cholinesterase activity has been localised ultrastructurally to the spaces between the axon and Schwann cells, but the function of the enzyme is unknown (YOKOTA et al. 1983).

The most prominent features of the varicosities of sensory fibres is the presence of many small mitochondria, large accumulations of glycogen particles, and a few electron-dense bodies (CHIBA and YAMAUCHI 1970; TRANUM-JENSEN 1975, 1979) (Fig. 51 A). Smooth endoplasmic reticulum arranged in parallel lamellae, closely stacked one on top of the other with glycogen particles often dispersed between, is another common feature (Fig. 51 B). Similar glycogen-lamellar membrane complexes have been observed in spinal ganglion cells (PANNESE 1969), photoreceptor cells (YAMADA 1960; COHEN 1963; PETIT 1968), and in skeletal muscle (GARANT 1968). There is still controversy about whether the lamellar stacks of membrane cisternae play a significant role in glycogen metabolism. Evidence for a relationship between tubules of endoplasmic reticulum, ribosomes and glycogen synthesis can be found in hepatocytes although the lamellar complexes of endoplasmic reticulum are not observed (VRENSEN and KUYPER 1969; VRENSEN 1970; CARDELL 1971).

Lamellar membrane complexes without glycogen particles but with associated irregular electron-dense bodies also occur, as do autophagic vacuoles sometimes containing recognizable mitochondria (CHIBA and YAMAUCHI 1970; TRANUM-JENSEN 1979).

c) Chemoreceptors

YAMAUCHI (1979) has described the ultrastructure of presumed chemoreceptors in the sinoatrial region of the dog heart near cardiac ganglia. These chemoreceptors show morphological similarities to those in the aortic and carotid bodies (HANSEN and YATES 1975; YAMAUCHI 1976). They are composed of granule-containing cells forming groups intimately in contact with chemosensory-type axons and closely associated with capillaries. In carotid bodies, the granule-containing cells of the paraganglia may be contacted by chemosensory-type fibres or efferent fibres. Such connections have been observed amongst chemoreceptors in the dog aorta but not in the dog heart (YAMAUCHI 1979).

Fig. 50. A Cholinesterase-stained whole-mount preparation of the atrial wall of the mini-pig. Two end-organs (E) and nerve fibres (N) are stained. × 120. B Low-power electron micrograph of a sensory end-organ. The large varicosities of nerve fibres (V) contain numerous small mitochondria and some also contain numerous dense bodies (varicosity at upper right). Varicosities are in contact with both the modified Schwann cells (S) and dense bundles of collagen (C). Very few elastin fibres (E) are present. × 5000. (From TRANUM-JENSEN 1979)

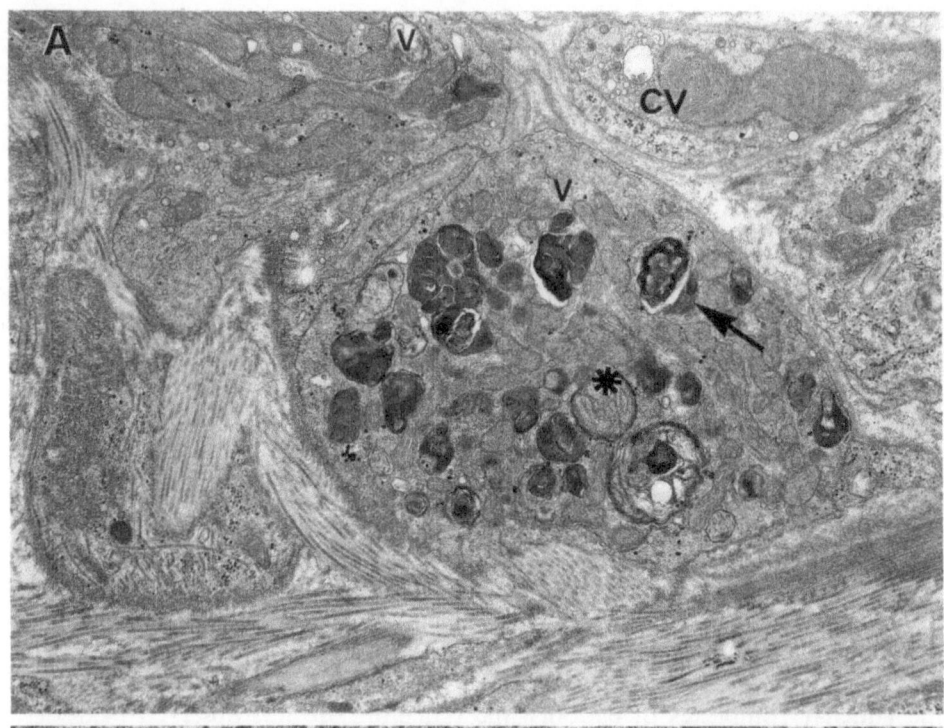

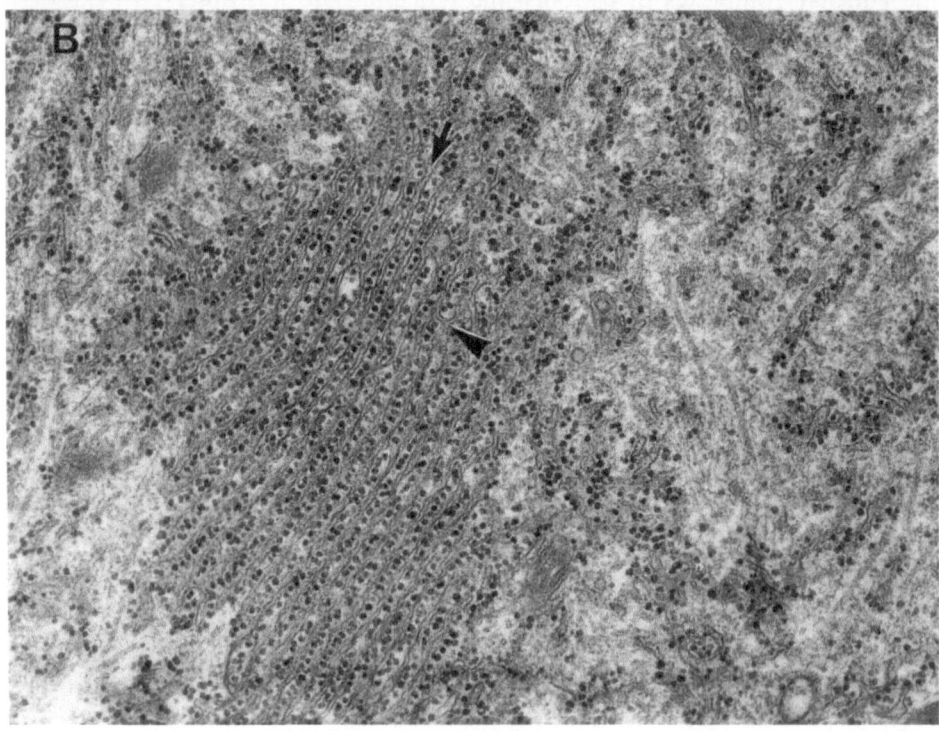

III. Histochemical and Chemical Studies of Efferent Fibres

1. Distribution of Parasympathetic Nerves

Quantitative estimations of the density of cholinergic nerves in the heart rely on chemical measurements of acetylcholine (ACh) or choline acetyltransferase. This is largely because acetylcholinesterase (AChE), which is commonly used as a histochemical marker for cholinergic nerves, is produced by many non-neural tissues. In particular, cells of the heart's conduction system contain AChE and it is therefore not an accurate measure of the innervation of conduction tissue. In histochemical studies, however, clear differentiation between AChE-positive nerves and conduction tissue is possible (BOJSEN-MØLLER and TRANUM-JENSEN 1972).

a) Comparison of the Four Chambers

Parasympathetic innervation is non-uniform throughout the heart in a variety of mammals, such as rabbit (VLK 1958; LEWARTOWSKI and BIELECKI 1963; KILBINGER 1973), cat (BROWN 1976), and guinea-pig (SCHMID et al. 1978). The SA node has the highest density of innervation and the left ventricular wall the lowest. The innervation density from highest to lowest is as follows: SA node, right atrial appendage, right atrium, left atrium, papillary muscle, right ventricular wall, left ventricular wall.

The absolute values of right atrial ACh expressed in nmol/g ranges from 7.6 to 9.1, and of left atrial ACh from 4.3 to 6.7 (VLK 1958; LEWARTOWSKI and BIELECKI 1963; KILBINGER 1973). ACh is higher in the rabbit interventricular septum (2.4 nmol/g) than in the left (1.0 nmol/g) and right (2.2 nmol/g) ventricular free wall (KILBINGER 1973). However, in cat heart ACh is lower in the interventricular septum (1.5 nmol/g) than in the left and right ventricular walls (2 and 5 nmol/g respectively) (BROWN 1976). SCHMID et al. (1978) measured choline acetyltransferase activity in the guinea-pig heart and found that the activity of the enzyme in the inferior part of the interventricular septum was between the values of the right and left ventricular free walls. Since they also compared enzyme activities in the conduction system their sampling technique for the interventricular septum purposefully avoided the inclusion of Purkinje tissue. Both BROWN (1976) and SCHMID et al. (1978) have reported very high measurements of ACh and choline acetyltransferase in the conduction system,

Fig. 51. **A** Two varicosities (*V*) are shown which contain many small mitochondria, but only the lower one has many dense bodies (*arrow*), and autophagic vacuoles containing mitochondria (*asterisk*) which appear to be part of degenerative processes. Note the presence of large dense-cored vesicles and small clear vesicles (*CV*) in a third varicosity with a large mitochondrion. × 15100. (From TRANUM-JENSEN 1979) **B** Glycogen-cistern complex, often observed in nerve fibres of end-organs. The section has been tilted to demonstrate the cisterns of smooth endoplasmic reticulum (*arrow*), which intercommunicate (*arrowhead*), and contain a moderately electron-dense material. Glycogen particles are arranged in rows between the cisterns. × 32700. (From TRANUM-JENSEN 1975)

including the Purkinje tissue. Therefore the high value of ACh content in the interventricular septum recorded by KILBINGER (1973) may reflect the presence of a significant amount of Purkinje tissue which is much more densely innervated with parasympathetic nerves than working myocardium. This also raises the question of the effect of relatively well-innervated ramifications of Purkinje tissue present within the ventricular myocardium on measurements of ACh and choline acetyltransferase. If the contribution of the Purkinje fibres' innervation to these measurements is significant, then the real innervation density of working myocardium may be lower than presently believed.

b) Densely-Innervated Atrial Muscle Bundles

The SA ring bundle of the rabbit heart has a much greater density of innervation than the surrounding atrial myocardium (BOJSEN-MØLLER and TRANUM-JENSEN 1972). It is made up of myocardial cells and ChE-positive nerves, and is formed around the edge of the junction of the right atrium proper and the embryonic remnant of the sinus venosus. It forms a ring around the superior and inferior venae cavae. The right edge of the ring is associated with the tail of the SA node, and runs alongside the crista terminalis which forms a major preferential pathway of conduction to the AV node (GOODMAN et al. 1971).

The rat heart contains other specialized cells in ring formation with a dense innervation of AChE and catecholamine (CA)-containing nerves (ANDERSON 1972c). These are distributed around the base of the tricuspid orifice as a continuous ring and make no contact with the ventricular myocardium, being separated from it by the connective tissue of the valve base. The cells are in communication with atrial myocardium through interposing transitional cells. The ring of specialized tissue is continuous with the posterior portion of the AV node as a continuation of the bundle of lower nodal cells. The entire circumference of the base of the mitral valve also contains well innervated specialized ring tissue which is connected to the tricuspid ring by a second node-like structure, separated from and above the penetrating AV bundle. This structure has been called the "retro-aortic knot" since it is positioned between the aortic root and the mitral valve ring (ANDERSON 1972c). As well as being innervated with CA-containing and AChE-positive nerves, the cells of the retro-aortic knot and ring are AChE-positive. The ultrastructure of the ring tissue is typical of specialized cells, being smaller than working atrial myocardial cells and poor in myofibrils which appear disorganized (HIBBS and ELLISON 1973). Large numbers of nerve axons with varicosities are present in the interstitium of the ring tissue.

A similar arrangement exists in the rabbit heart (ANDERSON 1972c), but the ring tissue here is less well formed and remains discontinuous with the retro-aortic knot although the cells are still well supplied with AChE- and CA-containing nerves. Guinea-pig ring tissue (ANDERSON 1972b) is less well formed than in rabbit. However, the retro-aortic knot of specialized cells is well developed, with extensions to the mitral and tricuspid rings. Unlike the rabbit and rat, no direct connection to the posterior portion of the AV node exists in the guinea-pig.

Only 15% of adult human hearts contain evidence of specialized ring tissue of the tricuspid valve base (ANDERSON et al. 1974). However, in fetal human hearts, a complete ring can be identified around the tricuspid orifice. The myocardial tissue in this ring is considered specialized due to its histology and AChE content. Continuous with the tricuspid ring is a segment of identical specialized tissue applied to part of the mitral valve base. This only reaches the posterior of the mitral opening and does not form a complete ring. It is joined to the tricuspid ring by a node-like structure reminiscent of the retro-aortic knot of rabbit, rat, and guinea-pig. However, in human fetuses this knot has a more anterior position and has been named the anterolateral knot. In those adult hearts in which portions of specialized ring tissue are found, the most frequent site corresponds to the position of the anterolateral knot of the fetus (ANDERSON et al. 1974, 1979).

These observations of densely-innervated specialized tissue leaves open the possibility that the high measurements of overall ACh reported in the right atrium may not be due entirely to direct innervation of the right atrial working myocardium. The right atrial working myocardium may be no more densely innervated with cholinergic nerves than is the left. It is also not known whether the very high measures of choline acetyltransferase in the atrial appendage (SCHMID et al. 1978) are due to the presence of a densely innervated muscular branch of the SA ring bundle which enters the appendage (BOJSEN-MØLLER and TRANUM-JENSEN 1972).

c) Functional Implications of Densely-Innervated Atrial Muscle Bundles

Many questions about the innervation of the atria remain unanswered. For instance, the functional basis of the dense innervation of the SA ring bundle or the extranodal ring tissue (ANDERSON 1972b, c) is not well understood. Cells within the SA ring bundle possess some pacemaker-like properties, such as spontaneous depolarisations, low amplitude of the action potential and plateau as well as the ability to conduct at high K^+ concentrations (PAES DE CARVALHO et al. 1959; DE MELLO and HOFFMAN 1960; TAKAYASU et al. 1969). These observations suggest that parts of the SA ring bundle may function as subsidiary pacemakers or O-points (BOINEAU et al. 1980) but this is not known for certain (see D.I.7). The density of innervation around the veno-atrial junction may be related to sensory function since many sensory end-formations occur in this region (see E.II.2). The possibility of similarly well-defined muscle bundles with dense innervation residing in the left atrium of mammalian heart has not been thoroughly investigated.

ANDERSON et al. (1974) have discussed the probability that remnants of specialized rings in humans may correspond to extra nodes (KENT 1913, 1914; SHANER 1929). They also suggest that the anterolateral knot may become significant in malformed hearts where the Purkinje system communicates with it instead of the true AV node. The basal two thirds of mitral and tricuspid valve cusps contain myocytes and nerves with efferent and sensory endings as well as highly-organised connective tissue (MILLER and KASAHARA 1964; COOPER et al. 1966; ELLISON and HIBBS 1973; HIBBS and ELLISON 1973). During electrical stimulation of mitral valve cusps from cat heart, noradrenaline increases and acetylcholine decreases the amount of tension developed (SONNENBLICK et al. 1967). These observations suggest that specialised ring tissue can influence valve closure and may receive sensory feedback from the valves. Control of valve closure may be one function of the integrated intrinsic nervous system of the heart.

2. Distribution of Sympathetic Nerves

a) Catecholamine Fluorescence Studies

Sympathetic nerves show a different distribution from parasympathetics in the heart of mammals. Left and right ventricles receive similar amounts of catecholamine-containing nerves and are well innervated. Atria are more densely innervated than the ventricles and the SA node is the most densely-innervated cardiac structure, followed by the AV node and some other parts of the right atrium (DAHLSTRÖM et al. 1965; JACOBOWITZ et al. 1967; NIELSSON and OWMAN 1968; DOLEŽEL et al. 1978; SCHMID et al. 1979; LEVY et al. 1981; MANGER 1982).

Because of the very high density of innervation of the SA node it has been suggested that the majority of the SA nodal cells are individually innervated (DAHLSTRÖM et al. 1965; NIELSSON and OWMAN 1968). Sympathetic innervation of the AV node is also high. In contrast, sympathetic innervation of Purkinje cells is sparse (DAHLSTRÖM et al. 1965), although the fibres are well supplied with cholinergic nerves (KENT et al. 1974; BOJSEN-MØLLER and TRANUM-JENSEN 1971).

The functional activity of the autonomic nervous system can also show regional variation. In dogs with experimentally-induced right-heart failure, bio-chemical studies show a nonuniform alteration to sympathetic and parasympa-thetic nerve function (LUND et al. 1982). Furthermore, there is some evidence that central neural activity may selectively influence cardiac regions nonuniform-ly (SCHMID et al. 1979).

b) Comparisons with Chemical Studies

Chemical estimations of noradrenaline content do not correlate well with catecholamine fluorescence observations. Low concentrations of adrenaline have been found in all parts of mammalian myocardium (SHORE et al. 1958; ANGE-LAKOS et al. 1963; SHINDLER et al. 1968) and there is no significant difference in noradrenaline content of the SA node and the surrounding atrium (SHORE et al. 1958; ANGELAKOS et al. 1963; LEVY et al. 1981). However, dopamine is in much higher concentration in the SA node than in the right atrium (ANGE-LAKOS et al. 1963). Similarly, noradrenaline turnover and tyrosine hydroxylase and dopamine beta-hydroxylase activities are greater in the SA node and right atrial appendage than in the right atrium (SCHMID et al. 1979; DICKSON et al. 1981). Noradrenaline turnover is higher in the ventricular Purkinje system than in the left or right ventricular working myocardium (SCHMID et al. 1979), al-though the activities of tyrosine hydroxylase and dopamine beta-hydroxylase are similarly low in the ventricular conduction system and working myocardium (DICKSON et al. 1981).

Apart from the Purkinje system, which has few catecholamine-containing nerves, measurements of noradrenaline turnover correlate well with the density of innervation observed with fluorescence histochemistry in the different heart regions, but not with the tissue concentration of noradrenaline. This suggests that noradrenaline content is not related to its turnover or the density of sympathetic innervation.

c) Evidence for Different Types of Sympathetic Nerves in the Heart

Morphological and histochemical evidence suggests that adrenergic fibres supplying the coronary circulation and the myocardium are from two different types of neurons (DOLEŽEL et al. 1978) in agreement with functional studies (SZENTIVANYI and JUHASZ-NAGY 1959; SZENTIVANYI et al. 1976; NIELSSON and OWMAN 1967). More recently, DOLEŽEL et al. (1980) have postulated that three distinct types of sympathetic neurons innervate the heart: (1) neurons innervating the whole coronary circulation; (2) neurons innervating the ventricles; and (3) neurons innervating the atria. As well, two kinds of intrinsic ganglia have been distinguished in the interatrial septum and AV junction of rat heart (MORAVEC and MORAVEC 1984).

3. Species Differences and Regional Variation of AV Nodal Innervation

The use of histochemical techniques for the presence of cholinesterase (ChE) and AChE, as well as fluorescence for catecholamine-containing nerves, has led to an appreciation of both the variability between species and regional differences of innervation within the AV junction.

In rabbits (see D.II.8 c) the midnodal cell knot (ANDERSON 1972a) is itself AChE-negative and is sparsely supplied with both AChE and CA-containing nerves. Transitional cells are weakly AChE-positive and richly innervated with both AChE and CA-containing nerves. The lower nodal cell bundle is strongly AChE-positive and has dense innervation with both kinds of nerves, but CA-containing nerves are more numerous (ANDERSON 1972a, b, c; BOJSEN-MØLLER and TRANUM-JENSEN 1972).

The rat AV node is similar, being profusely innervated with both AChE and CA-containing nerves. Cells of the conduction system are AChE-positive and the innervation of the AV node appears richer than that of the SA node (FINLAY and ANDERSON 1974).

In guinea-pig numerous ganglion cells are present in the connective tissue near transitional cells and surround the compact part of the node (ANDERSON 1972a). Transitional cells are weakly AChE-positive, similar to the rabbit and rat. The compact node is intensely AChE-positive and AChE-positive nerves are abundant in the compact node, transitional cells and AV bundle. Few CA-containing nerves are present, which differs from the rabbit and rat heart. The density of CA-containing nerves in the guinea-pig transitional cell zone and compact node is similar to the atrial myocardium but the AV bundle is even more poorly innervated.

The superficial part and transitional cell zone of the dog AV node is well innervated with ChE-positive nerves (DAVIES et al. 1983).

A dense plexus of AChE-positive nerves is present in all parts of the conduction system of sheep and nonspecific ChE activity is only present in large preganglionic nerve fibres (MUNNELL 1982). CA-containing nerves form a dense plexus around the SA and AV nodes but not the Purkinje system (MUNNELL 1982).

A complete study of the regional distribution of AChE and CA within the AV junctional areas is lacking for human heart. Strong positive reactions for AChE have been observed in the human AV node, which is less densely innervated than the SA node (JAMES and SPENCE 1966). Nodal cells themselves are also AChE-positive, as in other species. BECKER and ANDERSON (1976) have expressed some reservations about the accuracy and reliability of histochemical methods performed on post-mortem material, although histochemical methods for AChE activity on the myocardium of human postmortem material obtained less than 12 h after death has been compared favourably with fresh myocardium taken at operation (JAMES and SPENCE 1966; KENT et al. 1974).

IV. Ultrastructure of Neuromuscular Relationships in the Heart

1. Terminology

A wide variety of neuromuscular relationships has been reported in the heart and these can be classified into the following categories:

a) En Passant

These are the most common neuromuscular associations seen in the heart (Figs. 52, 53, 55). This term is widely used and synonymous with "loose association". The nerve varicosity in this type of neuromuscular association is usually some distance from the nearest myocardial cell surface and connective tissue and fibroblasts may be present between the nerve varicosity and the myocardial cell (JENSEN et al. 1978).

b) Close Associations

Close associations are neuromuscular junctions where the axon varicosity is separated from the myocardial cell membrane by less than 100 nm and only basal-lamina material occupies the space between. This probably constitutes a direct relationship between the nerve fibre and a specific myocardial cell (Fig. 56).

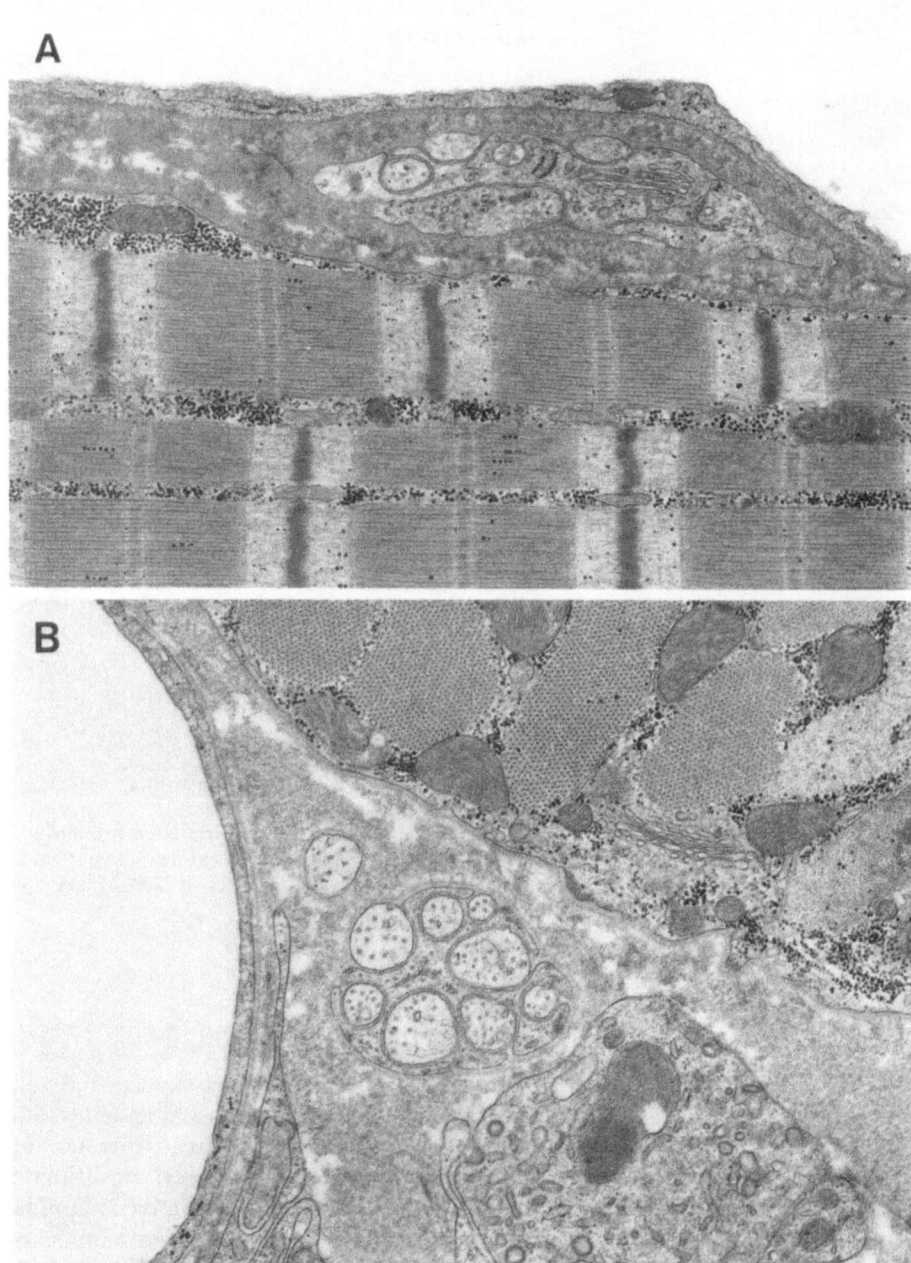

Fig. 52 A, B. Innervation of working myocytes in sheep heart. **A** A small bundle of axons can be seen between the endothelial cell of a capillary and a myocyte. The axon varicosity facing the myocyte is not ensheathed by the Schwann cell and contains small granular vesicles. × 15000. **B** A single naked axon, together with eight others all enclosed by a single Schwann cell, lie between a capillary and a myocyte. × 15800

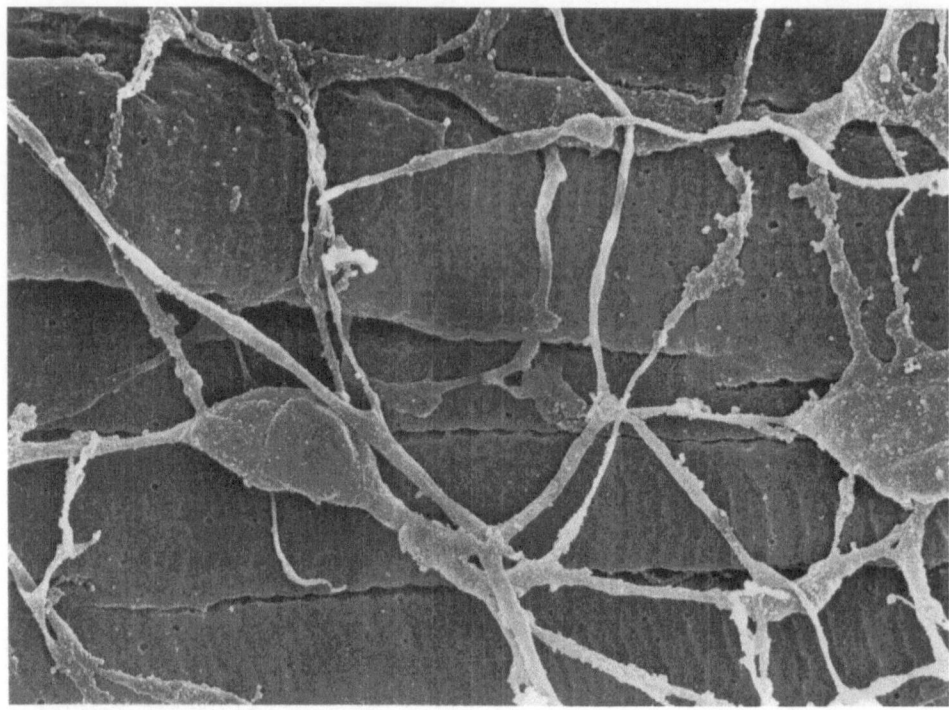

Fig. 53. Scanning electron micrograph of myocyte surfaces from rat ventricle after fixation and removal of surrounding connective tissue by hydrolysis in 8 N HCl. Here neuronal elements have not been removed demonstrating their relationship to the underlying myocytes. × 2200. Micrograph courtesy of Dr. TAKASHI FUJIWARA

c) Intimate Contacts

Intimate contacts or close contacts occur where the nerve-myocardial cell separation is 25 nm or less and no intervening basal lamina is visible (Figs. 54, 57 B). The intimacy of this type of relationship suggests a direct influence of the nerve on the specific myocardial cell. THAEMERT (1970) defined an intimate contact as one where the separation is less than 40 nm and lacks a basal lamina interposed between the membranes. Most of the intimate contacts with myocardial cells so far reported have a separation of 20 nm or less (THAEMERT 1966, 1969, 1970, 1973; HAYASHI 1971; TRANUM-JENSEN 1976, 1978; TRAUTWEIN and UCHIZONO 1963; CHENG 1971). The distinction between close associations and intimate contacts is entirely anatomical and whether these correlate with functional characteristics is unknown.

Although a 'close association'-type of sensory neuromuscular junction has not been reported, presumed sensory endings that fit the category of en passant have been reported in sheep Purkinje fibres (CANALE et al. 1983 c), and intimate contacts of sensory endings are present in mouse and rat AV node (THAEMERT 1973; MORAVEC-MOCHET et al. 1977).

d) Direct and Diffuse Innervation

Electron-microscopic observations of neuromuscular relationships suggest that cholinergic and adrenergic innervation of the working myocardium can be either generalized or directed to a specific cell (THAEMERT 1966; HARTZELL 1980). The accepted limit for separation between nerve and effector cell enabling a direct functional effect is approximately 100 nm (BENNETT and MERRILLEES 1966; MERRILLEES 1968; ROBINSON 1969). Close associations and intimate contacts fall within this limit. En passant nerve endings are, however, separated from their effectors by a greater distance (KISCH 1960) and therefore must employ other mechanisms. Since the fibres have many varicosities filled with vesicles along their length (Fig. 55) and ACh receptors are randomly distributed over the entire surface of the muscle cells (HARTZELL 1980), it has been suggested that the nerves exert their action by a mass release of transmitter which diffuses across the extracellular space, affecting a number of muscle cells (RICHARDSON 1964).

e) Cholinergic and Adrenergic Varicosities

Vesicle-filled varicosities of nerves within the heart usually contain a predominance of either small clear vesicles (Figs. 55, 56) or small dense-cored vesicles (Figs. 54, 57) 30–50 nm in diameter (YAMAUCHI 1973), as well as a few large dense-cored vesicles 80–140 nm in diameter (YAMAUCHI 1973). Small dense-cored vesicles are indicative of adrenergic nerves and the dense core is the site of storage of biogenic monoamines (TRANZER et al. 1969; DUFFY and MAR-KESBERG 1970; GEFFEN and LIVETT 1971). However, the preservation of the electron-dense cores is inconsistent during routine fixation and processing for electron microscopy (RICHARDSON 1966), and special fixation procedures (see YAMAUCHI 1973) are needed to distinguish cholinergic and adrenergic nerve varicosities on the basis of vesicles present. Therfore accumulations of small clear vesicles seen with standard fixation are not necessarily indicative of a cholinergic axon.

2. Innervation of Working Myocardium

The most common neuromuscular association in the working myocardium is of the en passant type. These have been reported in rat papillary muscles (NOVI 1968), rat ventricle (VIRÁGH and PORTE 1961), mouse ventricle (THAEMERT 1969), cat and rabbit papillary muscle (HADEK and TALSO 1967), and human ventricle (KISCH 1960; BATTIG and LOW 1961; CHIBA and YAMAUCHI 1970). Axons often run alongside blood vessels before branching out to course between myocardial cells (NOVI 1968) (Figs. 52, 53) and vesicles are present in varicosities along the length of the axon as well as at their terminations (NOVI 1968; THAE-MERT 1969).

Close associations (VIRÁGH and PORTE 1961; NOVI 1968) and intimate contacts (THAEMERT 1966, 1969) between working myocardial cells and nerve vari-

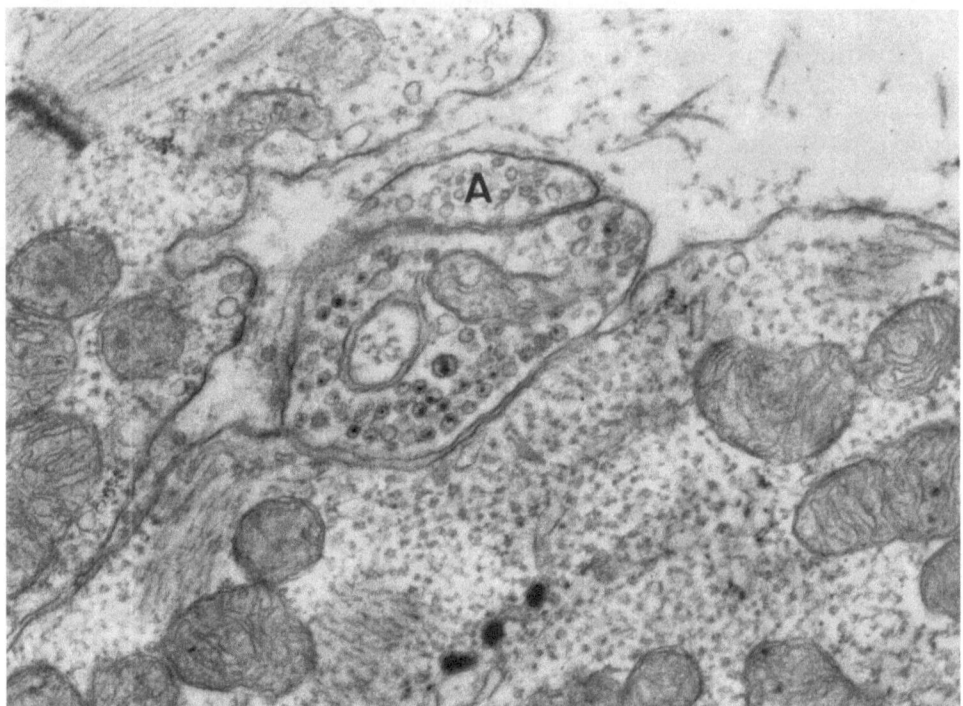

Fig. 54. A nerve varicosity containing predominantly granular vesicles makes intimate contact with the surface of a cardiac muscle cell. The adjacent nerve axon (*A*) contains only agranular vesicles. × 32900. (From THAEMERT 1966)

cosities also occur (Fig. 54). VIRÁGH and PORTE (1961) reported close associations where the gap between the myocardial cell and the varicosity was 50 nm. NOVI (1968) reported a gap of 66 nm in her study of rat papillary muscle. THAEMERT (1969) observed a number of examples of intimate contacts where the separation between nerve axon and myocardial sarcolemma was only 20 nm and lacked an intervening basal lamina. In many cases the axon varicosities lie in a depression on the surface of the myocardial cell.

Innervation density appears to vary according to species. Few axons have been found in the myocardium of cat and rabbit (HADEK and TALSO 1967) but many are present in mouse heart (THAEMERT 1966). Human ventricular myocardium contains many adrenergic axons, but few have been reported in the atria (CHIBA and YAMAUCHI 1970).

Apart from vesicle-filled varicosities, large varicosities containing numerous mitochondria have also been described within the working myocardium (CHIBA and YAMAUCHI 1970). These presumptive sensory endings are predominantly located in the pericapillary space near the myocardium. They are approximately 1.5 to 3 µm in diameter and also contain some large dense-cored vesicles, glycogen, and lysosome-like bodies.

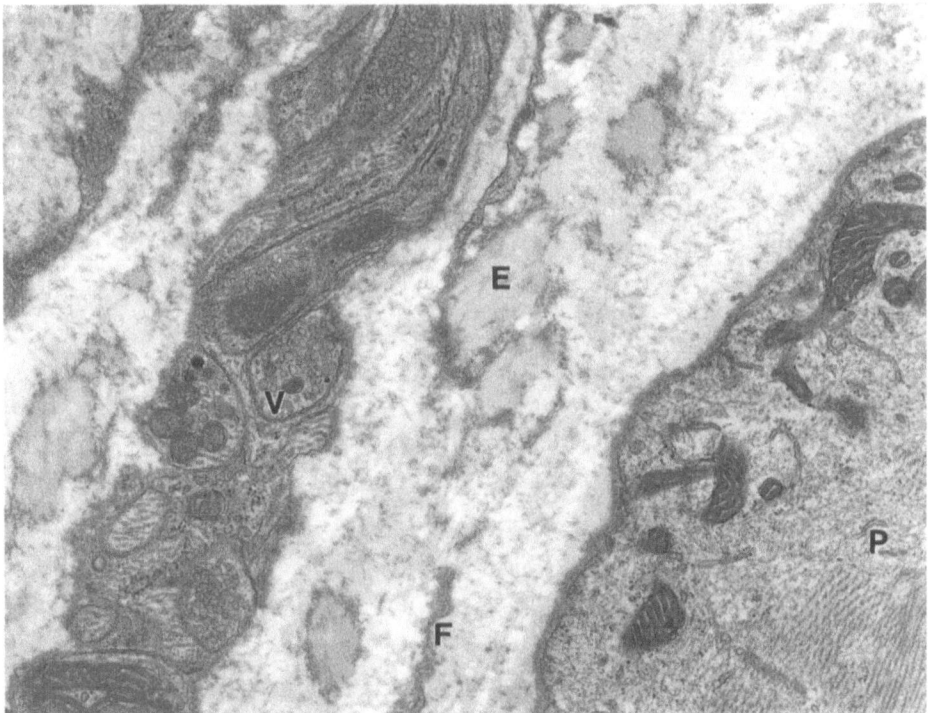

Fig. 55. Diffuse innervation is mediated by many en passant varicosities (*V*) which are distant to the effector cell surface, in this case a Purkinje cell (*P*). Elastin (*E*) and fibroblasts (*F*). × 20 000

3. Innervation of the SA Node

Innervation of the SA node of mammals is very extensive with vesicle-containing varicosities commonly observed in association with typical nodal cells and transitional cells. As well as ganglia in the connective tissue near the SA node, nerve fibres infiltrate the connective tissue framework of the node. NIELSON and OWMAN (1968) suggest that the innervation of the SA node in mammals is so dense that every nodal cell may receive at least one varicosity. The innervation of the guinea-pig SA node has been studied by enzyme histochemistry for AChE activity, and formaldehyde-induced fluorescence for the presence of catecholamines (ANDERSON 1972 a, b, c). Nodal cells are more densely innervated than transitional cells with both AChE-containing nerves and CA-containing nerves, but atrial working myocardium is only sparsely innervated with AChE-containing nerves. The pattern of innervation of the AChE and CA-containing nerves is very similar suggesting that both types run together.

Many ultrastructural studies have reported on the innervation of the SA node in a variety of mammals including man (KAWAMURA and HAYASHI 1966; JAMES 1977, 1980), monkey (KAWAMURA and HAYASHI 1966), dog (KAWAMURA 1961 b; HAYASHI et al. 1970; HAYASHI 1971; TRANUM-JENSEN 1978), rabbit

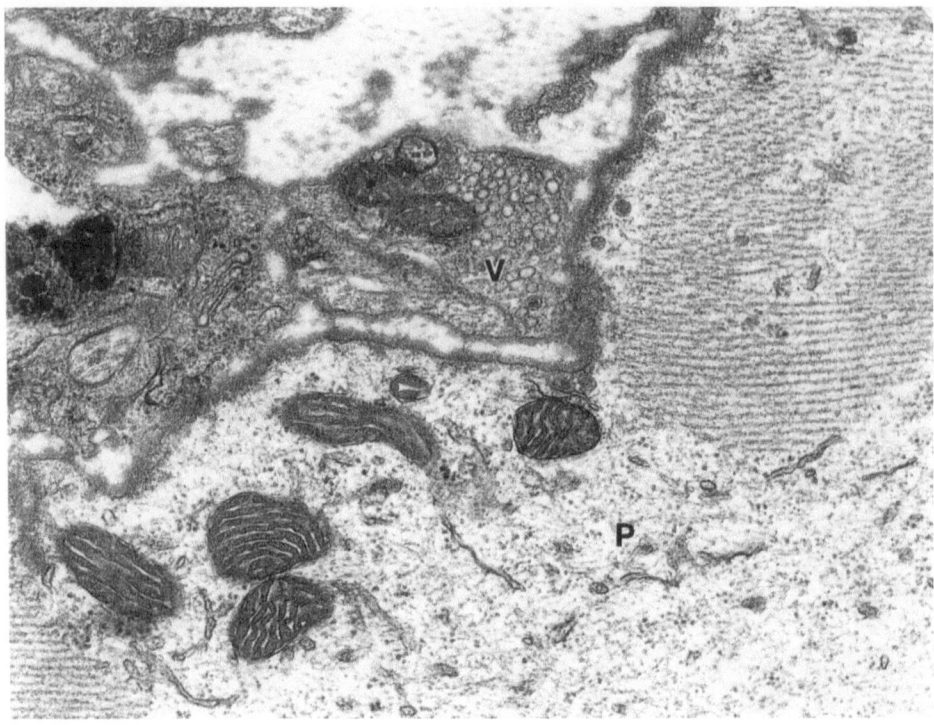

Fig. 56. One form of direct innervation is the close association. Only basal-lamina material and less than 100 nm separate the nerve varicosity (*V*) and Purkinje cell (*P*). × 32500

(Torii 1962; Trautwein and Uchizono 1963; Nilsson and Sporrong 1970; Tranum-Jensen 1976, 1978), rat (Virágh and Porte 1961; Cheng 1971), mouse (Maekawa et al. 1967), cow (Hayashi 1962), squirrel monkey (Colborn and Carsey 1972), sheep (Munnell 1982), pig (Tranum-Jensen 1978), bat (Kawamura et al. 1978), and mole (Kikuchi 1976). All these studies demonstrate an abundance of nonmyelinated nerve axons forming numerous en passant associations with the nodal cells.

Myelinated nerves have been observed in SA nodes of cow (Hayashi 1962) and mole (Kikuchi 1976), but not in other mammals (Tranum-Jensen 1978). Colborn and Carsey (1972) observed myelinated fibres only at the periphery of the SA node in squirrel monkey.

The en passant type of neuromuscular relationship is the most common, however close associations of less than 100 nm with only basal lamina material between the axon varicosity and nodal cell sarcolemma do occur (Hayashi 1971; Kikuchi 1976). Indeed, direct individual innervation of SA nodal cells has been reported with increasing regularity in recent studies (Fig. 57). Hayashi (1962) observed close neuromuscular contacts in cow SA node while Trautwein and Uchizono (1963) noted them in electrophysiologically-determined SA nodal tissue of the rabbit. Intimate contacts have since been described in moles (Kiku-

CHI 1976), rats (CHENG 1971), squirrel monkeys (COLBORN and CARSEY 1971), rabbits and pigs (TRANUM-JENSEN 1976, 1978) and sheep (MUNNELL 1982). The scarcity of intimate contacts visible in routine sampling of the SA node may be considered a consequence of both the small volume of nerve varicosities compared to nodal cells and the small volume of thin sections used in transmission electron microscopy. Hence there is a high probability that sections will not include many examples of intimate neuromuscular contacts (YAMAUCHI 1973). Although this is certainly a valid point, species variation also exists (TRANUM-JENSEN 1978). Intimate contacts are common in the mole SA node (KIKUCHI 1976) but are rare in the dog SA node (HAYASHI et al. 1970; HAYASHI 1971; TRANUM-JENSEN 1978). In the SA node of rabbits and pigs (TRANUM-JENSEN 1976, 1978) intimate contacts appear more common than in the dog but less common than in the mole SA node. Although the innervation of the SA node of the bat is particularly rich, neuromuscular contacts are rarely formed (KAWAMURA et al. 1978).

4. Innervation of the AV Junctional Tissues

AV junctional tissues are well innervated by the large number of nerves in the interstitium of the node (TRUEX 1961; HAYASHI 1962; TORII 1962; TRUEX and SMYTHE 1965 b; JAMES and SPENCE 1966; YAMAUCHI 1969, 1973; VIRÁGH and PORTE 1973 b; VIRÁGH and CHALLICE 1973 a; TRUEX et al. 1974; MOCHET et al. 1975; TRANUM-JENSEN 1976; BECKER and ANDERSON 1976; KAWAMURA et al. 1978; MORAVEC-MOCHET et al. 1978; JAMES 1980; MORAVEC and MORAVEC 1982, 1984). With routine histology, many large nerve bundles and ganglion cells can be observed in the connective tissue around the AV node. Although by far the majority of nerves are unmyelinated, some myelinated nerves can be found near or in the AV node. Myelinated nerves appear to be more common in some animals (e.g. cow, HAYASHI 1962). Within the node and transitional cell zones surrounding it, many small nerve bundles are associated with the conduction cells. Associated with the node are chromaffin cells, parasympathetic and sympathetic ganglion cells and sensory endings (MORAVEC and MORAVEC 1982, 1984). The importance of neural function within the heart has been reviewed by JAMES (1980).

a) Efferent Fibres

Most associations between axon varicosities and AV nodal or transitional cells are of the en passant type with the separation between the myocardial cell and the vesicle-filled varicosity well over 100 nm. This type of nerve ending is the only one reported in the AV node by many authors in a range of animals, such as dog (KAWAMURA 1961 a, b), rabbit (TORII 1962), human (JAMES and SHERF 1968), monkey (VIRÁGH and PORTE 1973 b), rat (KAWAMURA et al. 1978), and cow (HAYASHI 1962).

However, there are some reports of intimate contacts between axons and nodal cells. In the mouse AV node, a variety of arrangements of nerve processes

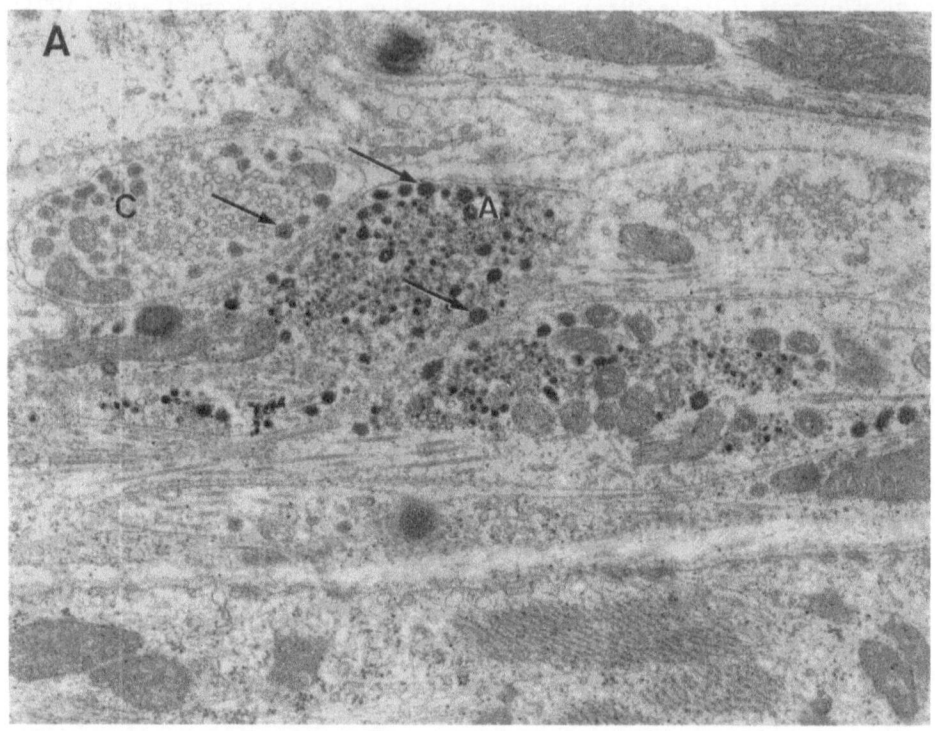

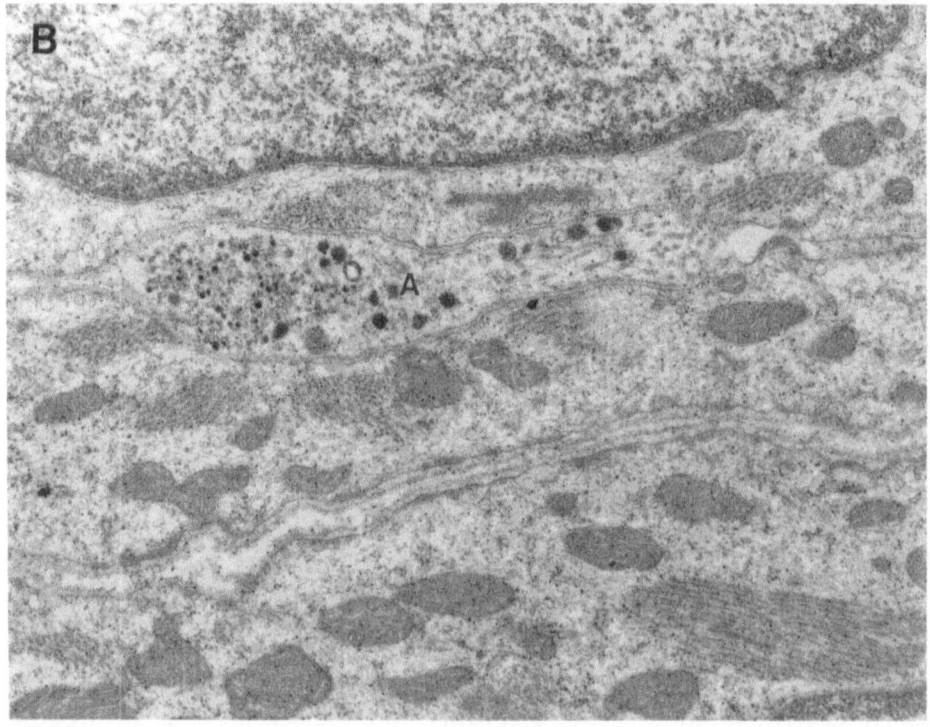

and nodal cells produce intimate associations (THAEMERT 1970, 1973; THAEMERT and EMMETT 1968). Single nerve processes are either surrounded by Schwann cells or without a Schwann cell sheath, residing in tunnels or a cul-de-sac within the nodal cells (Fig. 58). Small bundles or single nerve processes with Schwann cell investments take up positions within shallow or deep grooves on the surface of nodal cells. In all four arrangements, vesicle-filled varicosities form intimate contacts with the nodal cell surface. The vesicles are 30–100 nm in diameter and most are small and clear. In rat AV node, axons are separated by 75 nm or less from the surface of nodal cells (MOCHET et al. 1975; MORAVEC-MOCHET et al. 1977). Basal lamina is interposed between the nerve and the nodal cell membrane. In rabbit heart intimate contacts in all portions of the AV junction, particularly in the transitional cell zone and including the AV bundle, are a common feature (TRANUM-JENSEN 1976). In his study of the innervation of sheep conduction system, MUNNELL (1982) found only one example of an intimate neuromuscular contact with an AV nodal cell. The membrane separation was less than 20 nm and the varicosities contained small clear vesicles.

b) Sensory Endings

The least common kind of nerve varicosity is filled with mitochondria and pleomorphic dense bodies. These varicosities have been interpreted as sensory endings and are AChE-positive and ChE-negative and have a neuromuscular separation of less than 30 nm (MUNNELL 1982; MORAVEC and MORAVEC 1982). They are larger than vesicle-filled varicosities and the mitochondria are densely packed, elongated and smaller than other mitochondria in nerve processes (Fig. 59). Sometimes the mitochondria are outnumbered by pleomorphic dense bodies. A few clear vesicles are also sometimes noted. Similar large varicosities have been reported in the mouse (THAEMERT 1973) and in rat AV node (MORAVEC-MOCHET et al. 1977, 1978). MORAVEC-MOCHET et al. (1977, 1978) identified some nodal cells which formed intimate contacts with both efferent and sensory endings. After vagotomy or thoracic sympathectomy there is no degeneration of the sensory endings, suggesting that the cell bodies of the sensory fibres are located in intrinsic cardiac ganglia (MUNNELL 1982). The apparent complexity and duality of the intrinsic innervation (effector and sensory) has led to a revision of the "relay station concept" for intracardiac ganglia (see E.6).

c) Species Differences and Regional Variation

Inconsistency exists between authors on the incidence of close neuromuscular contacts in the AV nodal region. Two reasons for this have been briefly discussed

Fig. 57 A, B. Innervation of SA nodal cells. A Adrenergic (A) and cholinergic (C) varicosities may occur in close apposition and form en passant association with nodal cells. Both types of varicosity may contain large dense-cored vesicles (arrows) but only adrenergic varicosities contain predominantly small dense-cored vesicles. Varicosities with predominantly small clear vesicles are probably cholinergic. × 24700. B Intimate contact between an adrenergic axon varicosity (A) and two SA nodal cells. The gap from axon to nodal cell membrane is 15 nm. × 28900. (From TRANUM-JENSEN 1976)

Fig. 58. This illustration represents a three-dimensional reconstruction of serial thin sections showing the path of a nerve process within a sarcolemma-lined tunnel inside an AV nodal cell of mouse heart. The nerve process branches and forms a number of varicosities before emerging from the nodal cell. (From THAEMERT 1970)

in Chap. E.IV.3; these are species variation and the low probability of observing neuromuscular contacts in relatively few sections. TORII (1962) found none in the rabbit, but TRANUM-JENSEN (1976) claimed they occurred there regularly. Since the study by THAEMERT (1970, 1973) is the only one employing serial sectioning techniques, it perhaps provides the best indication of how common the intimate neuromuscular junction is in nodal tissue. He suggests that in the inferior portion of the mouse AV node every nodal cell forms at least one intimate contact with a nerve process.

Fig. 59. A sensory ending (*S*) with many small, elongate mitochondria and a few dense bodies lies in an invagination of this AV nodal cell (*NC*) of mouse heart. This varicosity is much larger than the more common vesicle-filled varicosities (*V*), and is in intimate contact with the nodal cell sarcolemma. × 11 130. (From THAEMERT 1973)

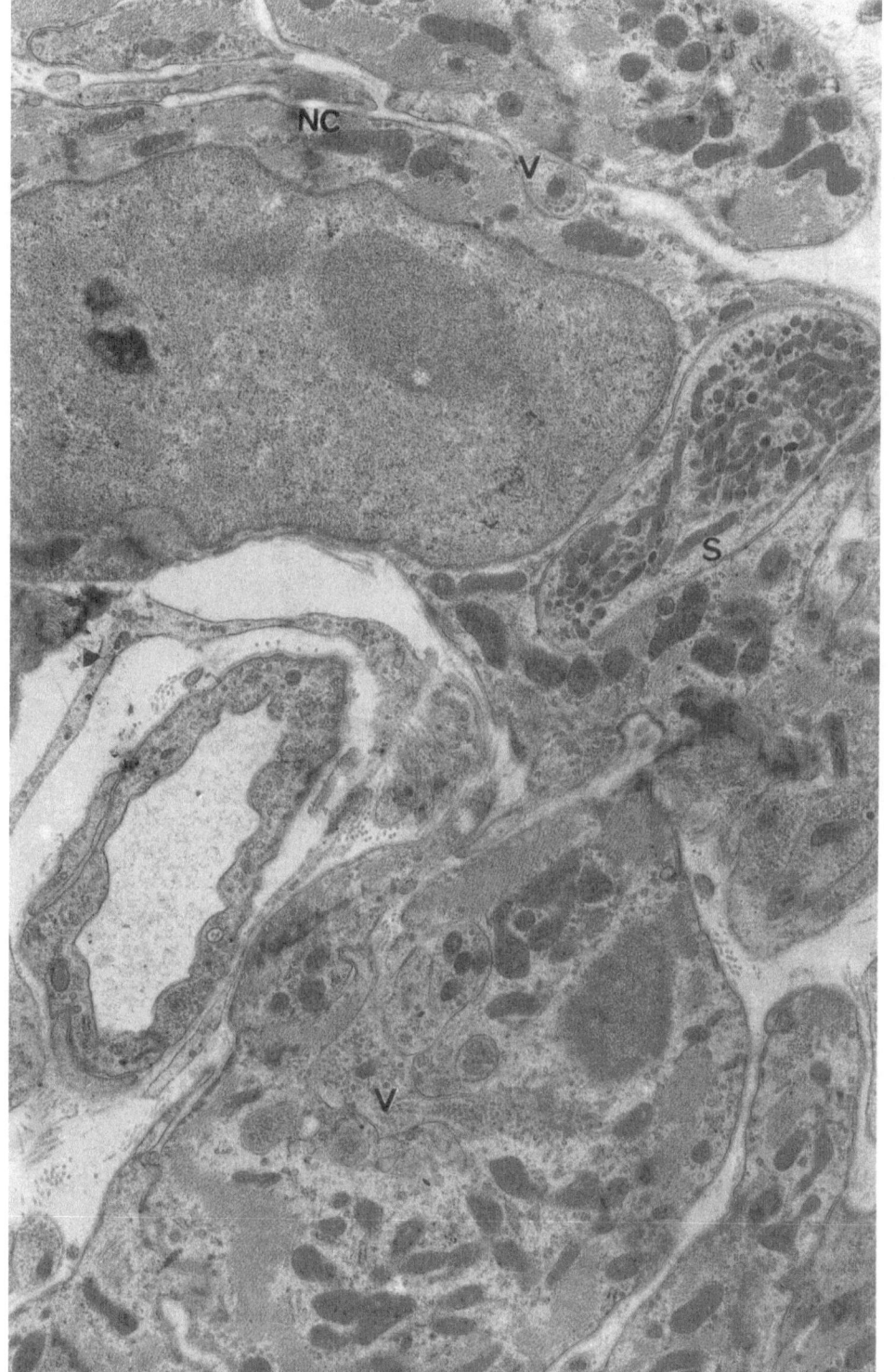

Although regional variations in innervation of the AV junctional tissues have been mapped histochemically in some animals (ANDERSON 1972a, b, c), the ultrastructure of regional variations in nerve-nodal associations has not been examined in detail. THAEMERT (1973) found variations in the frequency of intimate contacts in the mouse AV node. Intimate neuromuscular contacts were common in the lateral portion of the posterior part of the node. This contrasted with the anterior portion which had fewer intimate contacts with nerve processes, whereas the central region was intermediate between the two.

Because of the marked variation between mammals in the distribution of AChE and CA-containing nerves and differences in the structural plan of the AV junctional region, extrapolation between species should be avoided. The difficulty in obtaining information is that serial thin sectioning is a painstaking and laborious procedure. THAEMERT (1970, 1973) was able to fit the width and breadth of the AV node of young mice onto one block-face for thin sectioning. In larger animals this is not possible. The only conclusions that can be reached at the present time are that intimate neuromuscular associations are more common than is apparent from routine sampling for transmission electron microscopy of the AV node and regional variation exists in the incidence of intimate associations, and in the predominant form of innervation.

5. Innervation of the AV Bundle and Purkinje Fibres

A discussion of the innervation of the Purkinje system as determined by histochemical techniques can be found in Chaps. E.III.1a and E.III.2. Briefly, histochemical evidence demonstrates that Purkinje fibres are similarly or less densely innervated with cholinergic nerves than SA node and AV node (KENT et al. 1974; SCHMID et al. 1978). The density of adrenergic innervation seems to be similar to ventricular working myocardium (ANGELAKOS et al. 1963; DICKSON et al. 1981; MUNNELL 1982), although the activity of sympathetic nerves in the conduction system may be greater than in the working myocardium (SCHMID et al. 1979).

a) Nerve Fibre Density

Ultrastructural studies demonstrate a high density of nerve fibres surrounding the AV bundle and Purkinje fibres. No obvious differences in the density of innervation of the SA node, AV node, AV bundle, and right and left bundle branches have been reported in the cow heart (HAYASHI 1962). All parts of the conduction system appear to be very well innervated (Fig. 60) except the peripheral branches of the Purkinje system where fewer nerves are present. Ganglion cells along the penetrating bundle and its branches are also present in the cow and albatross hearts (TRUEX et al. 1961). In canine heart the AV node and false tendons are better innervated than the AV bundle (HAYASHI 1971). TORII (1962) found nerves to be sparse in the Purkinje tissue of the rabbit heart, but TRANUM-JENSEN (1976) reported a high density of nerve fibres

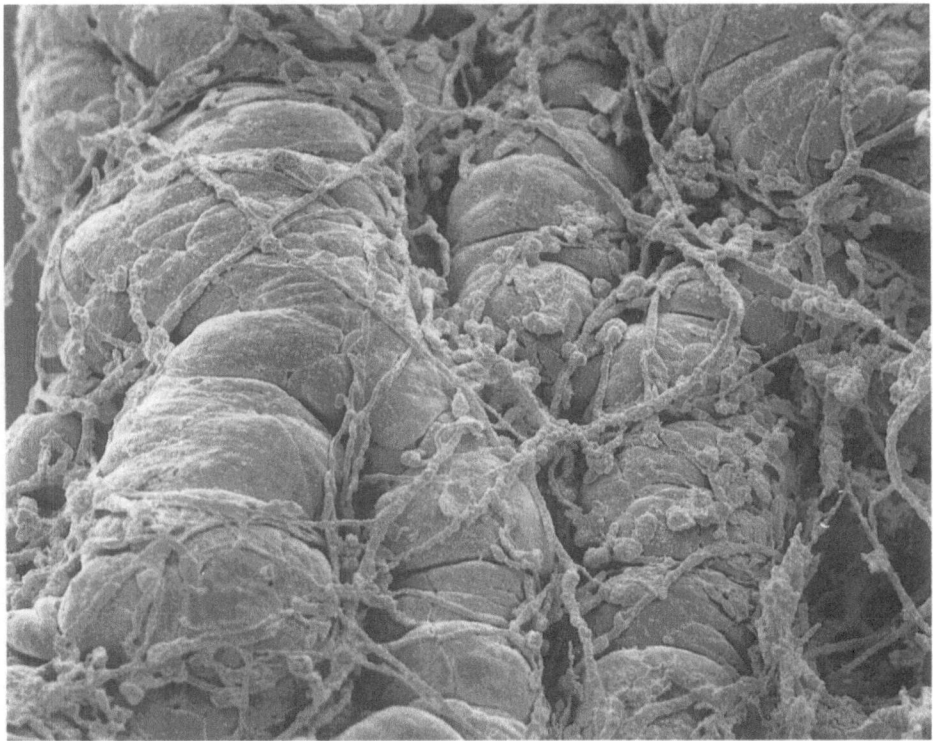

Fig. 60. Scanning electron micrograph illustrating a nerve plexus with en passant endings over Purkinje strands from false tendon, after removal of connective tissue with HCl. × 550. (From CANALE et al. 1983c)

in the AV bundle of this species. Purkinje fibres of the sheep (MUNNELL 1982; CANALE et al. 1983c), pig (BOJSEN-MØLLER and TRANUM-JENSEN 1971), ox, and goat (JENSEN et al. 1978) are surrounded by a dense plexus of nerves.

b) Efferent Fibres

By far the most common form of neuromuscular association is the en passant type (Fig. 55). The large majority are formed by varicosities filled with small clear vesicles although some with predominantly dense-cored vesicles are encountered (JENSEN et al. 1978; CANALE et al. 1983c). The nerve plexus surrounding the Purkinje fibres has been demonstrated in whole mounts of moderator band after cholinesterase staining (BOJSEN-MØLLER and TRANUM-JENSEN 1971) (Fig. 61). This plexus is revealed in detail by scanning electron microscopy after hydrolytic removal of the connective tissue in sheep false tendons (Fig. 60).

Intimate nerve contacts and close associations with Purkinje cells occur in rabbit AV bundle (TRANUM-JENSEN 1976), guinea-pig interventricular septum (HIRANO and OGAWA 1967), and sheep false-tendons (CANALE et al. 1983c).

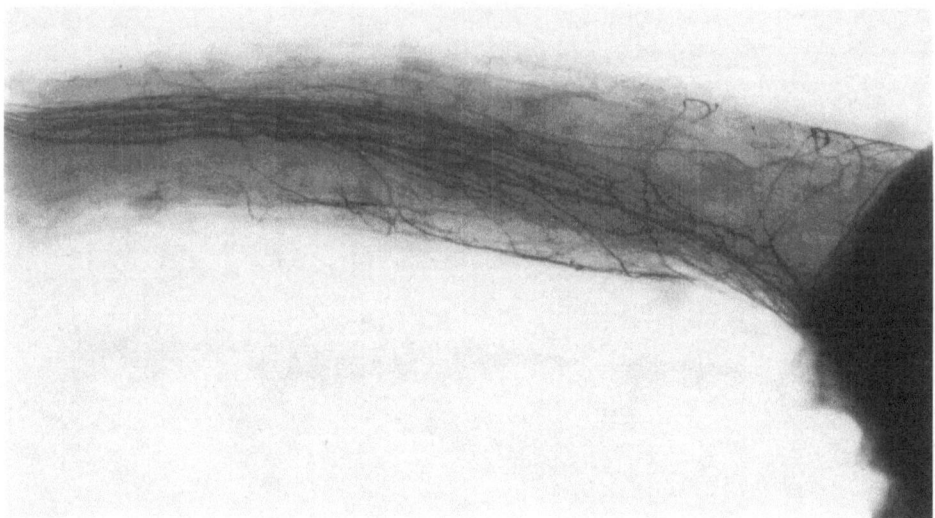

Fig. 61. Whole mount of septomarginal band of the pig showing cholinesterase-containing nerves which run with the main Purkinje fibre bundle before spreading out near the junction with the anterior papillary muscle (on the right). × 16. (From BOJSEN-MØLLER and TRANUM-JENSEN 1971)

In guinea-pig the ultrastructural relationships between nerve varicosities and Purkinje cells (HIRANO and OGAWA 1967) are similar to those in mouse AV node (THAEMERT 1970, 1973). As well as contacts on the outer surface of the cells, the nerve fibres sometimes enter shallow or deep invaginations of the myocardial cell sarcolemma and are AChE-positive (HIRANO and OGAWA 1967). In contrast, nerve axons and their varicosities in sheep false tendons are found in deep clefts, in tunnels formed between two or more coupled Purkinje cells (Fig. 62) and embedded within individual Purkinje cells. In the last type the Purkinje cell wraps around the nerve bundle, and where opposite extensions of Purkinje sarcolemma meet, gap junctions, fasciae adherentes and desmosomes are present (Fig. 63). The observation of a myocardial cell forming "intercellular" junctions between its own cytoplasmic extensions is unusual. Elaborate sarcolemmal outfoldings, many of which contain multivesicular bodies (PAGE et al. 1969; RYBICKA 1978b), are frequently observed near nerve bundles and varicosities. Whether these are involved in Purkinje cell-nerve interaction is unknown but would significantly increase surface area for membrane-transmitter interaction (Fig. 63).

c) Sensory Endings

Mitochondria-filled sensory terminals occur in the sheep Purkinje system (MUNNEL 1982; CANALE et al. 1983c), but no intimate contacts or close associations involving sensory terminals have been observed (Fig. 62).

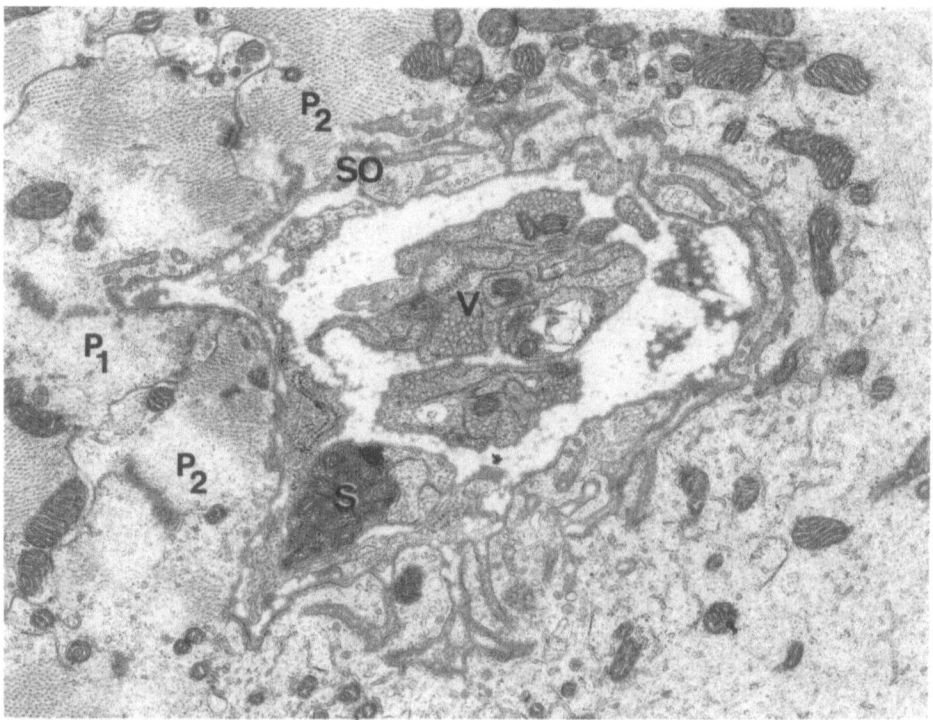

Fig. 62. En passant associations within a tunnel formed by two coupled Purkinje cells (P_1 and P_2). A mitochondria-filled sensory ending (S) is present as well as a few vesicle-filled varicosities (V), and sarcolemmal outfoldings (SO). × 15000

d) Species Differences and Regional Variation

Species variation in the morphology of the Purkinje system and innervation of the AV node suggests there may be species differences in the neuromuscular relationships of the Purkinje system. Regional variation within the Purkinje system is implied by the observation of either no direct innervation or various degrees of innervation (HIRANO and OGAWA 1967; TRANUM-JENSEN 1976; CANALE et al. 1983c). There are no close associations between nerves and Purkinje sarcolemma in the septomarginal band of goat and ox heart (JENSEN et al. 1978), but they are present in large false tendons of sheep heart (CANALE et al. 1983c). Thorough ultrastructural investigations of the innervation of the ventricular Pukinje system have not been made.

Functional relationships between Purkinje fibres and nerves or neural transmitters have been demonstrated (GADSBY et al. 1978; SCHMID et al. 1979; LIPSIUS and GIBBONS 1980; MUBAGWA and CARMELIET 1983; RARDON and BAILEY 1983). However, our knowledge of the functional effects of intimate contacts, close associations and en passant endings is incomplete.

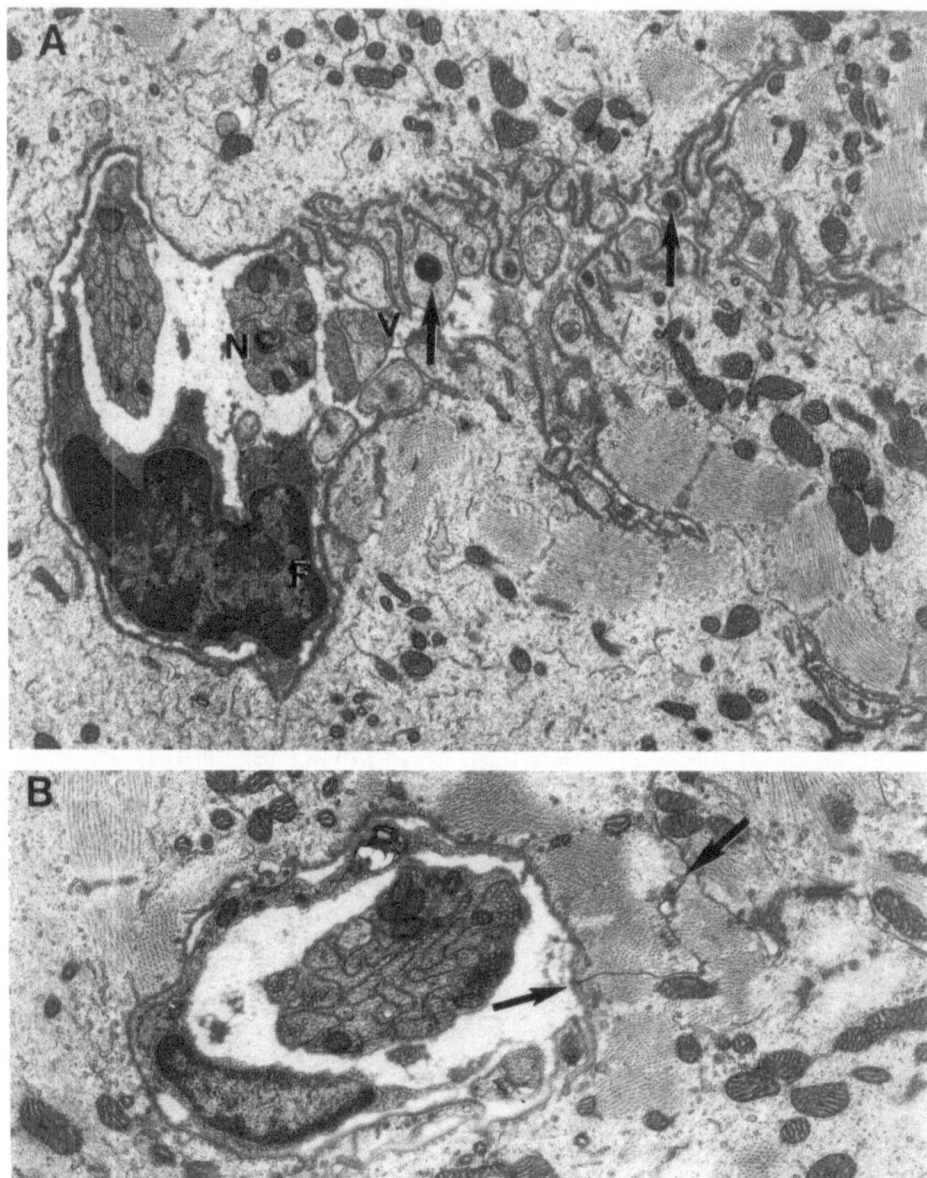

Fig. 63. A In some regions of Purkinje cells of the sheep, nerves are associated with sarcolemmal outfoldings containing multivesicular bodies (*arrowed*). Nerve bundle (*N*), varicosity (*V*), and fibroblast (*F*). × 10000. **B** In some instances nerve bundles are enclosed within a Purkinje cell which forms junctions with its own sarcolemma (*between arrows*). × 16000. (Micrographs from CANALE et al. 1983c)

6. An Alternative Concept for the Role of the Intrinsic Innervation of the Heart

A number of categories of sensory endings have been discussed and the innervation of the specialised AV rings, SA ring bundle and the conduction system have been outlined. An inter-relationship between these various parts of the heart is not clear, particularly since the specialised AV rings are not always well represented (e.g. in man). However, all of these anatomically-defined parts of the heart are well-innervated and have special electrophysiological and/ or ultrastructural properties (E.III.1 b,c). The classic concept of the role of the intracardiac ganglia is that they are simple relay stations for parasympathetic outflow. Sensory information is processed in the extrinsic autonomic nervous system. However, many sensory endings of the free fibre type appear to have their cell bodies located within the heart (PRIOLA 1980; MUNNELL 1982), and many other intracardiac neurons are adrenergic (MORAVEC and MORAVEC 1984).

Based on the fact that the conduction system is in a unique position as both a source of information to the autonomic nervous system and a means of influencing some parameters of heart function, an alternative concept is that the intrinsic innervation of the heart is sufficiently integrated with local sensory feedback to modulate cardiac function on a beat-to-beat basis (MORAVEC and MORAVEC 1982, 1984). In this concept, the extrinsic autonomic nervous component integrates sensory information from intra- and extracardiac sources, modifying the activity of the intrinsic nervous component.

V. Innervation of Non-Mammalian Cardiac Muscle

1. Fishes

Cardiac innervation in fish has been the subject of considerable study, aimed at increasing our understanding of the evolution of neural regulation in the heart. An excellent review by LAURENT et al. (1983) is recommended.

a) Parasympathetic Innervation

Except for the cyclostomes all fishes so far studied possess cardiac inhibitory vagal nerve fibres (McWILLIAM 1884, 1885a; GREENE 1902; IZQUIERDO 1930; YOUNG 1931, 1933, 1936; HIATT 1943; NICOL 1952; JOHANSEN and MARTIN 1965; JOHANSEN et al. 1966; RANDALL 1966; BURNSTOCK 1969; SANTER 1977; ANTHONIOZ et al. 1978). In one group of the cyclostome class, the myxinoids (hagfishes), branches of the vagus nerve join a plexus near the heart but have no effect on heart function when stimulated (AUGUSTINSSON et al. 1956). An early electron-microscopic study of the hagfish heart concurs with physiological evidence that it is aneural (HOFFMEISTER et al. 1961). More recently, a ganglion

cell and some nerve fibres have been observed in the sinus venosus of one
species of hagfish but no associations were observed between the nerves and
the myocardial cells (YAMAUCHI 1980).

The other group of cyclostomes, the lampetroids, possess cardiac innervating
fibres which branch from the vagus and reach the sinus venosus (BERINGER
and HADEK 1972) and bulbus cordis (NAKAO et al. 1981). Their stimulation
produces cardioacceleration which is mimicked by acetylcholine administration,
which also has a negative inotropic effect (AUGUSTINSSON et al. 1956; BURN-
STOCK 1969). The cholinergic receptors in lamprey heart are blocked by tubocu-
rarine but not atropine and so are of the nicotinic type (AUGUSTINSSON et al.
1956; FALCK et al. 1966). The cardioacceleratory action of vagal stimulation
may either be a direct effect of the cholinergic innervation or due to the release
of catecholamines from chromaffin cells mediated by the cholinergic nerve fibres.
However, no axon varicosities associated with chromaffin cells have been found
in the lamprey heart (BERINGER and HADEK 1973).

In elasmobranchs and teleosts, as in mammals, the cholinergic innervation
of the heart is inhibitory and receptors are of the muscarinic type, blocked
by atropine (IZQUIERDO 1930; YOUNG 1936; BURGER and BRADLEY 1951; FALCK
et al. 1966; HOLMGREN 1977).

b) Sympathetic Innervation and Circulating Catecholamines

Paravertebral ganglia are lacking in cyclostome fish and are only poorly
developed in elasmobranchs. Species in both classes lack sympathetic innervation to the heart (IZQUIERDO 1930; LUTZ 1930; YOUNG 1933; HIATT 1943;
OSTLUND 1954; AUGUSTINSSON et al. 1956; OSTLUND et al. 1960; FALCK et al.
1966). However, there is some evidence of a minor sympathetic innervation
of the sinus venosus in a few species of elasmobranch fish (GANNON et al. 1972;
SÆTERSDAL et al. 1975). Lungfish (Dipnoids) also apparently lack cardiac sympathetic innervation (ANTHONIOZ et al. 1978; ABRAHAMSON et al. 1979).

It was originally considered that myxinoids were without cardiac responses
to exogenous catecholamines (OSTLUND 1954; AUGUSTINSSON et al. 1956). However, the isolated hagfish heart becomes responsive to treatment with reserpine.
This is thought to be due to depletion of endogenous stores (BLOOM et al. 1961).
The lamprey and all elasmobranch hearts have positive inotropic and chronotropic responses to catecholamines (IZQUIERDO 1930; HIATT 1943; BURGER and
BRADLEY 1951; OSTLUND 1954; FALCK et al. 1966; MACEY et al. 1984), and
this effect is similarly enhanced after pre-treatment with reserpine (LIGNON
1979). The action of adrenaline and noradrenaline in these hearts is via beta-
receptors (FALCK et al. 1966; LIGNON 1979; ASK 1983), and possibly alpha-
receptors (LIGNON 1979).

It is now firmly established that the hearts of the majority of teleosts receive
sympathetic nerves (GANNON and BURNSTOCK 1969; HOLMGREN 1977; CAMERON
1979; DONALD and CAMPBELL 1982). Cardiac sympathetics are absent in two
closely-related species, the plaice and greenback flounder (FALCK et al. 1966;
SANTER and COBB 1972; SANTER 1977; DONALD and CAMPBELL 1982). In other
teleosts the sympathetic nerves reach the heart with the vagus, forming a vago-

sympathetic trunk. Together with the vagal fibres, the adrenergic fibres innervate the sinus venosus and atrium. In the ventricle, the adrenergic fibres enter along the blood vessels of the compact layer of myocardium. They may penetrate slightly into the spongy myocardium in some species or remain in the compact layer only (GANNON and BURNSTOCK 1969; DONALD and CAMPBELL 1982). In contrast, the heart of the plaice which is without sympathetic innervation, consists of spongy myocardium only (SANTER 1977). These distinguishing features may be related to the relatively inactive lifestyle of this teleost (DONALD and CAMPBELL 1982).

Observations from fluorescence histochemical studies suggest that the nerve fibres observed in the spongy myocardium of the trout ventricle are not adrenergic (YAMAUCHI and BURNSTOCK 1968b; GANNON and BURNSTOCK 1969). Since administration of acetylcholine has no effect on the ventricle of teleost fishes (GANNON 1971; FALCK et al. 1966; COBB and SANTER 1973; HOLMGREN 1977; DONALD and CAMPBELL 1982), the nature of these nerves is still in question.

c) Chromaffin Tissue

In the absence of an adrenal medulla, the source of circulating catecholamines in fish is clusters of chromaffin cells dispersed throughout various tissues (PAIEMENT and McMILLAN 1975). Chromaffin cells synthesize and release catecholamines and thus fluoresce brightly in formaldehyde-induced fluorescence preparations, and are stained brown by dichromate solution (COUPLAND 1972). These cells have been assigned a variety of names, including specific interstitial cells (SIC), and small intensely fluorescent cells (SIF) (see YAMAUCHI 1980).

Chromaffin cells are distributed within the hearts of cyclostomes directly under fenestrated vascular endothelium (LEAK 1969; SAETERSDAL et al. 1975; YAMAUCHI 1980). In lampreys they are most abundant in the sinus venosus and atrium but in hagfish are most numerous in the ventricle (DAHL et al. 1971; BERINGER and HADEK 1972; YAMAUCHI 1980). In lungfish they are absent from the ventricle but line the inner surface of the atrium (ABRAHAMSON et al. 1979).

Cyclostome hearts have a very high catecholamine content (AUGUSTINSSON et al. 1956; EULER and FANGE 1961; STABROVSKY 1967; DASHOW et al. 1982), the source of which is the intracardiac chromaffin tissue (BLOOM et al. 1961; LIGNON 1979). As well, handling markedly enhances circulating catecholamine levels in the lamprey (DASHOW et al. 1982), and this should be taken into consideration in experiments (MACEY et al. 1984).

The normal mechanism of release of catecholamines from the chromaffin cells is generally assumed to be mediated by nerve fibres (LAURENT et al. 1983). However, except for one report (SÆTERSDAL et al. 1975), ultrastructural evidence for the innervation of intracardiac chromaffin cells is lacking (BERINGER and HADEK 1973; PAIEMENT and McMILLAN 1975; OTSUKA et al. 1977; LIGNON and LE DOUARIN 1978). Blood pressure and sensitivity to oxygen tension have been suggested as probable regulatory influences on chromaffin cell catecholamine release in fish hearts (LIGNON and LE DOUARIN 1978).

d) Morphology of Innervation

Nerve fibres are distributed throughout the fish heart in the following order of decreasing density: sinus venosus and pacemaking region, atrium, ventricle (McWILLIAM 1884, 1885a; YAMAUCHI and BURNSTOCK 1968b; GANNON and BURNSTOCK 1969; SANTER and COBB 1972). In more primitive fish (e.g. cyclostomes), nerves may only reach as far as the sinus venosus (BERINGER and HADEK 1972, 1973; YAMAUCHI 1980). Ganglion cells, some of which exhibit catecholamine fluorescence, are present in the region of the SA junction and the walls of the sinus venosus in teleost hearts (YAMAUCHI and BURNSTOCK 1968b; SANTER and COBB 1972; YAMAUCHI et al. 1973; SÆTERSDAL et al. 1974; HOLMGREN 1977).

The fine structure of the nerve fibres and their varicosities is similar to that described in mammals. The varicosities contain round clear vesicles, 20 to 60 nm in diameter (YAMAUCHI 1969). Close contacts (less than 20 nm separation and no intervening basal lamina) between nerve varicosities and myocardial cells are far more numerous in the SA nodal region of teleosts than mammals. Every myocardial cell in the SA node of the trout heart forms at least one close contact with nerves (YAMAUCHI and BURNSTOCK 1968b). The SA nodal region of loach heart is also densely innervated (YAMAUCHI et al. 1973). As well, intimate contacts have been observed in the sinus venosus of lamprey heart (YAMAUCHI 1980).

In the trout atrium, most of the myocardial cells have close neuromuscular contacts, while nerve fibres in the spongy and compact layers of the ventricular myocardium are very sparse (YAMAUCHI and BURNSTOCK 1968b).

The cardiac internuncial cell seems to be a peculiarity of the loach. These cells are interposed between the autonomic nerve fibres and SA nodal cells and are thought to modulate autonomic nerve action (YAMAUCHI et al. 1973). The internuncial cell is devoid of myofibrils, has few mitochondria or ribosomes but has a well-developed smooth endoplasmic reticulum and Golgi apparatus (YAMAUCHI et al. 1973). They form apparently specialised junctions with many axon varicosities surrounding them. These close contacts differ from nerve-myocardial cell contacts in that aggregations of electron-dense material may occur under the presynaptic and post-synaptic membranes. YAMAUCHI et al. (1973) measured an interspace of less than 5 nm in some junctions between internuncial cells and nodal cells but could not be certain that these were gap junctions. In contrast, no specialised junctions were observed where internuncial cells abutted each other.

2. Amphibians and Reptiles

a) Functional Anatomy

All amphibians and reptiles possess inhibitory vagal fibres innervating the heart. In general, the cardiac branches of the vagus enter the heart along the great veins to the sinus venosus (TAXI 1976). Acetylcholine administration mimics the inhibitory action of vagal stimulation, and both are blocked by atropine

(GASKELL 1884a; MILLS 1884, 1885; McWILLIAM 1885b; GARREY and BASS 1937; GARREY and CHASTAIN 1937b; DUFOUR et al. 1956; DE LA LANDE et al. 1962; KIRBY and BURNSTOCK 1969). At low concentrations acetylcholine can cause a negative inotropic effect without a change in rate (GARREY and CHASTAIN 1937a).

The arrangement of paravertebral ganglia and sympathetic nerves is subject to considerable individual and species variation among the lower vertebrates. Sometimes two normally separate ganglia can be fused into one (GASKELL and GADOW 1884). In the frog, sympathetic spinal nerves run cranially to a large ganglion situated just outside the cranium, which is a conglomerate of vagal, glossopharyngeal and sympathetic ganglia (GASKELL and GADOW 1884). Thereafter, the sympathetics accompany the vagus. At least some of the cardiac sympathetic fibres join the vagosympathetic trunk while passing near the ganglion but not entering it (GASKELL 1884a, b).

In tortoises the main cardiac sympathetic nerve comes from the paravertebral ganglion at the level of the tenth spinal nerve. The vagal and sympathetic nerves are usually separate for the greater part of their course but enter the heart together (MILLS 1884). In some animals the sympathetic nerve runs to the vagal ganglion and accompanies the vagus to the heart (GASKELL and GADOW 1884). In alligators the cardiac sympathetic nerve arises at the level of the 11th spinal nerve (2nd or 3rd thoracic ganglia) and remains distinct from the vagus all the way to the heart (GASKELL and GADOW 1884).

Within the frog heart the vago-sympathetic fibres join a series of paired ganglia, the first being REMAK's ganglia in the sinus venosus. From there, connecting fibres reach LUDWIG's ganglia in the interatrial septum and BIDDER's ganglia, also in the interatrial septum but disposed near the AV junction (TAXI 1976). The frog interatrial septum is transparent and this characteristic has been successfully exploited for the study of the ganglia within it. Thus, histological studies suggest that most of the ganglion cells are innervated by a single axon about 1 µm in diameter, which may form up to 27 synaptic boutons (McMAHAN and KUFFLER 1971). Electrophysiological studies suggest that only about 45% receive a single axon and most of the remainder are doubly innervated (DENNIS and SARGENT 1978). The pre-synaptic axon coils around the axon and axon hillock of the innervated ganglion cell body, covering an average 3% of the cell surface. Less frequently, any other part of the nerve cell body may receive a synapse and some are innervated by 2 or 3 axons (McMAHAN and KUFFLER 1971). Recently scanning electron microscopy of these ganglia has demonstrated that varicosities on the axon hillocks sometimes occur in clusters and are often sites of branching of the preganglionic fibre (BALUK and FUJIWARA 1984). None of the ganglion cells exhibit catecholamine fluorescence (WOODS 1970a). Ganglion cells measure 25 to 50 µm along their major axis and 15 to 35 µm along their minor axis (McMAHAN and KUFFLER 1971). In normally-innervated parasympathetic ganglia only the nerve cell surface in the immediate vicinity of synaptic boutons is particularly sensitive to acetylcholine, but after denervation sensitivity becomes uniformly distributed (HARRIS et al. 1971; KUFFLER et al. 1971). Re-innervation leads to many more synapses than in the normal cardiac ganglion (DENNIS and SARGENT 1978).

Adrenaline and noradrenaline administration mimics the positive inotropic and chronotropic responses of the heart to sympathetic nervous stimulation (MILLS 1884; DUFOUR et al. 1956; DE LA LANDE 1962; KIRBY and BURNSTOCK 1969). Adrenaline also enhances relaxation of the myocardium (EHARA 1974). However, cardiac chronotropic responses show seasonal variation in the frog heart. There is a relatively minor response in spring and early summer but this becomes stronger as summer progresses (BURNSTOCK 1969; TAXI 1976). The seasonal variation is not related to normal environmental temperature changes, but artificially low temperatures for 7 to 15 days (cold acclimation) has a marked effect (NICKERSON and NOMAGUCHI 1950). The myocardium's sensitivity to sympathomimetics is lowered in cold-acclimated frogs (TIRRI et al. 1974). It has been hypothesised that alpha- and beta receptors are interconvertible, subject to the influence of temperature, however this remains a controversial issue (KUNOS and NICKERSON 1976; KUNOS 1978; STENE-LARSEN and HELLE 1978b, 1979).

It is now well established that adrenaline, not noradrenaline, is the neurotransmitter in frog heart (FALCK et al. 1963; ANGELAKOS et al. 1965; AZUMA et al. 1965; BANISTER and MANN 1965; ANGELAKOS and KING 1967). The cardiac action of adrenaline is mediated by beta$_2$ receptors and vascular effects by alpha receptors (ERLIJ et al. 1965; STENE-LARSEN and HELLE 1978a). In reptiles, as in mammals, noradrenaline is the neurotransmitter (AZUMA et al. 1965; COOPER et al. 1966).

As in other vertebrates, neural influences in amphibian and reptilian hearts have unequal effects in different parts of the heart. Vagal influence is greater on the atria than on the ventricle (MILLS 1884; GARREY and BASS 1937; DUFOUR et al. 1956). Catecholamine content is greatest in the sinus venosus, then atria and ventricle (ANGELAKOS et al. 1965). Enzyme histochemistry and catecholamine fluorescence demonstrate that sympathetic and parasympathetic nerve fibres are distributed similarly, and are more extensive in the sinus venosus than in the atria (ANGELAKOS et al. 1965; ANGELAKOS and KING 1967; WOODS 1970a). It has been suggested that the myocardium of the turtle ventricle entirely lacks innervation because acetylcholine, in even very large doses, produces no response and large doses of catecholamines result in only a slight inotropic effect (GARREY and CHASTAIN 1937a; HIATT and GARREY 1943). In contrast, frog ventricle develops marked inotropic and chronotropic responses (GARREY and CHASTAIN 1937b; COOPER et al. 1966).

Fluorescence-histochemical studies have demonstrated sympathetic nerves in the ventricle as well as atria of the turtle heart (OTSUKA and TOMISAWA 1969; cited in YAMAUCHI 1969; CHIBA and YAMAUCHI 1973). The relationship of these fibres to the ventricular myocardium awaits further investigation by transmission electron miroscopy.

Afferent activity of cardiac nerve fibres has been described in frogs and turtles and is believed to come from ventricular and atrial mechanoreceptors which are sensitive to the amount of venous return (KAMENSKAYA et al. 1976, 1977a, b). As well, increases in the "tone" of cardiac muscle induced by increasing the concentration of calcium ions, increases afferent discharges from mechanoreceptors of the isolated frog heart (KONTANI and KOSHIURA 1981).

b) Ultrastructure

Axons without a Schwann cell covering (naked axons) commonly run between myocardial cells and form en passant associations (FAWCETT and SELBY 1958a; GRIMLEY and EDWARDS 1960; STALEY and BENSON 1968; HARTZELL 1980). In turtle atrium, axons are usually between 0.3 and 0.6 µm in diameter and may contain numerous vesicles 20 to 30 nm in diameter (FAWCETT and SELBY 1958a). More rarely, intimate contacts are observed in the frog heart where 20 nm or less separates the nerve and myocardial cell membranes, with no intervening basal lamina visible (THAEMERT 1966). However, there are no regions of widespread intimate myo-neural contact in amphibian and reptilian hearts comparable to the SA nodal region of the teleost fish (YAMAUCHI and BURNSTOCK 1968b; YAMAUCHI et al. 1973).

Two morphological types of nerve varicosities have been observed in amphibians and reptiles. These are similar to other vertebrates and contain predominantly small clear vesicles or small dense-cored vesicles (THAEMERT 1966; McMAHAN and KUFFLER 1971). Either type may also contain large dense-cored vesicles. In Amphibia the dense-cored vesicles are slightly larger than those of their mammalian counterparts (YAMAUCHI 1969; WOODS 1970b). In response to pharmacological treatments affecting catecholamine content, small dense-cored vesicles may undergo ultrastructural changes. Administration of alpha-methyl-DOPA or alpha-methyl-tyrosine (inhibitors of noradrenaline synthesis) causes a reduction in the size of the electron-dense core. A greater reduction is achieved with reserpine, and treatment with 5-hydroxydopamine causes an increase in both the size of the cores and vesicles. Variation in the density of the cores is also observed (WOODS 1977).

SIF, or granule-containing cells are present in the sinus venosus and around the base of the great vessels of the turtle heart. They are 10–20 µm in diameter and occur in groups of 8 to 40 cells (CHIBA and YAMAUCHI 1973). The granules are 40 to 200 nm in diameter and the outer membrane of the SIF cell is often in close contact with smooth muscle cells and nerve fibres but bears no relationship to endothelium of blood vessels (in contrast to the cardiac chromaffin cells of fish). In the sinus venosus they are also in intimate contact with ganglion cells (CHIBA and YAMAUCHI 1973). No nerve varicosities with predominantly small dense-cored vesicles have yet been observed in contact with the SIF cells (YAMAUCHI et al. 1975).

Reciprocal synapses (i.e. where an axon receives a synapse from the cell it innervates), occur between cholinergic axons and the granule-containing cells. In this case, where the polarity of the synapse is from cholinergic axon to SIF cell, aggregations of small clear vesicles are observed under the presynaptic membrane of the axon, but no vesicles are present under the post-synaptic membrane of the SIF cell. The reverse synaptic polarity is interpreted as occurring where SIF cell granules aggregate over the presynaptic membrane (in this case the SIF cell plasmalemma), but no vesicles are accumulated under the cholinergic axon's post-synaptic membrane (YAMAUCHI et al. 1975).

The axon-to-SIF cell synapses are more common than those of SIF cell-to-axon polarity, but the latter cover a larger surface area (YAMAUCHI et al. 1975).

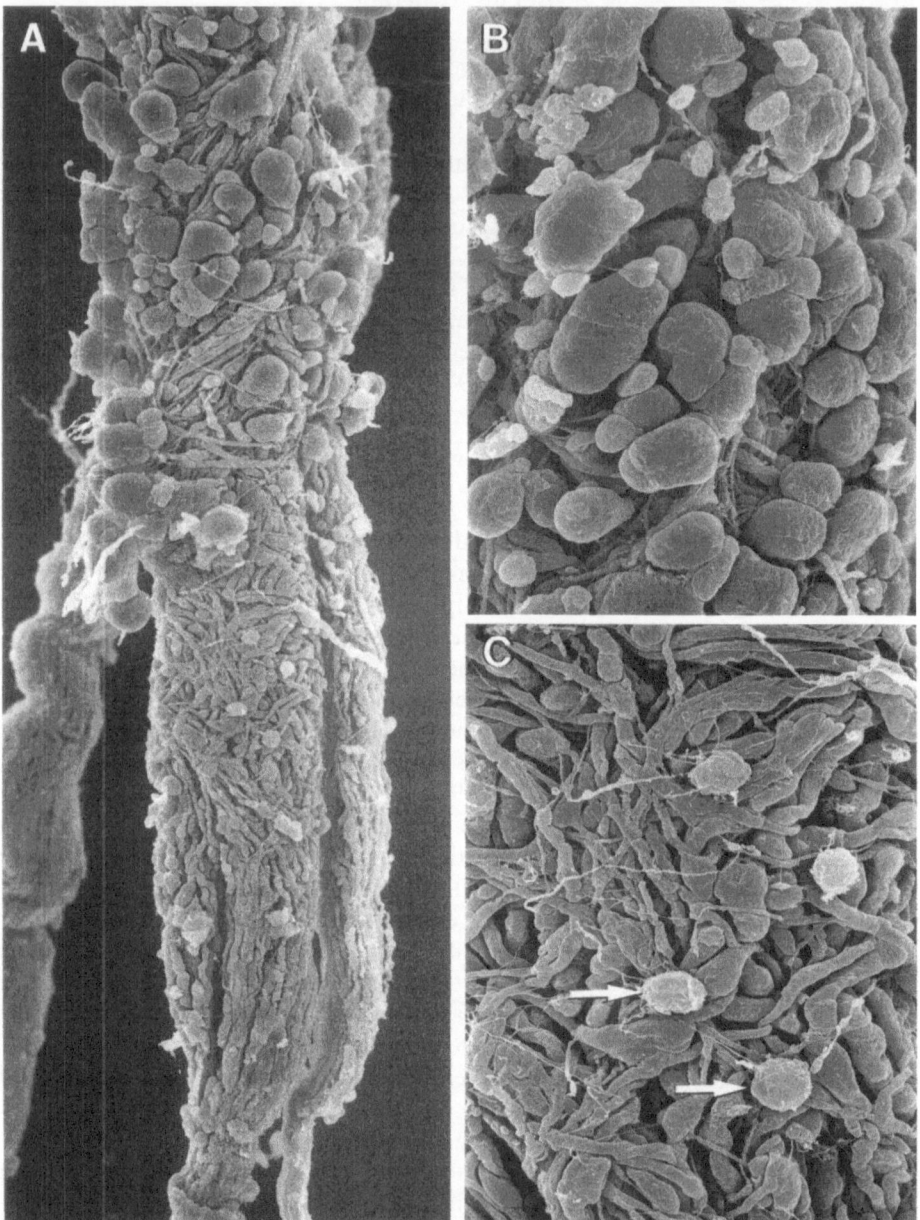

Fig. 64 A–C. Scanning electron micrographs of bullfrog cardiac ganglion from the atrioventricular junction after removal of connective tissue with 8N HCl. **A** Two regions are distinguishable. The upper part is the ganglion and the lower part contains presumed specialised myocardium which has remained in association with the neural elements. × 360. **B** Ganglion cells are clearly distinguishable by their large size. × 700. **C** Cells in the lower part are long, slender and irregularly intertwined. Fibroblasts which have rounded up and nearly removed by the digestion procedure can be observed (*arrowed*). × 1200. Micrographs courtesy of Dr. PETER BALUK and DR. TAKASHI FUJIWARA

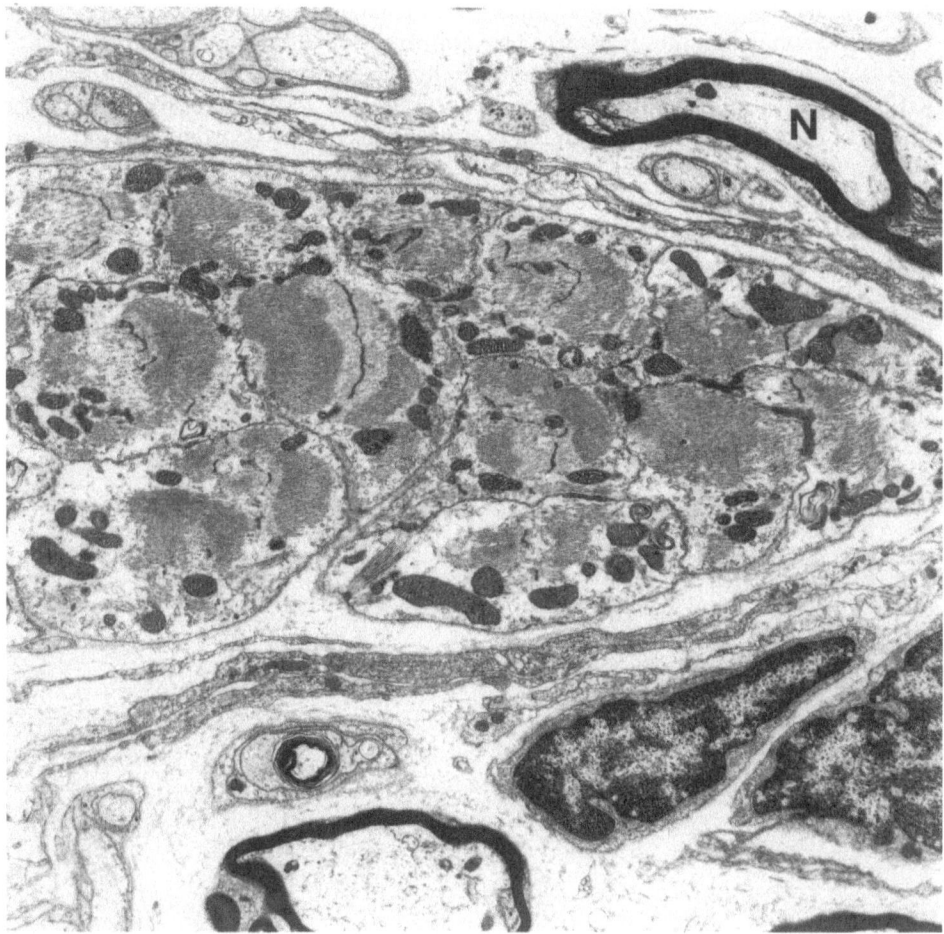

Fig. 65. Transmission electron micrograph from a similar area of Bidder's ganglion as in Fig. 64C, confirming the presence of cardiac myocytes in the ganglion tissue. Myelinated nerve fibres are also present (*N*). ×6100. Tissue generously supplied by Dr. PETER BALUK and Dr. TAKASHI FUJIWARA

In both types of synapse, electron-dense material accumulates under the pre- and post-synaptic membranes but is thicker on the presynaptic side. Degeneration experiments indicate that the nerves associated with the SIF cells are nearly all postganglionic fibres (YAMAUCHI et al. 1975).

A peculiarity of at least some ganglia in amphibian and reptilian hearts is the intermingling of nerve fibres, nerve cell bodies and bundles of myocardial cells within the confines of a ganglion. In the sinus venosus of the turtle heart, this has been named a "somato-muscular complex". Within this complex, 27% of a ganglion cell surface is in intimate contact with cardiac myocytes, and axons are occasionally observed in contact with both a nerve cell body and a myocardial cell (YAMAUCHI et al. 1974). An apparent variant form of the somato-muscular complex has been observed in association with BIDDER's ganglion of the frog heart, using scanning electron microscopy (BALUK and FUJI-WARA, personal communication). The continuation of the main nerve bundle into the dorsal wall of the AV canal includes cardiac myocytes intermingled with the nerve fibres and ganglion cells (Figs. 64, 65), but unlike the turtle (YAMAUCHI et al. 1974), no intimate contacts between myocardial and ganglion cells have yet been observed.

The somato-muscular complex of turtle sinus venosus may exert neural influence on the pacemaker (YAMAUCHI et al. 1974), while the somato-muscular complex in BIDDER's ganglion may be involved in neural regulation of AV conduction. This is consistent with the scanning-electron-microscopic appearance of the myocytes in the complex, which appear larger than normal working myocytes (BALUK and FUJIWARA, personal communication), and with functional evidence that the integrity of the AV junction in the vicinity of BIDDER's ganglia is critical to successful AV conduction in toad hearts (KANNO 1963).

F. Development of Cardiac Muscle

I. Developmental Stages

Developmental studies require an accurate assessment of the progress of development within embryos. Two commonly used indices are embryo size and age.

Size, usually measured as crown-to-rump length (CRL), is often used as a measure of development in humans and other large animals. However, it can be inaccurate due to natural individual variation. In ectothermic embryos such as chick the rate of development is also influenced by environmental temperature (HAMBURGER and HAMILTON 1951).

Age is reliable for some species where the time of mating can be accurately determined. But in species such as chick this may be difficult or impossible to determine. In addition fertilization may occur many hours after mating and this lag may be significant in species of short gestation.

A more accurate means of determining developmental progress is by embryo staging. The stage of development of any normal embryo can be ascertained regardless of differences in size or age, by comparison of external and internal features of the embryo with known standards. Furthermore, staging allows correlation of the timing of developmental events in a single organ such as the heart (O'RAHILLY 1971) and across species (SISSMAN 1970; WITSCHI 1972). Embryos have been staged for the chick (HAMBURGER and HAMILTON 1951), mouse (THEILER 1972), and human (STREETER 1942, 1945, 1948, 1951; HEUSER and CORNER 1957). STREETER's classification ends at horizon 23 which is considered to be the end of the embryonic period. Later fetal stages are usually compared by size and weight.

II. Embryonic Heart Formation

1. Cardiogenic Crescent

The earliest stages of cardiac development have been studied most thoroughly in the chick (RAWLES 1943; DE HAAN 1963a, b; MANASEK 1970; ROSENQUIST and DE HAAN 1966; STALSBERG and DE HAAN 1969) and mouse (CHALLICE and VIRÁGH 1973a,b; VIRÁGH and CHALLICE 1973b; KAUFMAN and NAVARATNAM 1981) and supplemented by studies in other species such as man (DAVIS 1927) and guinea-pig (YOSHINAGA 1921).

In the mouse there is a single heart-forming primordium (KAUFMAN and NAVARATNAM 1981). In the chick, the cells which give rise to the primitive myocardium originate from two paired bilateral primordia which later fuse in the midline (RAWLES 1943; STALSBERG and DE HAAN 1969). The two heart-forming regions are in the splanchnic mesoderm and, although morphologically undifferentiated in the blastoderm at stage 5, their heart-forming capacity is known from transplantation studies (RAWLES 1943). By a process of migration in stages 6 to 12, the cardiogenic regions fuse at the midline to form the cardiogenic crescent. This layer migrates and folds as a cohesive sheet during the formation of the tubular heart (STALSBERG and DE HAAN 1969; ROSENQUIST and DE HAAN 1966)

A small part of the myocardium is not derived from the splanchnic mesoderm but from migrating mesenchymal cells in the septum transversum (VIRÁGH and CHALLICE 1973 b; CHALLICE and VIRÁGH 1973 b). In the mouse heart this involves the ventral side of the developing sinus venosus. Mesenchymal cells extend along sinus endothelium, gradually forming a continuous sheet and develop myofilaments, becoming indistinguishable from other developing myocytes (VIRÁGH and CHALLICE 1973 b). It seems likely that mesenchymal cells from the same area are the source of myocardium lining pulmonary veins in man, dogs, and rodents (see B.IV).

2. Development of the Epicardium

Although PATTEN (1953) and ROMANOFF (1960) considered that the epicardium originated from the outer layer of developing myocardium, the earlier suggestion of KURKIEWICZ (1909) that it was derived from cells of the somatic mesoderm of the septum transversum near the sinus venosus is now firmly established (MANASEK 1969 b, 1980; VIRÁGH and CHALLICE 1973 b, 1981). The cells of the somatic mesoderm rapidly proliferate to form vesicle-like structures which, upon reaching the atria and ventricle, spread out over the surface of the developing myocardium (VIRÁGH and CHALLICE 1973 b, 1981), covering it incompletely at first (MANASEK 1969 b). The proliferating epicardial cells, which contain many ribosomes and a moderate amount of granular endoplasmic reticulum, form desmosomes between themselves and also with underlying myocardial cells. A subepicardial layer of ground substance, collagen fibres and mesenchymal cells develops when the covering is complete.

3. Development of the Endocardium

In chick the endocardium migrates with the precardiac mesoderm in the presomitic embryo to form initially paired endothelial tubes within the tubular heart. The paired tubes soon fuse in the midline in a cranial-to-caudal direction to form a single tube which bifurcates at the developing atrial end of the heart. The endothelial cells migrate as clusters which show some intermingling during formation of the tubular heart (ROSENQUIST and DE HAAN 1966; STALSBERG and DE HAAN 1969; MANASEK 1980).

III. Formation of the Four-Chambered Heart

After formation of the tubular heart, marked growth and configurational changes occur culminating in the formation of a highly efficient four-chambered muscular pump with its own conduction system. This is a rapid process which occurs between the 8th and 10th day in mouse embryo (CHALLICE and VIRÁGH 1973a), stages 11 to 25 in the chick (SISSMAN 1970; MANASEK et al. 1984), and STREETER horizons 10 to 15 in human embryos (O'RAHILLY 1971).

The formation of a four-chambered heart involves looping and septation. These aspects of cardiac morphogenesis have been extensively reviewed by VAN MIEROP (1979) and ICARDO (1984), and only a brief overview will be given here.

1. Looping

The beginning of the looping process is heralded by the bending of the heart tube into a "C" shape which also produces a right-sided bulge. As looping continues the forming venous end comes to lie dorsal to the truncus and cranial to the ventricles so that the heart forms a twisted "S"-shape.

The inital stages of looping occur over a period of hours and are not due to differential cell growth but result from deforming forces within the heart. The start of the looping process coincides with the onset of contraction. STALSBERG (1970) suggested that asymmetry in the first contractions determined the subsequent asymmetrical heart shape. However, hearts arrested by incubation in isotonic medium containing high potassium levels also develop a loop comparable to controls allowed to beat (MANASEK and MONROE 1972), suggesting that the initiation of normal looping is independent of cardiac contraction and of blood flow. The looping process appears to be a complex event involving cytodifferentiation, matrix maturation and fibrillogenesis (MANASEK et al. 1978; MANASEK 1981). The interactions of these factors and their individual contribution to the looping process have been considered in detail by MANASEK et al. (1984), and incorporated into a new model for the looping process. This model is based on the combined effects of (1) a helical arrangement of the developing myofibrils in the tubular heart, (2) inflation of the heart tube by the production of cardiac jelly, with subsequent rotation by virtue of the arrangement of the myofibrils, and (3) different physical properties along one side of the heart tube (dorsal mesocardium) leading to the formation of a bend in the heart tube as a consequence of inflation.

2. Septation

On completion of the looping process the heart has an outward appearance much like its definitive form. However, it is still basically a hollow twisted tube. Haemodynamic and mechanical forces influence septation of the outflow

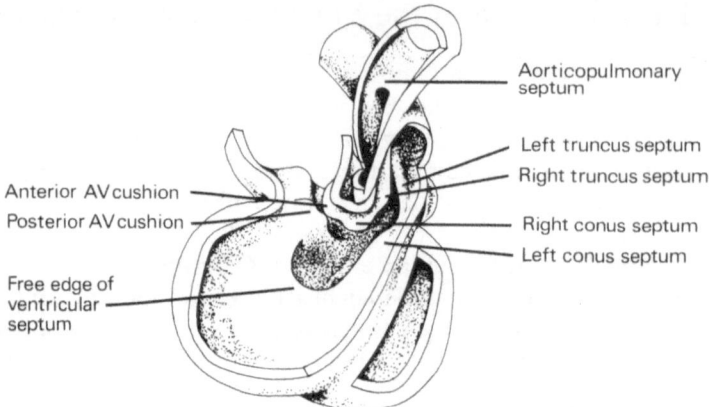

Anterior AV cushion

Posterior AV cushion

Free edge of
ventricular
septum

Aorticopulmonary
septum

Left truncus septum

Right truncus septum

Right conus septum

Left conus septum

Fig. 66. Schematic representation of the continuity of cardiac septa. The figure includes different developmental stages and the atrial septa have been omitted. (From STEDING and SEIDL 1980)

tract and development of the atrioventricular orifices (MCBRIDE et al. 1981; HUTCHINS et al. 1979; STREETER 1948; DE VRIES and SAUNDERS 1962; MOORE et al. 1980; SWEENEY et al. 1980). During septation the ventricle and atrium are divided into left and right chambers, while the atrioventricular canal is transformed into mitral and tricuspid valvular orifices and the truncus into aortic and pulmonary arteries. The process of septation shows regional variation. The formation of the interventricular and interatrial septa involves growth of surrounding myocardial tissue, while the septa have a more passive role. However in septation of the truncus and atrioventricular canal local connective tissue growth plays a major role (ICARDO 1984).

A new hypothesis about the factors influencing septation in the heart has recently been proposed (STEDING et al. 1981; STEDING and SEIDL 1980, 1981), based on the observation that "the anlagen of the cardiac septa are not isolated formations, but rather represent a continuous system of protrusions" (Fig. 66). The protrusions are thought to arise at the sites of depressions in the tubular heart as a consequence of correct looping. The depressions locally alter the originally round lumen to an oval form. MCBRIDE et al. (1981) independently made the same observation in the septation of the outflow tract in staged human embryos. However, ORTS-LLORCA et al. (1982) suggest that the bulbus cordis (conus) and truncus arteriosus are independent portions of the heart without continuity between the bulbar ridges and the septation of the truncus. Others consider that septa originate separately in the bulbus cordis, ventricle and atrium and then unite (see ICARDO 1984). It is now clear that bulbar rotation to re-align the aortic outlow tract on the left side of the interventricular septum does not occur (STEDING et al. 1981; ICARDO 1984). The apparent change in position of the bulbus relative to the atria (BORN 1889) is due to growth of the atria rather than movement of the bulbus (ICARDO 1984). Re-alignment of the aortic outflow tract is not necessary since the dorsal part of the interventricular septum is always to the right of the aorta (STEDING et al. 1981).

IV. Development of the Coronary Circulation

An important component of cardiac morphogenesis is the formation of a coronary circulation. This aspect of development is intimately associated with changes in the characteristics of the myocardium, and will therefore be covered in some detail.

The early embryonic heart is devoid of a definitive coronary circulation. Instead, its metabolic requirements are provided directly from the cardiac chambers and possibly, before formation of the epicardium, from the pericardial fluid.

The wall of the primitive heart tube is lined by endocardium, which in turn is separated from the myocardium by cardiac jelly (DAVIS 1924), an extracellular matrix consisting mainly of glycosaminoglycans and glycoproteins (MANASEK 1975, 1976). After looping of the heart tube, the extent of the cardiac jelly is reduced, and the endocardium extends into intercellular spaces between the myocytes, thereby forming trabeculae. Trabeculae are not the result of new growths or protrusions (HARH and PAUL 1975), but appear to arise secondary to preferential growth of the outer layers of the myocardial wall (STEDING and SEIDL 1980). In the atria, the trabeculae remain as simple branched ridges, but in the ventricles, trabeculation of the myocardium results in a labyrinthine and complex system of intramural blind-ending channels termed intertrabecular spaces or sinusoids, which communicate directly with the heart chambers (Figs. 67, 68). The sinusoids may extend as far as the epicardium, and are lined by a continuous layer of endocardium closely applied to the musculature (BENNET 1936). The endocardial cells are firmly connected by interlocking of their free margins and by desmosomes (OBRUCNIK et al. 1972). Trabeculation is also important in the development of the Purkinje system (see G.IV.2).

The combination of trabeculae and sinusoids results in a spongy myocardium. The development of a compact myocardium is associated with the formation of a definitive coronary circulation. This is evident from the progress of normal development in the chick and mammals (DBALY et al. 1968; LICHNOVSKY et al. 1978; OSTADAL et al. 1975), from phylogenetic studies (BRADY and DUBKIN 1964; MINOT 1900; OSTADAL and SCHIEBLER 1971), and from the rare instances of persistent, postnatal spongy myocardium in man (CHENARD et al. 1965; DUSEK et al. 1975; FELDT et al. 1969).

Current understanding of the development of the coronary circulation is largely based on studies in chick and rat embryos supplemented by observations in other mammals, including man. There is broad agreement about the general pattern of development of the coronary arterial and venous vessels, but two crucial questions still await unequivocal clarification. These relate firstly to the mode of formation of myocardial capillaries, and secondly, to the fate of sinusoids.

In most studies, the development of the coronary circulation has been heralded by the appearance of budding, canalized venous sprouts from the region of the sinus venosus (BENNETT 1936; VIRÁGH and CHALLICE 1981) or the coronary sinus (GOLDSMITH and BUTLER 1937; GRANT 1926), with the anlagen of

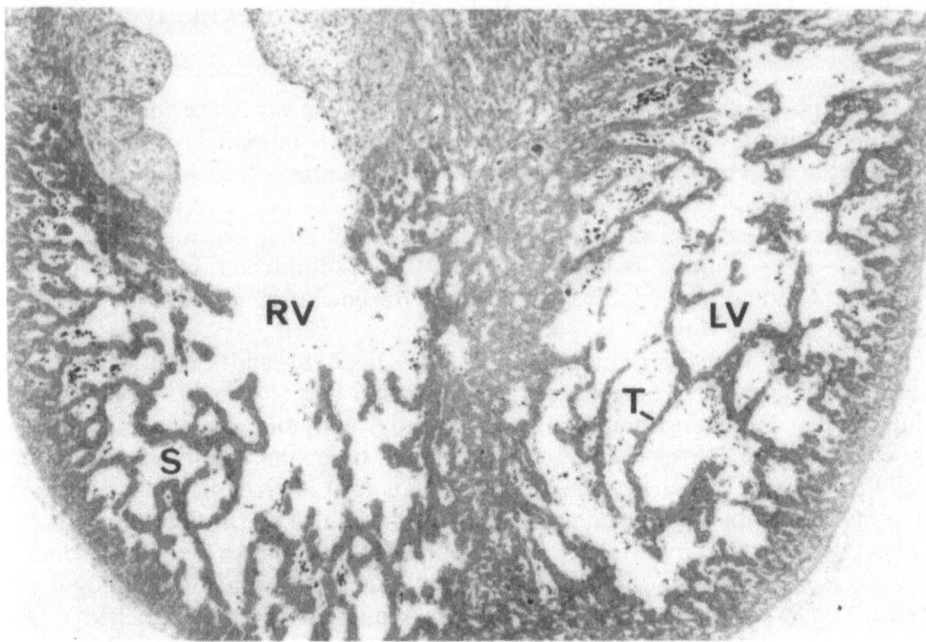

Fig. 67. A section through the heart of an 8-day chick embryo (stage 30 Hamilton Hamburger). The walls of the left (*LV*) and right (*RV*) ventricles contain many muscular trabeculae (*T*) separated by sinusoids (*S*). × 70

the coronary arteries appearing later as buds from the left and right sinuses of Valsalva at the root of the aorta. However, a recent report (CONTE and PELLEGRINI 1984) suggests that in human embryos the cardiac veins and arteries develop simultaneously, but independently. The left coronary artery bud usually appears before the right (BENNETT 1936; HIRAKOW 1983). In pig and human embryos, these buds may show a transient narrowing of their lumen at their commencement (BENNETT 1936; RAO 1958). The developing arteries branch and spread over the epicardial surface of the heart. Ventricular vascularization proceeds from base to apex, dorsal to ventral, and left to right (RYCHTER et al. 1972) so that vascularization usually terminates on the ventral wall of the right ventricle (RYCHTER et al. 1975). The arteries join onto an already formed terminal vascular bed (DBALY et al. 1968; GRANT 1926; RYCHTER and OSTADAL 1971).

The point in gestation at which the coronary circulation develops is species-dependent, and related to the length of gestation. Development begins in the first quarter of gestation in large mammals such as pig (BENNETT 1936; GOLD-SMITH and BUTLER 1937) and man (COOPER and O'RAHILLY 1971; HIRAKOW 1983; LICATA 1954; LICHNOVSKY et al. 1978; RAO 1958; VERNALL 1962), the first third of gestation in chick (OSTADAL et al. 1975; RYCHTER and OSTADAL 1968), and around mid gestation in rabbit (GRANT 1926; LEWIS 1904) and mouse (VIRÁGH and CHALLICE 1973b). However in the rat, coronary vascularization commences in the last quarter of gestation (DBALY et al. 1968; OSTADAL et al.

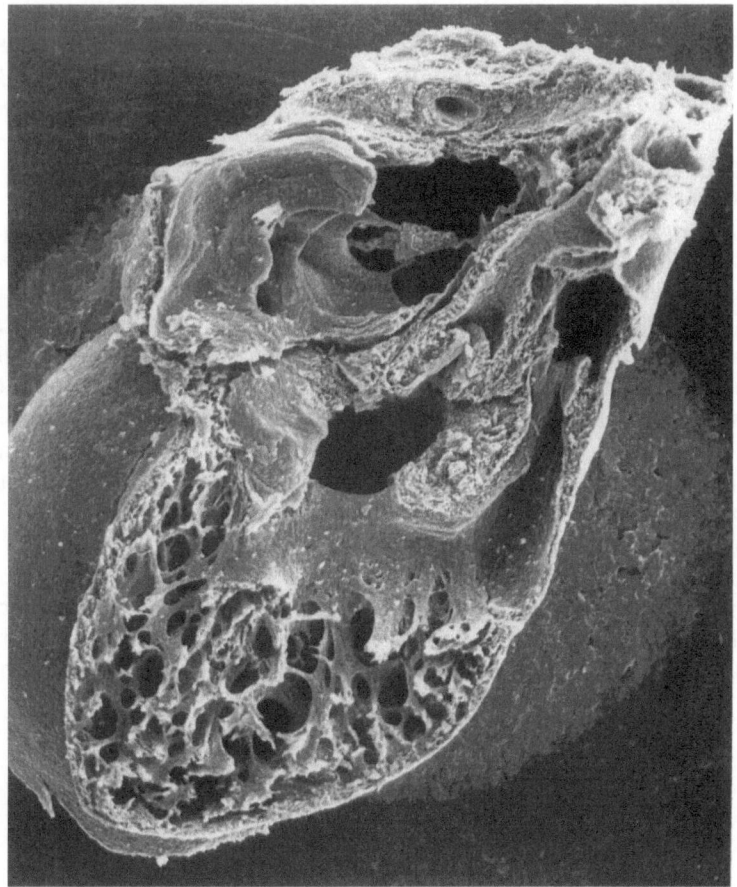

Fig. 68. A scanning electron micrograph of an embryonic heart, illustrating the extensive trabecular system which develops in the ventricles. (From Pexieder 1982)

1975), is completed over the epicardial surface by term (Rychter et al. 1971), and across the myocardial wall by the third or fourth postnatal day (Ostadal et al. 1968). There is less variation between species in the time taken for the spread of vascularization over the epicardium. This process is completed over 6 days in chick, 5 days in rat and 10 days in human embryos (Rychter et al. 1975).

Investigations on the origin of the myocardial capillaries have yielded divergent and conflicting results. Dbaly et al. (1968) and Grant (1926) considered that the capillaries arose by transformation of the sinusoids, whilst Bennett (1936), Goldsmith and Butler (1937) and Rychter and Ostadal (1968) contended that the capillaries originated by differentiation from the growing coronary vasculature. It is difficult to reconcile the latter with the results of Licata (1956) who observed the development of capillaries in the myocardial wall prior to the appearance of coronary arterial or venous buds. Furthermore, Hirakow

(1983) described blood islands in the subepicardial space of the apical incisure and interventricular groove of human embryos as the first sign of coronary vascularization. These blood islands were composed of clusters of primitive erythroblasts surrounded by a thin layer of endothelium. These blood islands appeared to be forming *de novo* because no connection with a sinusoid could be found.

The differing results may in part be due to inherent difficulties of discrimination within the tissue. RYCHTER and OSTADAL (1971) noted little difference in size between the growing coronary vessels and their branches, capillaries and sinusoids, whilst VIRÁGH and CHALLICE (1981) observed that the lumina of developing sinusoids or capillaries could be very small. On the available information, it seems reasonable to infer that the development of the myocardial capillaries is a complex phenomenon in which the capillary anlagen within the subepicardial space may be derived from three sources: a portion of the sinusoids, the growing coronary vessels, and in situ differentiation presumably from mesenchyme. These subepicardial capillaries then proliferate and vascularize the myocardial wall as cords penetrating between myocytes (MANASEK 1971).

Communications between sinusoids and developing capillaries appear to be infrequent (BENNETT 1936; VIRÁGH and CHALLICE 1981) and of little functional significance. These communications are presumably the basis of direct communications between the coronary vasculature and cardiac chambers in the adult. It would seem that many of the sinusoids disappear (BENNETT 1936; GOLDSMITH and BUTLER 1937; LEWIS 1904).

V. Ultrastructure of Developing Cardiac Muscle

1. Cytodifferentiation

The cells of the precardiac splanchnic mesoderm, termed "myoblasts" (MANASEK 1970), are epithelial in character with a cuboidal form, scattered microvilli at their free apical surface, and primitive junctional complexes (CHALLICE and VIRÁGH 1973a, b; MANASEK 1968a). The cytoplasm contains abundant free ribosomes and a few tubules of rough endoplasmic reticulum. The Golgi apparatus is extensive and usually located in the lateral cytoplasm. In the chick, these cells may also contain a few glycogen particles (MANASEK 1968a).

The transformation into definitive cardiac cells or "myocytes" is apparent with the appearance of thick and thin myofilaments within the cytoplasm, and their combination into myofibrils. Simultaneously in the chick, there is an increase in the amount of glycogen which forms large intracellular pools (MANASEK 1968a, 1970). In the mouse, however, VIRÁGH et al. (1982) first observed glycogen particles after the formation of thick and thin filaments. Around the time of transformation, a primary cilium is commonly seen in myocytes and myoblasts, projecting from the free surface or into the intercellular space (MANA-

SEK 1968a). Compared to myoblasts, myocytes also have an elongated appearance (MELAX and LEESON 1969).

Further ultrastructural development of myocytes is an asymmetric process which is associated with a heterogeneous appearance of cells. Adjacent cells, and indeed different regions within the same cell may show differing degrees of development of cytoplasmic contents and membrane specializations.

2. Ventricular Myocardium

In the following section, the development of the major constituents of myocytes only will be considered.

a) Myofibrils

Although HIBBS (1956) described contractions in the chick heart when no myofibrils were visible, it is now firmly established that myocytes contain myofibrils when beating of the heart commences (MANASEK 1968a), but that these may be short or primitive in appearance (MEYER and QUERIOGA 1961).

There is no constant pattern in the sequence of appearance of myofilaments. In guinea-pig (MARKWALD 1973) and mouse embryos (CHALLICE and VIRÁGH 1973a, b), thick filaments appear first whilst in human embryos thin filaments are first (MYKLEBUST et al. 1978b). In chick embryos, thick and thin filaments are evident at the same time (MANASEK 1968a).

The formation of myofibrils occurs by the attachment of thin filaments into amorphous Z substance associated with free ribosomes and rough endoplasmic reticulum (Fig. 69). This Z substance is present in sarcolemmal plaques situated at the sides and ends of the myocyte (Figs. 69, 80) and in sarcoplasmic condensations. In chick embryos, the formation of sarcoplasmic Z discs is preceded by the appearance of ovoid or circular electron-dense bodies 0.15–0.2 μm in diameter, which are not associated with myofilaments (GROSS 1977). Myofibrillogenesis at sarcolemmal plaques results in myofibrils located at the periphery of myocytes and aligned parallel to its long axis (Fig. 70). This pattern is commonly seen in the developing heart of man (LEAK and BURKE 1964), dog (LEGATO 1975), cat (MAYLIE 1982), rat (ANVERSA et al. 1975), sheep (SHELDON 1976a, b), and rabbit (MUIR 1957a). The sarcoplasmic condensations are commonly seen in chick embryos (MANASEK 1968a; WAINRACH and SOTELO 1961) but also in fetal rat (ANVERSA et al. 1975; CHACKO 1976) and newborn dog (LEGATO 1975). These condensations often form "Z centres" (Fig. 71) which radiate myofibrils in many directions, and appear to be the basis of myofibrillar branching seen in development. With ongoing development, myofibrils become distributed throughout the cytoplasm (Fig. 72).

An important role for the Z substance in myofibrillogenesis is suggested by autoradiography which demonstrates newly synthesized protein within the developing myocyte preferentially located at the Z disc (HAGOPIAN et al. 1975).

The rate of myofibrillogenesis is presumably related to functional demands. Morphologically, the presence of thick filaments appears to be a limiting factor

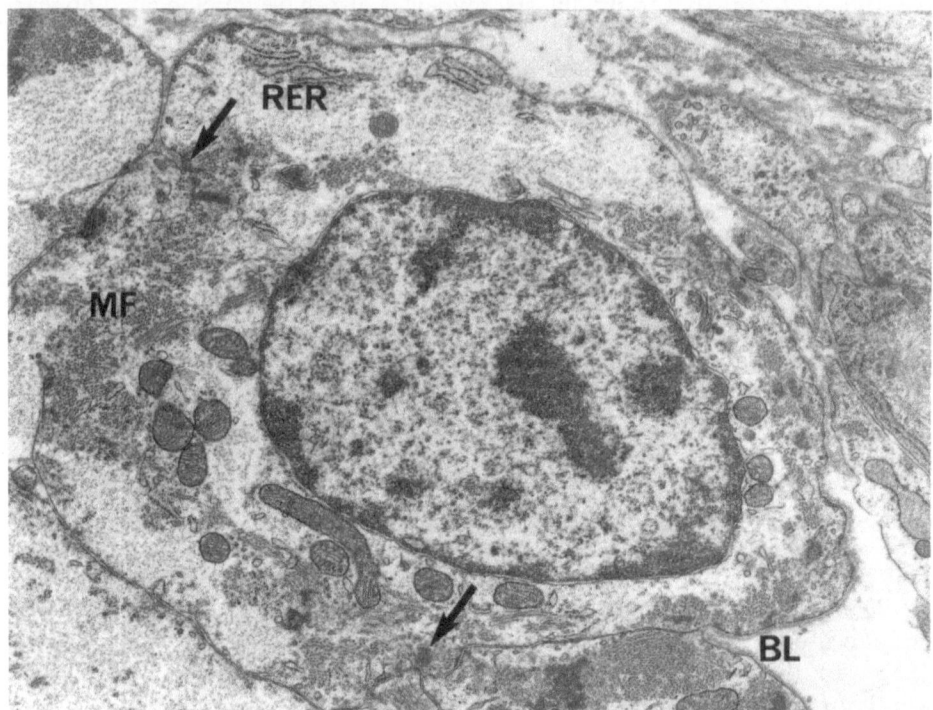

Fig. 69. Developing myocyte from 60-day fetal lamb showing a central nucleus, scattered and sparse myofibrils (*MF*) in various orientations, a few mitochondria, widespread rough endoplasmic reticulum (*RER*), adjacent to myofibrils in many places and abundant glycogen. The basal lamina (*BL*) is reduced at sites of close apposition. Note the sarcolemmal plaques of Z substance (*arrow*). × 13 200

in this process, as isolated bundles of thick filaments are very uncommon, whilst the converse is true for thin filaments (MANASEK 1968a).

Early myofibrils are often random in orientation (CHACKO 1976; CHALLICE and EDWARDS 1961; MANASEK 1970) and short, being only a few sarcomeres in length. However, in microscopic sections it may be difficult to determine whether the short length of these myofibrils is a real phenomenon or a reflection of myofibrillar disarray. Maturation of the myofibrillar apparatus during development involves an increase in organization of the myofibrillar bundles and the myofilaments.

In developing myofibrils containing few filaments, A or I bands are not clearly apparent (Fig. 73). With thickening of the myofibrils, A and I bands are discernible but their margins appear irregular and indistinct (Figs. 74, 75). Transverse sections of such myofibrils, however, show a typical lattice of thick and thin filaments.

In developing myocytes, the Z disc has an irregular shape and a variable width (Fig. 73). With maturation the Z disc narrows, and concomitantly the thin filaments become more closely packed and ordered (MARKWALD 1973).

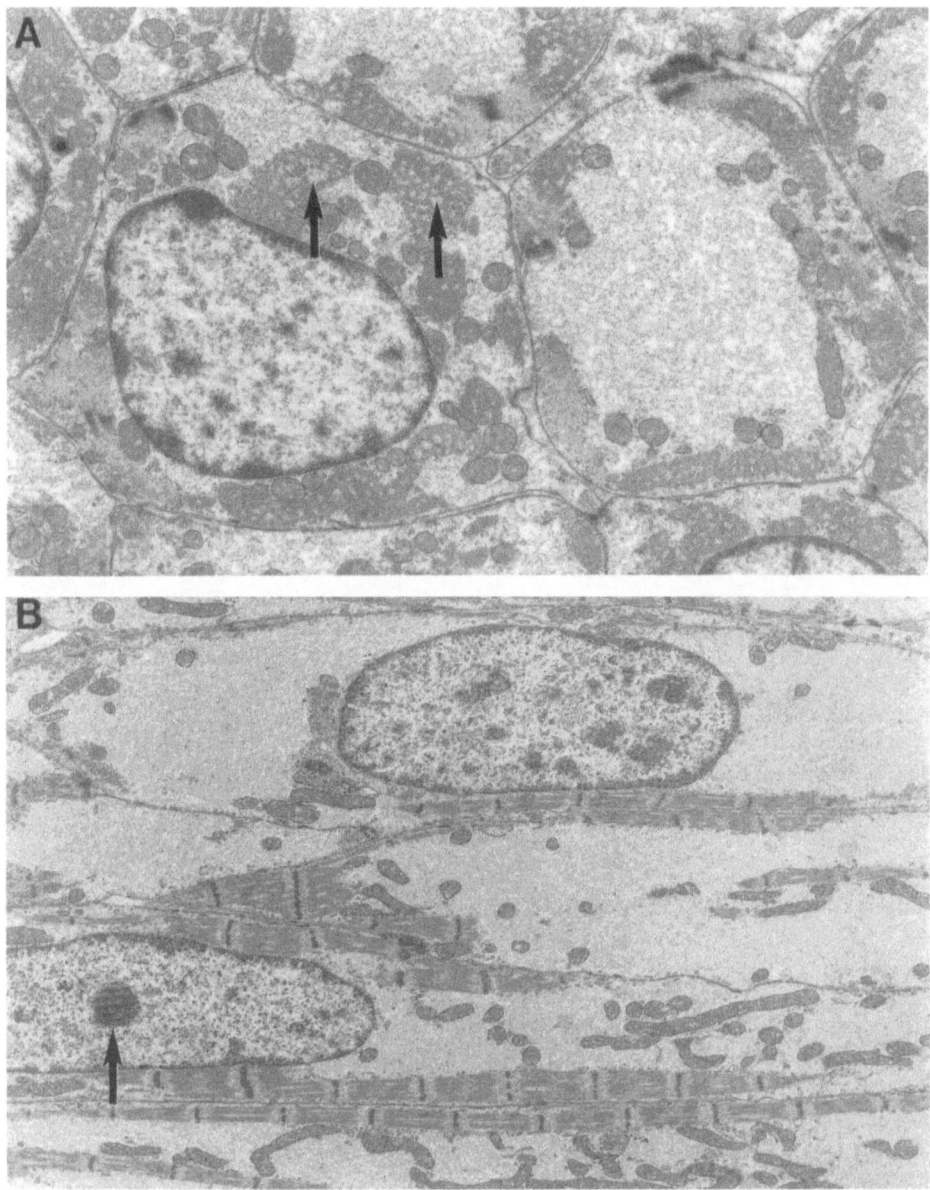

Fig. 70. A Transverse section of myocytes from 60-day fetal lamb showing prominent peripheral myofibrils punctuated by many interfibrillary spaces (*arrows*) containing glycogen. Glycogen is also widespread throughout the remainder of the cells. Note the oval nucleus and scattered mitochondria. × 10000. **B** Longitudinal section of comparable myocytes showing the extent of the glycogen. The nuclei are elongated and may have a prominent nucleolus (*arrow*). Most myofibrils occupy a subsarcolemmal position, but can also be found in the central region of the cells. The mitochondria are of variable shapes and sizes. × 6000

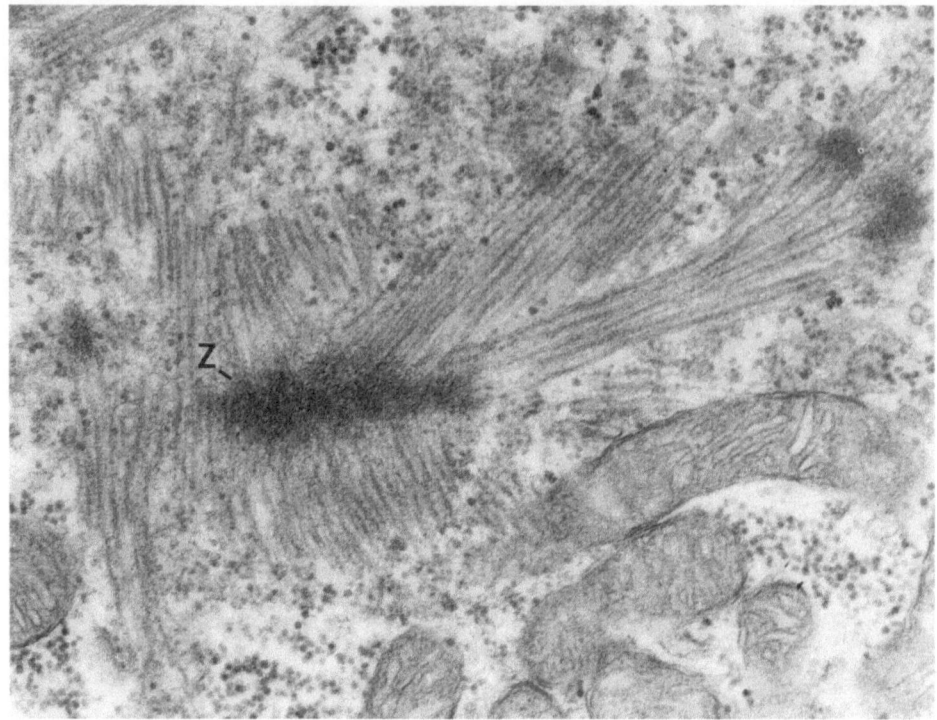

Fig. 71. A Z centre in a 45-day fetal lamb consisting of an irregular Z disc (Z) attached to myofibrils from 4 different directions. × 45000

The Z disc also increases in length with thickening of the myofibrils (MANASEK 1968a).

Although occasional M bands have been described early in development in sheep (BROOK et al. 1984) and man (LEAK and BURKE 1964), they are generally regarded as an endpoint of myofibrillar maturation (FORSGREN et al. 1982). In the cow, M bands appear in the late fetal period (FORSGREN and THORNELL 1981) whilst in the rat, their formation is a postnatal event (ANVERSA et al. 1981). The development of M bands is asynchronous within cells, and even in adjacent sarcomeres. The M band matures gradually with a progressive increase in density associated with an increase in the number of cross-bridges between the thick filaments. After the formation of M bands, the thick filaments display a more regular arrangement.

b) Mitochondria

In the myoblast, mitochondria are small, relatively few in number, and sparse in cristae (CHALLICE and VIRÁGH 1973a, b). In myocytes, the mitochondria are larger in size, elongated and contain a greater number of cristae. Their shape is variable, with cylindrical, dumbell, horseshoe and doughnut forms

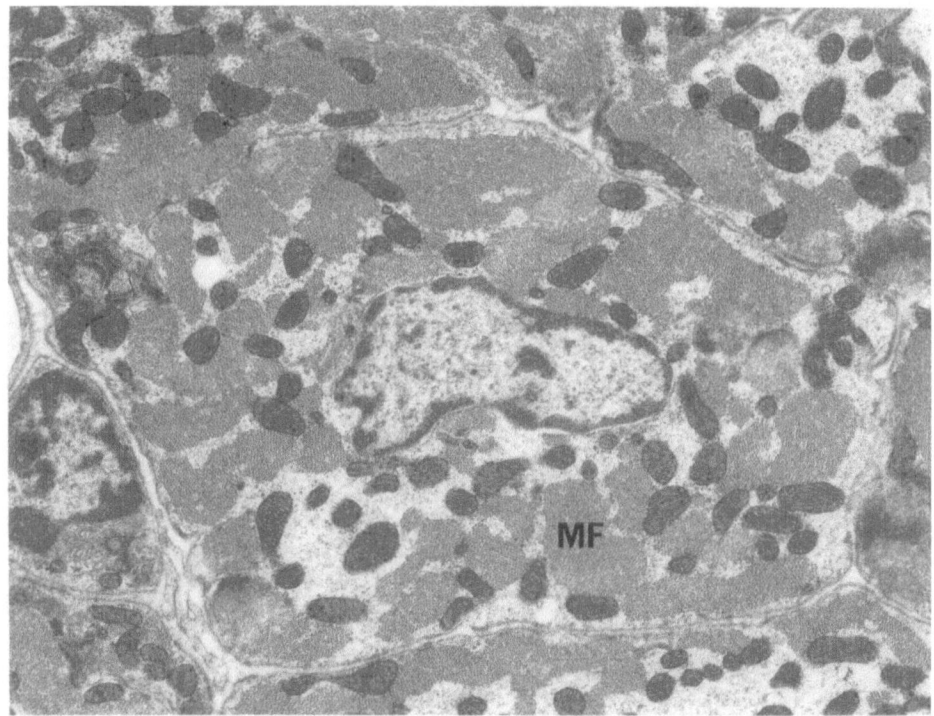

Fig. 72. Transverse section of myocyte from a 142-day fetal lamb (term = 147 days). Myofibrils (*MF*) are distributed throughout the sarcoplasm, and the central nucleus is less prominent. Many mitochondria occupy interfibrillary positions and the extent of the glycogen is diminished compared to earlier ages. × 13 200

being common. Branching is also prevalent. The distribution of mitochondria is initially scattered, but by term in most species many are found in rows between well-aligned myofibrils (ANVERSA et al. 1975a; CHACKO 1976; CHALLICE and VIRÁGH 1973a, b; FORSGREN and THORNELL 1981; SHELDON et al. 1976a, b).

c) Glycogen

Glycogen is abundant in developing myocytes, particularly those constituting the trabeculae (CHALLICE and VIRÁGH 1973a, b; VIRÁGH and CHALLICE 1973b; VIRÁGH et al. 1982). A relative decrease in the amount of glycogen is observed as the myocardial wall changes from the trabecular to the compact form (CHALLICE and VIRÁGH 1973a, b). A further decrease is evident with ongoing development (LEAK and BURKE 1964). Glycogen is initally distributed in large, central pools (Fig. 70), but is also found as discrete particles. Either form may be associated with endoplasmic reticulum and lipid droplets (MANASEK 1968a). Glycogen particles are also evident in elongated spaces in the A band of the myofibrils (Fig. 75). Glycogen-containing lysosomes also occur in developing myocytes, especially in areas of cell death (VIRÁGH et al. 1982). In the mouse,

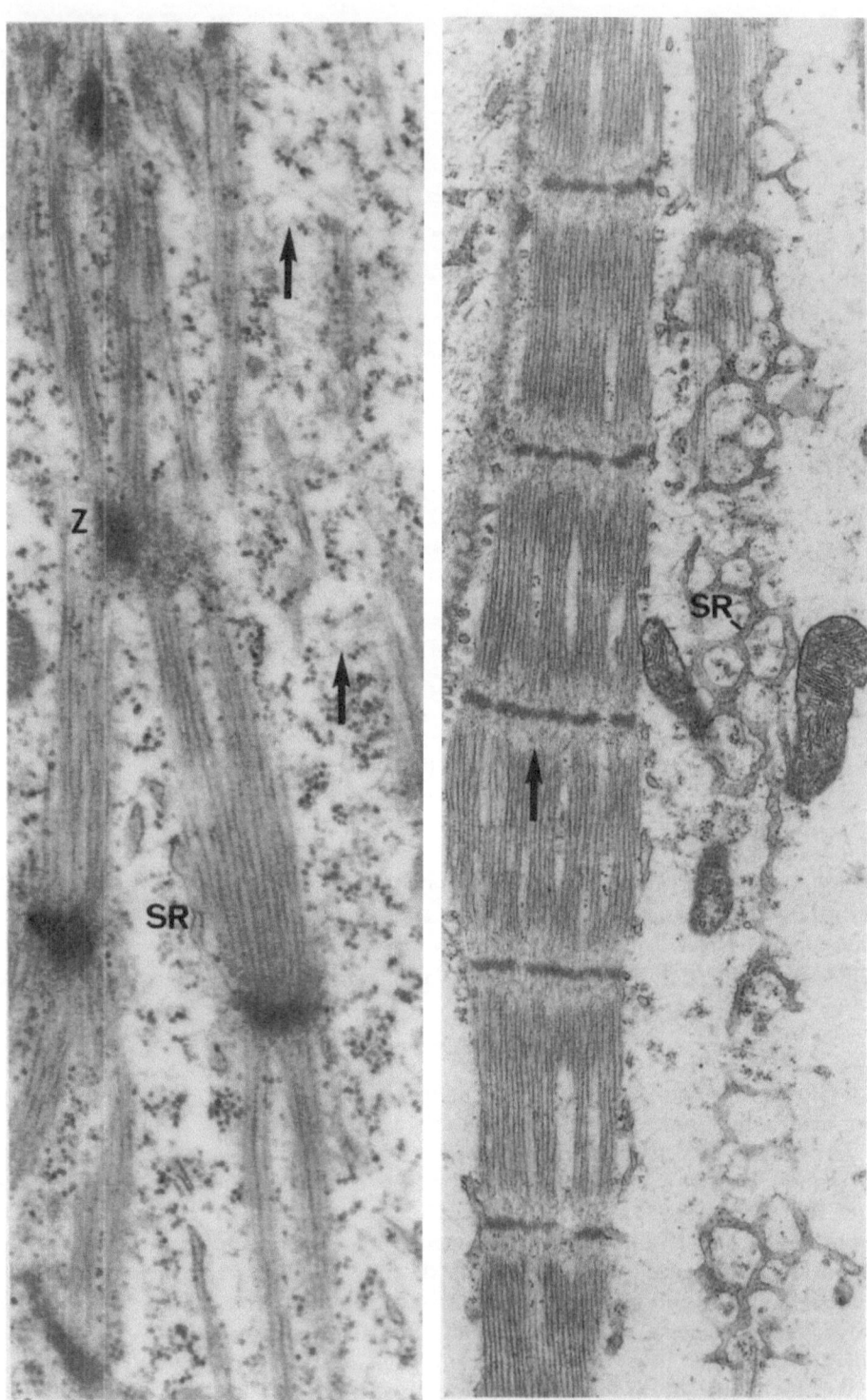

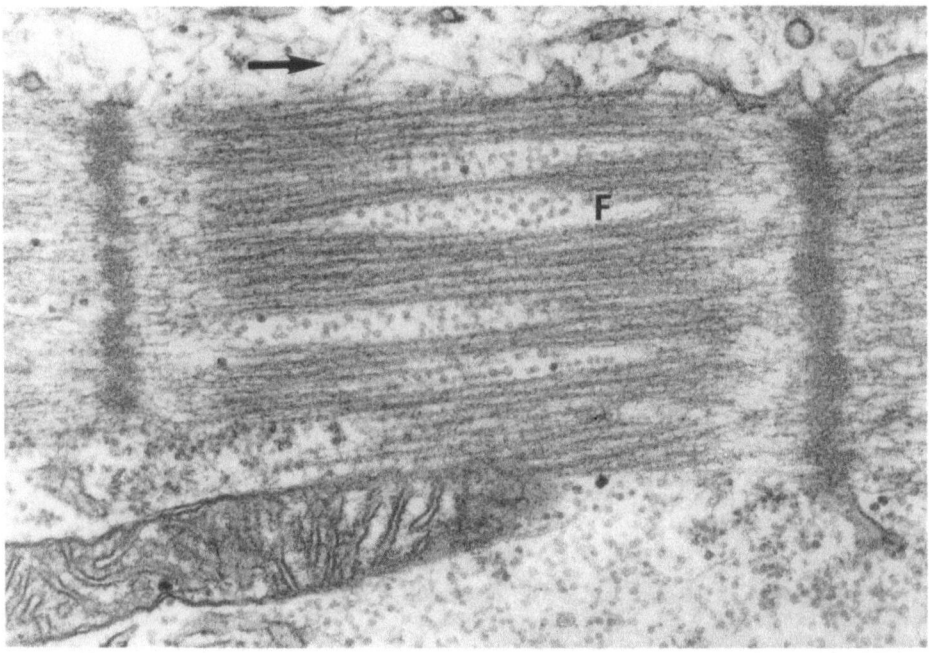

Fig. 75. A sarcomere within the myocardium of a 60-day fetal lamb. The Z discs are irregular and the M band absent. The thick filaments appear to be shortened (*F*) in the region of the intrafibrillary glycogen particles, but in reality are probably bowed. Note the intermediate filaments (*arrow*) and the mitochondrion with sparse cristae. × 51 000

these lysosomes are present throughout the fetal period, but their number increases with establishment of the coronary circulation.

d) Nucleus and Perinuclear Region

The nucleus of the developing myocytes is ovoid in shape and central in location (Figs. 69, 70, 72). A prominent nucleolus is often present. Ribosomes may be attached to the cytoplasmic side of the nuclear envelope, particularly in the neonatal period (LEGATO 1975). The appearance of the nucleus is not dramatically altered throughout development, but its size early in development is quite large relative to that of the myocyte.

The perinuclear region contains many organelles including a prominent Golgi apparatus, mitochondria, and a centriole. The Golgi complex (Figs. 76,

Fig. 73. Branching and strand-like myofibrils from a 45-day fetal lamb with irregular Z discs (*Z*) and indistinct A and I bands. Note the scattered tubules of sarcoplasmic reticulum (*SR*), the widespread 10 nm filaments (*arrows*) and glycogen. × 41 300

Fig. 74. Sarcoplasmic reticulum (*SR*) with a honeycomb pattern in a 60-day fetal lamb, adjacent to a myofibril showing A and I bands with an ill-defined junctional zone (*arrow*). × 20 000

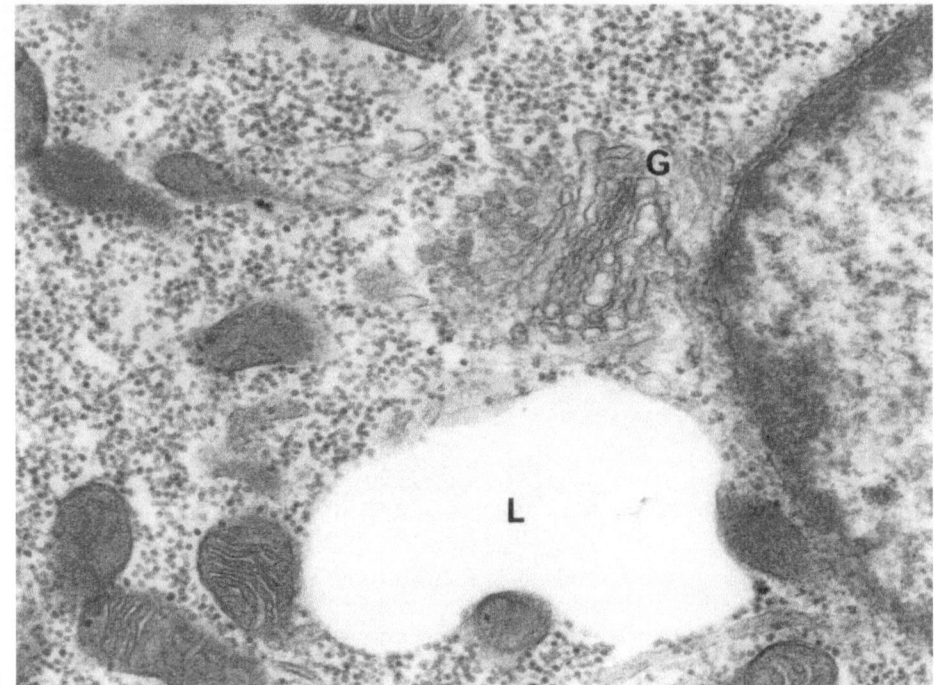

76

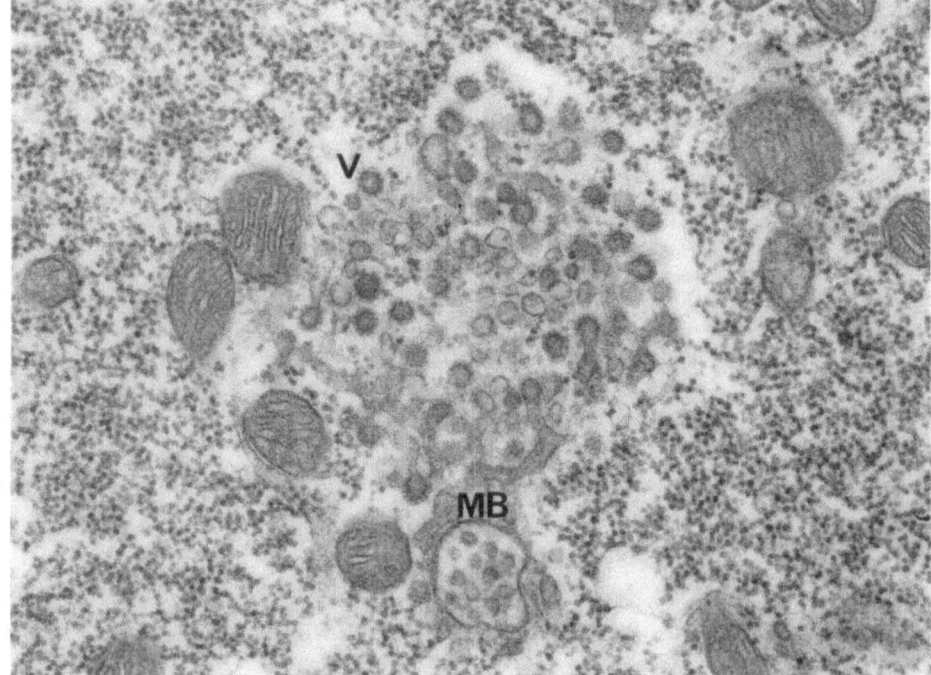

77

77) consists of many lamellae and vesicles, and is frequently associated with bristle-coated vesicles (CHALLICE and VIRÁGH 1973a, b), multivesicular bodies, and dense granules. These granules occur mainly in the atria, but are also found in the ventricular myocytes of chick (MANASEK 1969a), rat embryos (NAKATA 1977), and neonatal dog at 24–72 h of age (LEGATO 1979b).

e) T System

The formation of a transverse or T system is evident with the appearance of tubular invaginations of the plasmalemma opposite the Z disc of myofibrils (Fig. 78). It has been proposed by ISHIKAWA and YAMADA (1975) and FORBES and SPERELAKIS (1976) that the mechanism of formation consists of a proliferation of caveolae, which form rosettes under the plasmalemma and then join to form a tube. This tube is of small diameter, and initially has an opening the size of caveolae. The tip of the tube is closely associated with an aggregation of caveolae. This mode of formation of T tubules has been disputed by FERGUSON and LEESON (1983) in the atrial myocytes of neonatal rat. These investigators did not observe a proliferation of caveolae progressing to T tubule formation, but rather an abundance of caveolae occurring at the same time as tubular invaginations at the Z groove.

The developing T tubule is initially confined to the periphery of the cell, but with further extension, localized varicosities appear which usually form a coupling with the sarcoplasmic reticulum. The course of the growing T tubule with the myocyte may be irregular, oblique or longitudinal. FORBES and SPERELAKIS (1976) observed a basal lamina throughout the developing T tubule, whilst ISHIKAWA and YAMADA (1975) noted this feature within the dilatations, but not the slender portions of the tube.

The development of T tubules is a postnatal event in mammals (PENEFSKY 1983) such as the mouse (CHALLICE and VIRÁGH 1973a, b; ISHIKAWA and YAMADA 1975), cat (GOTOH 1983; MAYLIE 1982; SHERIDAN et al. 1979), rat (HIRAKOW and GOTOH 1975; HIRAKOW et al. 1980; NAKATA 1977), rabbit (HOERTER et al. 1981; PAGE and BUECKER 1981), dog (BISHOP 1973; LEGATO 1975, 1979b), hamster (COLGAN et al. 1978), and opossum (HIRAKOW and KRAUSE 1980). However, T tubule formation in the fetal period occurs in rhesus monkey (ALLEN and CARSTENS 1967), sheep (BROOK et al. 1984; SHELDON et al. 1976a, b), cow (FORSGREN and THORNELL 1981), and guinea pig (FORBES and SPERELAKIS 1976; HIRAKOW and GOTOH 1980).

In fetal shep, SHELDON et al. (1976a, b) first noted a T system with transmission electron microscopy at 110 days (term = 147 days), and transverse myofibrillar ridges at the Z disc with scanning electron microscopy at 90 days. Although

Fig. 76. The perinuclear region in a myocyte from a 115-day fetal lamb, showing a Golgi apparatus (*G*) consisting of tubular lamellae and peripheral vesicles. Note the large lipid droplet (*L*), scattered mitochondria and widespread glycogen. × 45000

Fig. 77. Portion of the Golgi apparatus from a myocyte in a 45-day fetal lamb showing abundant bristle-coated vesicles (*V*), and an adjacent multivesicular body (*MB*). × 45000

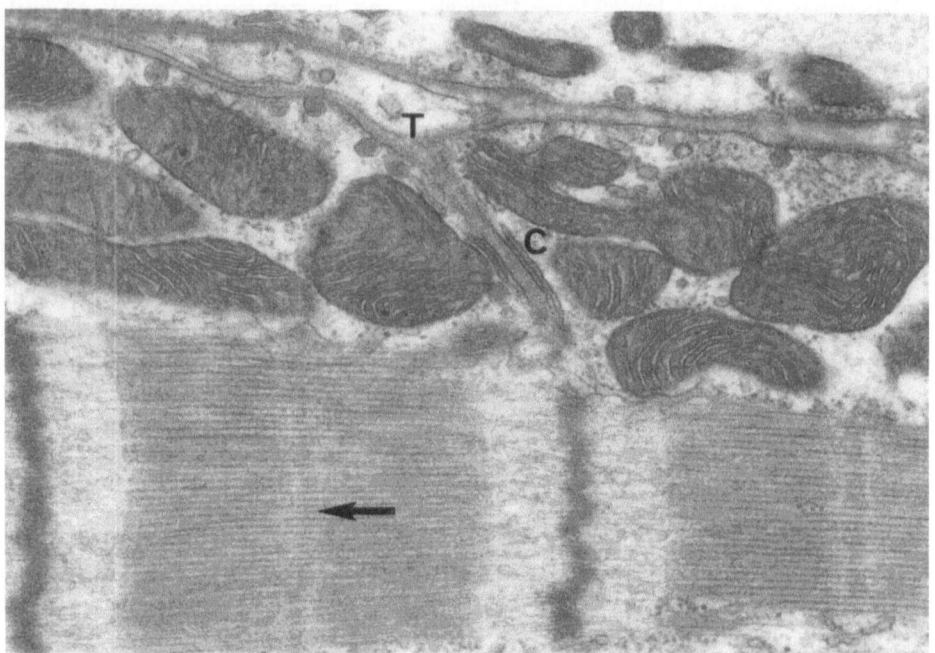

Fig. 78. A developing T tubule (*T*) in a myocyte from an 8-day lamb. The basal lamina is continued into the T tubule, which forms 2 central couplings (*C*) with tubules of sarcoplasmic reticulum. The myofibrils have M bands (*arrow*) and a well-defined A-I junction. × 37500

these ridges were presumed to be part of the T system, it is not clear that scanning electron microscopy can distinguish between a Z disc, T tubule or a tubule of sarcoplasmic reticulum. Furthermore, similar transverse ridges are present in avian cardiac muscle which does not have a T system (see SOMMER and JOHNSON 1979).

f) Sarcoplasmic Reticulum

The sarcoplasmic reticulum (SR) appears to be derived from the rough endoplasmic reticulum, because a direct communication is frequently seen in developing myocytes (ISHIKAWA and YAMADA 1975), particularly at the Z discs (CHALLICE and VIRÁGH 1973). The SR increases in extent with development, and initially appears as simple tubules. With the formation of myofibrils, the SR is seen as vesicles or tubules close to the Z disc (FORSGREN and THORNELL 1981; MANASEK 1968a). A honeycomb pattern subsequently develops (Fig. 74), but this is not frequent in the mouse until the postnatal period (ISHIKAWA and YAMADA 1975).

Typical peripheral couplings between the SR and sarcolemma are present very early in development (CHALLICE and VIRÁGH 1973; FORSGREN and THORNELL 1981; ISHIKAWA and YAMADA 1975), and their number increases with further development.

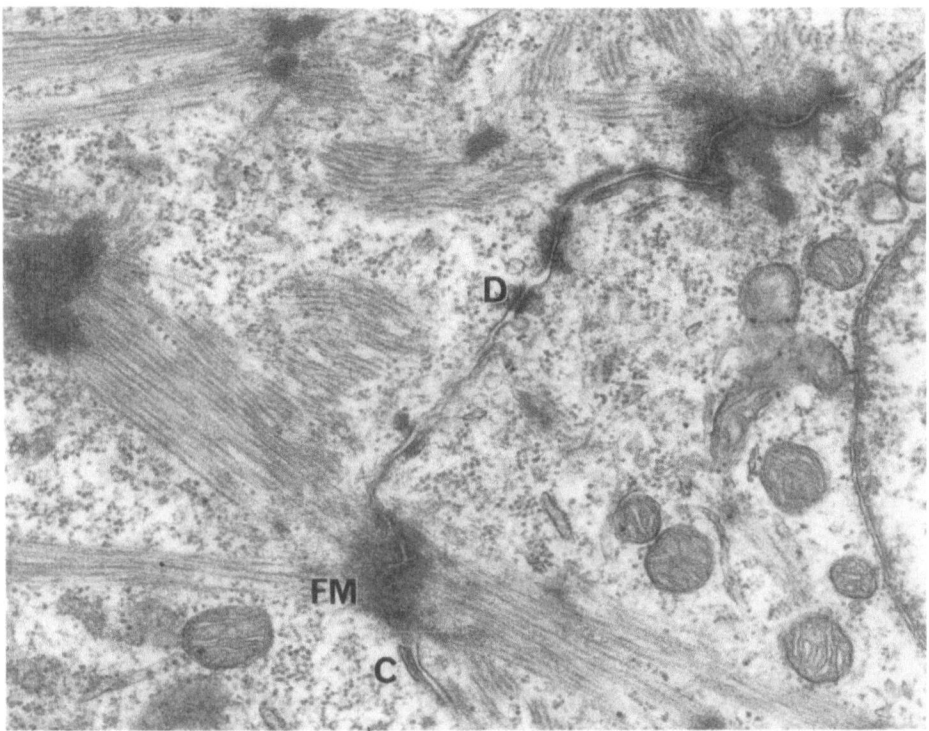

Fig. 79. Developing intercalated discs in a 45-day fetal lamb showing prominent filamentous mats (*FM*), and inserting myofibrils from many directions. Adjacent structures include a desmosome (*D*) and a peripheral coupling of sarcoplasmic reticulum (*C*). × 31 000

Immunofluorescence labelling of Ca^{++}-Mg^{++} ATPase of SR in developing chick myocardium suggests that the SR is present and perhaps functional in the regulation of cytoplasmic Ca^{++} concentration and hence the contraction-relaxation cycle of myocardial cells when the first contraction occurs (JORGENSEN and BASHIR 1984).

g) Intercalated Disc

The appearance of the intercalated disc shows marked changes during development, both in morphology and composition. The developing intercalated disc is initially oblique in rabbit (MUIR 1957a) and dog (LEGATO 1975) but wavy in rat (MELAX and LEESON 1969), sheep (BROOK et al. 1984), and man (LEAK and BURKE 1964). Immature discs are short, being usually less than 1–2 μm wide (Figs. 79, 81). Further development, which is presumably related to mechanical traction and an increase in the volume fraction of myofibrils, is characterized by an increase in complexity of the interdigitation (FORSGREN and THORNELL 1981) and the appearance of stepped end-to-end junctions (McNUTT 1970).

Temporal differences are evident in the development of the components of the intercalated disc. The earliest indication of a future intercalated disc

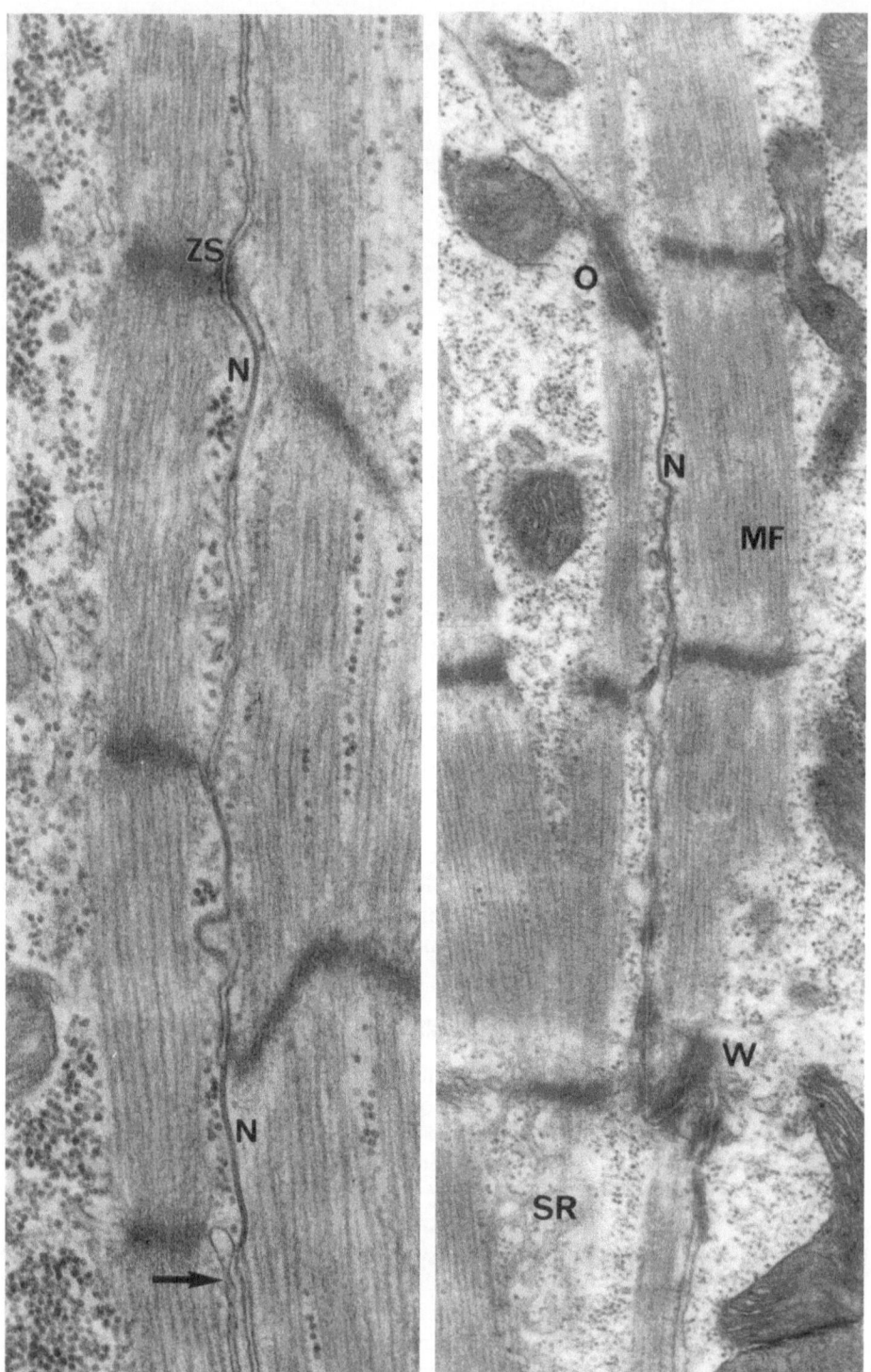

is the appearance of sarcolemmal plaques of electron-dense material, which herald the formation of a fascia adherens junction. These plaques occur at both end-to-end and side-to-side locations between myocytes. At these plaques, thin filaments insert obliquely or perpendicularly (MARKWALD 1973) into a dense filamentous mat with the same appearance as the adult (MUIR 1957a) (Figs. 79, 81). The side-to-side fascia adherens junctions migrate to the ends of the myocyte by differential cell growth (BISHOP and HINE 1975). Desmosomes appear with the increasing complexity of the intercalated disc, and also as isolated side-to-side junctions between myocytes (LEGATO 1979a). Desmosomes are not involved in myofibrillogenesis (see McNUTT 1970), despite such a suggestion by SPIRO and HAGOPIAN (1967) and HAGOPIAN and SPIRO (1970), based on a similarity between the electron density of Z disc material and desmosomes.

Nexal junctions are extremely rare in embryonic myocardium (CHALLICE and VIRÁGH 1973; VIRÁGH and CHALLICE 1973). During development, they are mainly located at the lateral contacts between cells as isolated junctions (Fig. 80) or adjacent to fascia adherens junctions (Fig. 81). Gap junctions often form a peg seen as a circular profile in cross-section. In the rat, before the tenth postnatal day, nexuses constitute a statistically-negligible component of end-to-end junctional areas, and thereafter increase progressively with a concomitant reduction in the amount of non-specialized junctional area (HIRAKOW and GOTOH 1975).

In the rabbit, the mean area of individual gap junctions doubles in the late fetal period, and redoubles between the first and third postnatal days (SHIBATA et al. 1980). These increases correspond to a rearrangement of junctional particles into an aisle configuration with an increase in interparticle distance.

h) Other Cytoplasmic Constituents

The cytoplasm of developing myocytes also contains abundant free ribosomes, rough endoplasmic reticulum, occasional lipid droplets and multivesicular bodies.

3. Mitosis

The increase in cell number within the developing heart is the result of division of myocytes, and not differentiation from a population of proliferating precursor cells (MANASEK 1968b). Unlike skeletal muscle, cardiac myocytes are able to undergo mitosis after the elaboration of myofibrils (CHACKO 1972; HAY

Fig. 80. Short nexuses (N) located close to irregular Z discs, and between lateral cell surfaces in a 45-day fetal lamb. Subsarcolemmal condensations of Z substance are prominent (ZS). Note the tubule of sarcoplasmic reticulum (arrow). × 50000

Fig. 81. Oblique (O) and wavy (W) appearances in a developing intercalated disc in a 115-day fetal lamb. A short nexus (N) is situated at the lateral cell surface. Note honeycombed sarcoplasmic reticulum (SR) and the irregular contour of the A-I junction in the myofibrils (MF). × 32000

and LOW 1972; MANASEK 1968b; MUIR 1957b). During the process of mitosis (Fig. 82), the nuclear envelope becomes discontinuous and disappears, whilst the myofibrils temporarily lose their Z discs, become disorganized and are often displaced to the cell periphery during metaphase, anaphase, and telophase (ER-OKHINA and RUMYANTSEV 1983). Myofibrils are gradually restored late in telophase. Junctions with adjoining cells, including the developing intercalated disc and desmosomes, are retained throughout the mitotic cycle as is the abundant glycogen.

A wide range of evidence indicates that DNA synthesis with mitotic division continues for several weeks after birth (see BISHOP 1973; OVERY and PRIEST 1966; ZAK 1973).

4. Cell Death

During cardiac morphogenesis, cell death is a feature of myocytes containing myofibrils or myofilaments. Cell death is rarely seen in ventricular trabeculae, but is randomly distributed in the remainder of the myocardial wall (MANASEK 1969a). Although the exact role of cell death within the myocardium is not clear, it appears to be a factor in the morphogenesis of the great arteries and outflow tracts of the ventricles in rat (OKAMOTO and SATOW 1976; PEXIEDER 1975), man (PEXIEDER 1975), and chick (HURLE and OJEDA 1979).

The main stages of myocyte death comprise nuclear condensation, rounding of the cell, loss of junctional contacts with adjacent cells, an increase in cytoplasmic density and disruption of myofibrils. These are followed by expulsion into the extracellular space and phagocytosis of the debris by neighbouring myocytes (HURLE et al. 1977, 1978; HURLE and OJEDA 1979) and macrophages.

5. Atrial Myocardium

The myoblasts of the atria are similar in appearance to their ventricular counterparts but they contain a greater number of organelles. In man, developmental events within the atria, including the development of the circulation are at least 2–3 weeks behind the ventricles (OBRUCNIK and LICHNOVSKY 1977). The myocytes also have a poorly developed myofibrillar system. The characteristic feature of atrial myocytes is the presence of atrial granules. These granules, which in man are apparently not found within the ventricles, appear as dense, round to oval membrane-bound structures, 0.25 to 0.5 nm in diameter. They are usually found within the Golgi region and often close to multivesicular bodies.

The development and complexity of the T system in the atria lags behind that of the ventricles (FORBES et al. 1984). Developing tubules are slender and tortuous, with infrequent dilatations. Couplings with junctional sarcoplasmic reticulum are present in both the slender and dilated portions (ISHIKAWA and YAMADA 1975).

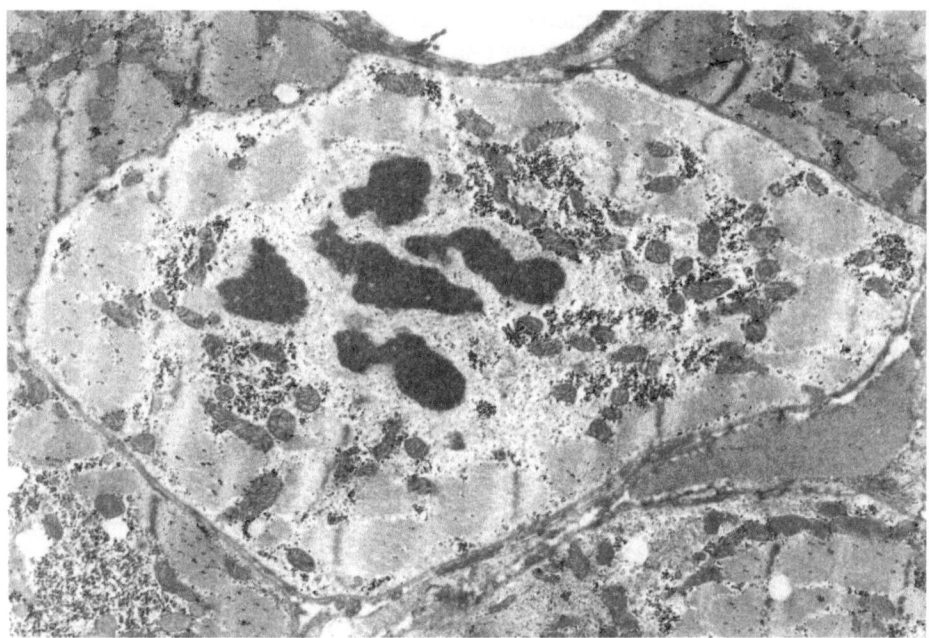

Fig. 82. A myocyte undergoing mitosis in a 4-day lamb. The cell is paler than those surrounding it, and the banding pattern in the myofibrils less distinct. × 7600

VI. Morphometric Changes During the Growth of Myocytes

1. General Features

The assessment of morphometric data from developing myocardium is complicated by inherent species differences in the level of developmental maturity in the perinatal period. For example the level of maturity in the late fetal guinea-pig is equivalent to that in the postnatal rat (see WITSCHI 1972). In effect, the morphometric picture is the summation of the differing levels of developmental maturity, the physiological changes resulting from reorganization of the circulation at birth, and the differing patterns of growth after birth.

During the fetal and early postnatal periods, growth of myocytes is achieved by hyperplasia with a variable amount of hypertrophy. With the subsequent loss of mitotic activity within myocytes, growth occurs by hypertrophy alone (see BISHOP and HINE 1975; ZAK 1973; CLUBB and BISHOP 1984).

During the fetal and postnatal periods an increase in heart weight is associated with a greater relative increase in the protein mass, resulting in an elevated percentage protein composition (LEWIS et al. 1984).

During development, an increasing contractile performance of the heart, in terms of absolute output, appears to be achieved by three main mechanisms.

Firstly, the amount of contractile material within the heart is increased by the cumulative effect of hyperplasia and hypertrophy of myocytes, as well as an increment in the myofibrillar volume fraction. Secondly, the organization of this contractile material is increased by improved alignment of myofibrils and myofilaments. Thirdly, the energy supply to the contractile material measured by the changes in morphology and morphometry of mitochondria is enhanced. Of these three mechanisms, only the first and third can be reliably quantified at present.

In this section, the main morphometric features of development will be described.

2. Dimensions

The dimensions of cardiac myocytes increase during development. After birth, the increase in myocyte diameter and length (ANVERSA et al. 1984) is associated with a decrease in myocyte density (RAKUSAN and POUPA 1963), and a decrease in the fibre-to-capillary ratio from 4–5 at birth, towards unity (BISHOP 1973; RAKUSAN et al. 1965; SHIPLEY et al. 1937).

In the rat (HIRAKOW and GOTOH 1975; HIRAKOW et al. 1980; NAKATA 1977), guinea-pig (HIRAKOW and GOTOH 1980), and cat (GOTOH 1983; SHERIDAN et al. 1979), a progressive increase in diameter occurs in the neonatal period. However, a different pattern has been described in the newborn dog. MUNNELL and GETTY (1968) found that between the ages of 2 days and 3 months, the myocytes of puppies averaged 6.5 μm in diameter. The growth of myocytes to a diameter of 13.3 μm occurred between the 4th and 6th month, in parallel with a period of rapid body growth. These findings were corroborated by BISHOP (1973), who although not providing figures, stated that up to the age of 8 weeks the myocytes in puppies increased little in cross-sectional area.

In rat, a difference in the dimensions of myocytes in the endomyocardium and epimyocardium of the left ventricle is not present in sectioned material from birth to 11 days of age (ANVERSA et al. 1980), or in isolated myocytes up to the age of 12 weeks (BISHOP et al. 1979a, b).

3. Volume Composition of the Myocardium

In dog the proportion of myocytes in myocardium fixed by immersion remains at 79% from birth to 5 months of age (LEGATO 1979a), but in rat myocardium fixed by perfusion this parameter decreases from 85% to 75% between the 1st and 11th postnatal days in both ventricles (ANVERSA et al. 1980).

4. Volume Composition of Myocytes

a) Myofibrils

There is a general increase in the myofibrillar volume fraction from the late fetal period to term. The increment is from 44% to 53.3% in guinea-pig

(HIRAKOW and GOTOH 1980), 21.1% to 46.4% in rat (HIRAKOW and GOTOH 1975) and 22.1% to 31.1% in rabbit (SMITH and PAGE 1977). The changes after birth are variable. Little change has been reported in cat (GOTOH 1983), guinea-pig (HIRAKOW and GOTOH 1980), rat (HIRAKOW et al. 1980), and rabbit (SMITH and PAGE 1977). However, an increase in myofibrillar volume fraction was noted in hamsters (COLGAN et al. 1978), rat (OLIVETTI et al. 1980) and cat (SHERIDAN et al. 1977). In puppies aged between 2 h and 25 days LEGATO (1975) reported values fluctuating between 39% and 52%, without an obvious trend. In the rat, DAVID et al. (1979) found a decrease from 52% on day 1 to 33% on day 5, followed by an increase to 55% betwcen 4–6 months.

There appears not to be a transmural gradient, although in the left ventricle of the hamster at birth the myofibrillar volume fraction is significantly lower in the midwall compared to the subendocardium (KIDD et al. 1981).

b) Mitochondria

The mitochondrial volume fraction increases from the late fetal period to term (HIRAKOW et al. 1980; SMITH and PAGE 1977). A further increase occurs after birth (COLGAN et al. 1978; DAVID et al. 1979; GOTOH 1983; HIRAKOW and GOTOH 1975, 1980; HIRAKOW et al. 1980; NAKATA 1977; OLIVETTI et al. 1980; SHERIDAN et al. 1979), although the time course of this increment appears to be shorter in rat than in the cat or guinea-pig.

c) Mitochondrial/Myofibrillar Ratio

The mitochondrial/myofibrillar ratio increases within 24 h of birth (LEGATO 1975), and this increase is mainly due to an elevation in the mitochondrial volume fraction (DAVID et al. 1979; HOERTER et al. 1981; LEGATO 1975; OLIVETTI et al. 1980). In later growth, the changes in the myofibrillar and mitochondrial volume fractions are proportionate, so that their ratio remains essentially constant (PAGE et al. 1971).

d) Matrix

In association with increasing myofibrillar and mitochondrial volume fractions, the matrix volume fraction characteristically decreases from 42% in newborn rats (OLIVETTI et al. 1980), to less than 10% in the adult. In guinea-pig (HIRAKOW and GOTOH 1980), and probably in cat (GOTOH 1983), the bulk of this decrease occurs before birth. However, in the rat (DAVID et al. 1979; HIRAKOW et al. 1980; OLIVETTI et al. 1980), it appears that a similar change occurs in the first postnatal weeks.

e) Nucleus

The nuclear volume fraction of 10–12% in the late fetus and at birth (AN-VERSA et al. 1975; COLGAN et al. 1978; SHERIDAN et al. 1977) decreases progressively to the adult value. HIRAKOW and GOTOH (1980) and NAKATA (1977) noted no change in the nuclear transverse dimensions with development, but OLIVETTI et al. (1980) found in the rat that absolute nuclear volume increased by a factor of 2.22 in the left ventricle and 2.31 in the right ventricle between the first and eleventh postnatal day, with most of the change occurring in the first 5 days. The percentage of binucleated myocytes increases dramatically in both ventricles after birth. ANVERSA et al. (1980) observed in sectioned material that in the left ventricle of the newborn rat the number of binucleate myocytes increased from 3% on day 1, to 15% on day 5, to 51% on day 11. BISHOP et al. (1979a, b), using isolated postnatal myocytes noted a similar pattern, progressing to binucleation of 85% of myocytes by 15 days of age. In isolated myocytes from both ventricles of puppies, the onset of binucleation within myocytes begins at 2 weeks of age and affects 85–90% of myocytes by the 6th–11th weeks (BISHOP and HINE 1975). Furthermore, a direct correlation is present between the number of nuclei and the cell volume. The most likely mechanism for the high percentage of binucleate myocytes in neonatal myocardium is nuclear division with a failure of cell division (ANVERSA et al. 1980; CLUBB and BISHOP 1984).

f) Sarcoplasmic Reticulum

HOERTER et al. (1981) found that the volume fraction of sarcoplasmic reticulum in rabbits increased progressively from 0.46% at 24 days postcoitum to 1.3% at 14 days postpartum. OLIVETTI et al. (1980) in rat and COLGAN et al. (1978) in hamster also observed an increase shortly after birth. However, PAGE et al. (1971) noted no further change in rat from the period of weaning to maturity.

The membrane area of the sarcoplasmic reticulum/unit myofibrillar volume increases in the early postnatal period (OLIVETTI et al. 1980), but remains constant during further development (PAGE et al. 1971).

g) T System

Both the volume fraction of the T system (HIRAKOW and GOTOH 1975, 1980) and the fraction of T system to cell surface area (SHERIDAN et al. 1979) increase after birth. There appear to be no morphometric data on the T system from animals in which this membrane specialisation is developed before birth.

Development of the T system slows but does not arrest the decline in the surface area-to-volume ratio of myocytes during development (see HOERTER et al. 1981; PAGE et al. 1971).

5. Correlation of Structure and Function in Developing Myocytes

Morphometric data from the developing heart need to be interpreted against a backdrop of the cardiovascular physiology in the fetal and postnatal periods. This has been most extensively studied in fetal and newborn lambs (see RUDOLPH 1974, 1976). In the fetus, the lungs serve no ventilatory purpose, as they are

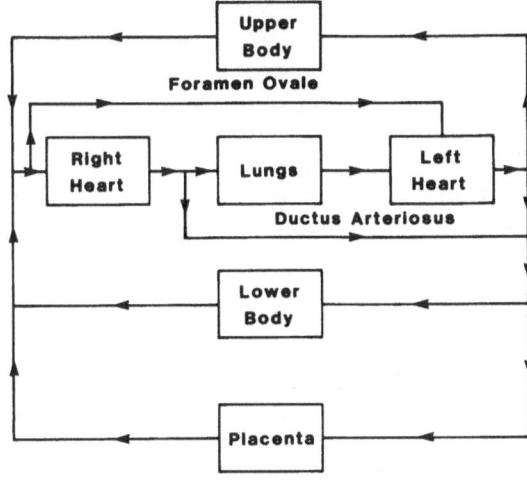

Fig. 83. Schematic diagram of the fetal circulation. The direction of blood flow is in the direction of the *arrowheads*

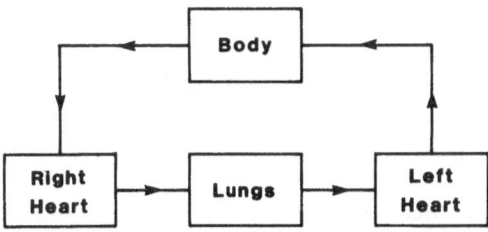

Fig. 84. Schematic diagram of the postnatal and adult circulation

collapsed and filled with lung liquid. All metabolic requirements are provided by the placenta. The fetus has two major vascular shunts which allow the movement of a large fraction of the circulating blood volume without passage through an intervening capillary bed. These shunts comprise the foramen ovale, between the termination of the inferior vena cava and the left atrium, and the ductus arteriosus, between the pulmonary trunk and the descending thoracic aorta (Fig. 83).

The arrangement of the fetal circulation is "in parallel". The significance of this is that the right and left ventricular outputs are both systemic in nature and potentially of unequal volume. Although conflicting results have been obtained in the past due to the acute nature of the experiments (see RUDOLPH 1976), it is now generally accepted that *in utero*, the right ventricular output is 1.5 to 2 times that of the left (ANDERSON et al. 1981; ASSALI et al. 1974; HEYMANN et al. 1973; PITLICK et al. 1976). Since these outputs occur against very similar pressures in the pulmonary and systemic vessels, the right ventricle can be considered to be the "dominant" ventricle in the sheep fetus.

After birth with ventilation of the lungs, loss of the placental circulation, closure of the foramen ovale, and constriction of the ductus arteriosus, the circulation becomes arranged "in series" (Fig. 84). All blood returning to the heart enters the right atrium and ventricle, and all the systemic output is derived from the left ventricle. In this situation, the left ventricle is "dominant" because although the volume outputs of the left and right ventricles are, on average, equal, the pressure in the systemic vessels is far greater than that in the pulmonary circulation.

The differences in the relative function of the right and left ventricles in the fetal, as opposed to the newborn and adult periods, is reflected in the pattern of myocardial blood flow. In fetal lambs the flow per 100 g of tissue to the right ventricular free wall and right side of the septum is 1.3 times the flow to the left ventricular free wall and left side of the septum (FISHER et al. 1982; SMOLICH, unpublished observations). This flow pattern is reversed in newborn lambs and adult ewes.

If a structure/function relationship is present in the developing heart, then it might be anticipated that in the fetus the right ventricle compared to the left would contain more myocytes, or larger myocytes, or myocytes with a higher myofibrillar volume fraction, or a variable combination of these. After birth, a reversal in the pattern of these parameters would be expected between the right and left ventricles.

Unfortunately, although the developmental physiology has been well documented in sheep, there has not been any comprehensive study of morphometric parameters in the same animal. Instead, most of the morphometric data have been derived from smaller laboratory animals. In addition, most investigations have been performed in the postnatal period, and relatively few at the fetal stage.

LEGATO (1979 b) observed in puppies aged between 1 day and 5 months, that the average myofibrillar volume fraction in the left ventricle (64%), was significantly greater than that in the right ventricle (59%). In addition, at birth, the right ventricular myocytes had a more mature appearance than those on the left. HIRAKOW and GOTOH (1980) did not find any significant differences in the volume fractions of myofibrils, mitochondria or matrix between left and right ventricular myocytes in fetal and neonatal guinea-pigs. Similar findings were obtained by HIRAKOW et al. (1980) and NAKATA (1977) in newborn rats, and by GOTOH (1983) in the neonatal cat.

With respect to the dimensions of right and left ventricular myocytes, no significant differences were found in fetal and neonatal guinea-pigs (HIRAKOW and GOTOH 1980), newborn rat (ANVERSA et al. 1980; HIRAKOW et al. 1980), or neonatal cat (GOTOH 1983). LEGATO (1979) observed that at birth myocytes in the right ventricle are greater in diameter than those in the left ventricle (7.7 μm vs 3.7 μm), but this result is difficult to interpret in the light of the fluctuating pattern of results obtained at older ages. Interestingly, ASHLEY (1945) noted that right ventricular myocytes were greater in diameter than the left in the human fetus, about the same in the infant, and that in children left ventricular diameters were larger than those on the right side.

The growth patterns of right and left ventricular myocytes appear to be different, at least in the rat. NAKATA (1977) found that whilst the diameter of right ventricular myocytes is unchanged between birth and 7 days of age and then increased, the diameters of left ventricular myocytes increase steadily from the time of birth. ANVERSA et al. (1980) concluded that the main difference in the postnatal growth of the right and left ventricles is the degree of myocyte proliferation. Thus, nuclear division with or without cell division is greater in extent and persists longer in the left ventricle. This results in a change in the ratio of numbers of myocytes in the left and right ventricles from 1:1 in the neonate, to 2:1 in the adult.

In puppies, BISHOP (1973) also found a differing pattern between the left and right ventricles. At birth 2–4% of myocyte nuclei are labelled with tritiated thymidine in both ventricles. This level persists in the left ventricle for 2 weeks then decreases progressively to less than 1% at 4–5 weeks. In the right ventricle, the labelling falls to less than 1% between days 2 to 10, then increases to 3% until day 15, followed by a decline similar to the left ventricle.

G. Development of the Conduction System

I. Conduction in Embryonic Heart

1. ECG of Embryonic Heart

In the mature heart, the conduction system plays a major role in determining the normal wave patterns of the ECG. Embryos with a four-chambered heart and an immature conduction system produce similar wave patterns. Cardiac electrical activity has been recorded from chick embryos as early as stage 24 (HOFF et al. 1939; HAMBURGER and HAMILTON 1951) and stage 16 (SEIDL et al. 1981) (Fig. 85), when no morphologically definable conduction system is present and the outward appearance of the heart is different from the adult. Perhaps the most significant result of electrocardiographic studies is the demonstration of an atrio-ventricular (AV) conduction delay as early as the 20-somite stage in the chick embryo (HOFF et al. 1939). This AV delay may be related to the development of specialised ring tissue (ANDERSON and TAYLOR 1972; WENINK 1976; ANDERSON et al. 1976). Alternatively, conduction delays may be due not to intrinsic properties of cell membranes, but to the effects of tissue geometry and cellular arrangement, which have been demonstrated to affect impulse propagation and action potential configuration in myocardium (VAN CAPELLE and JANSE 1976; SPACH 1982; SPACH and KOOTSEY 1983). The constriction of the AV canal first appears in the embryo of 19 somites (HAMBURGER and HAMILTON 1951) and may be implicated in providing sufficient local alterations in cell relationships to affect conduction at the 20-somite stage.

2. Preferential Conduction in the AV Canal

In 9- and 10-day-old mouse embryo hearts a group of large polygonal cells can be found in a triangular area on the dorsal wall of the AV canal (VIRÁGH and CHALLICE 1977a). It has been proposed that these cells are arranged so as to preferentially conduct impulses from the atria towards the still developing left ventricle (Fig. 86) before the development of a specialised conduction system (VIRÁGH and CHALLICE 1977a).

3. Preferential Conduction in the Ventricular Trabeculae

The branching network feature of trabeculae and their gradual extension during ventricular development have implicated these myocardial muscle bun-

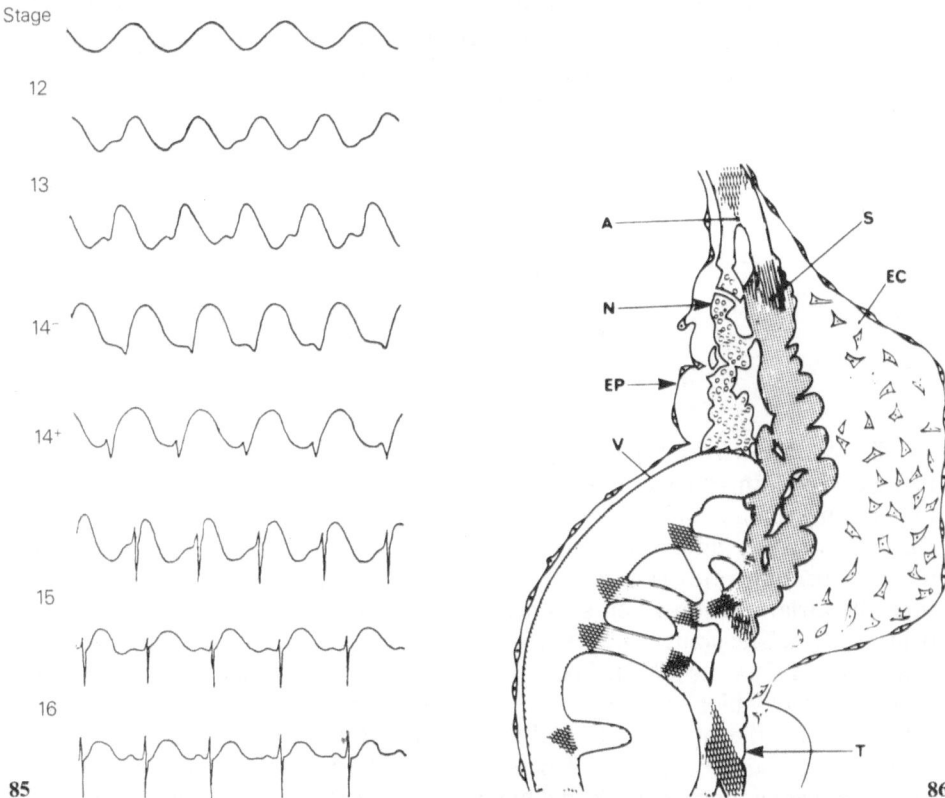

Fig. 85. Survey of development of shape of chick embryo electrocardiogram between stages 11 and 17. (From SEIDL et al. 1981)

Fig. 86. This diagram shows the atrioventricular interconnection along the sagittally sectioned dorsal wall of a 10-day-old mouse embryo heart. The dorsal atrial wall (*A*) splits into two layers in the atrioventricular canal. The inner specialized cell layer (*S*) facing the dorsal endocardial cushion (*EC*) is continuous with the ventricular trabecular system (*T*). The outer degenerated and necrotic cell layer (*N*) underneath the epicardium (*EP*) is in continuity with the outer left ventricular wall (*V*). (From VIRÁGH and CHALLICE 1977a)

dles as templates for the development of the Purkinje network (PATTEN 1956; ROBB and PETRI 1961; ROBB 1965; see G.IV.2). The trabeculae themselves also appear to play a special role in activation of the embryonic ventricular myocardium by preferential propagation of impulses in the absence of a specialised conduction system. In the dorsal wall of the AV canal, the inner layer of myocytes is in continuity with the ventricular trabeculae (Fig. 86). Since it is this inner layer of AV canal cells which conduct the atrial impulse, VIRÁGH and CHALLICE (1977a) have proposed that the trabeculae are responsible for conducting the impulse to the left and right ventricular chambers. Interestingly, trabecular myocytes are more advanced in their development of intercalated discs (including gap junctions) than those in the ventricular walls (CHALLICE and VIRÁGH 1973; VIRÁGH and CHALLICE 1977a). In addition, a short ventricular activation time is observed in the ECG of embryonic hearts (HOFF et al. 1939; SEIDL et al. 1981). Trabecular cells, morphologically indistinguishable from working myocardium, have been labelled as Purkinje cells in early embryonic dog heart on the basis of differences in action potential configurations (DANILO et al. 1983).

II. Development of the Sinoatrial Node

1. Development of the Pacemaker

The conoventricular part of the tubular heart, which is the first part of the heart to form, is also the first part to contract. As subsequent parts develop toward the venous end, they also gradually begin to contract (PATTEN 1949). Furthermore the different parts of the young tubular heart have their own intrinsic beating rates (DE HAAN 1965). The ventricular portion beats slower than the atrial portion when separated, but is normally entrained by the faster beating atria. The sinus venosus, when formed, has a still higher beat frequency (PATTEN and KRAMER 1933). These observations are in agreement with the pacemaker concept which predicts that the part of the heart with the greatest intrinsic rate overrides the other parts and acts as the pacemaker. PATTEN (1949) proposed a gradual progression of pacemaking regions as the caudal portions of the heart were added on, each one having a higher intrinsic rate until the formation of the sinus venosus, within which the SA node develops. This sequence of pacemaker progression in the developing tubular heart was accepted for many years until micro-electrode studies demonstrated that the pacemaker was already located in the sinoatrial portion of the heart at the time of the first ventricular contraction (VAN MIEROP 1967). However, this work has been largely overlooked. In a major review on the SA node (BROOKS and LU 1972) the hypothesis of PATTEN (1949) was still presented.

VAN MIEROP's observations have more recently been supported and extended by studies using voltage-sensitive dyes on early tubular hearts, including stages prior to the onset of contractions (FUJII et al. 1980, 1981 a, b, c; KAMINO et al. 1981; SAKAI et al. 1983; HIROTA et al. 1983). A summary of the sequence of events in pacemaker development in the chick heart, as revealed from these studies is seen in Table 9. As early as the 7 somite stage, when the forming atria still have fused and unfused portions, spontaneous and rhythmic action potentials are present (FUJII et al. 1980, 1981a). From this period until the beginning of the 9-somite stage the site of the pacemaker is highly variable and changeable. A switching phenomenon where the pacemaker shifts position between left and right pre-atrial regions, and double pacemakers where two independent pacemaking regions can be identified in the left and right pre-atrial portions of the heart, have been observed (SAKAI et al. 1983). As development proceeds a single pacemaking area becomes established on the left side. This

Table 9. Summary of sequence of events in development of pacemaker function in early embryonic chick heart. (From SAKAI et al. 1983)

Stage	Pacemaker function
7 somites	Generation of spontaneous and rhythmical action potential[a]
8 somites	Labile stage of regional pacemaking priority
	Organization of a single pacemaking area
	Localization of the pacemaking area at the left pre-atrial tissue[b]
9 somites	First spontaneous and rhythmical contraction[c]
10 somites	
	Translocation of the pacemaking area to the sinus primordium[d]
11 somites	

[a] From FUJII et al. (1980, 1981 a)   [b, d] KAMINO et al. (1981)   [c] FUJII et al. (1981 b)

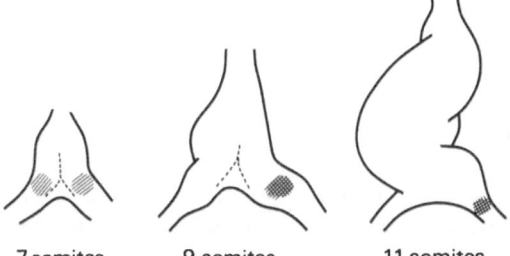

7 somites   9 somites   11 somites

Fig. 87. A summary of the location of pacemakers and the development of left-sided priority (*cross-hatch*) during early development in the chick heart. (Adapted from KAMINO et al. 1981)

occurs in the early 9 somite stage, when contractions usually begin (VAN MIEROP 1967; FUJII et al. 1981 b, c). During the 9- and 10-somite stages the heart's pacemaker is located in the left atrial region (Fig. 87), but becomes localized in the left sinoatrial region by the 11-somite stage (KAMINO et al. 1981).

2. Cell Coupling in the Absence of Gap Junctions

There is good evidence that ultrastructurally-defined gap junctions are not necessary for electrical coupling. Freeze fracture techniques reveal only a few

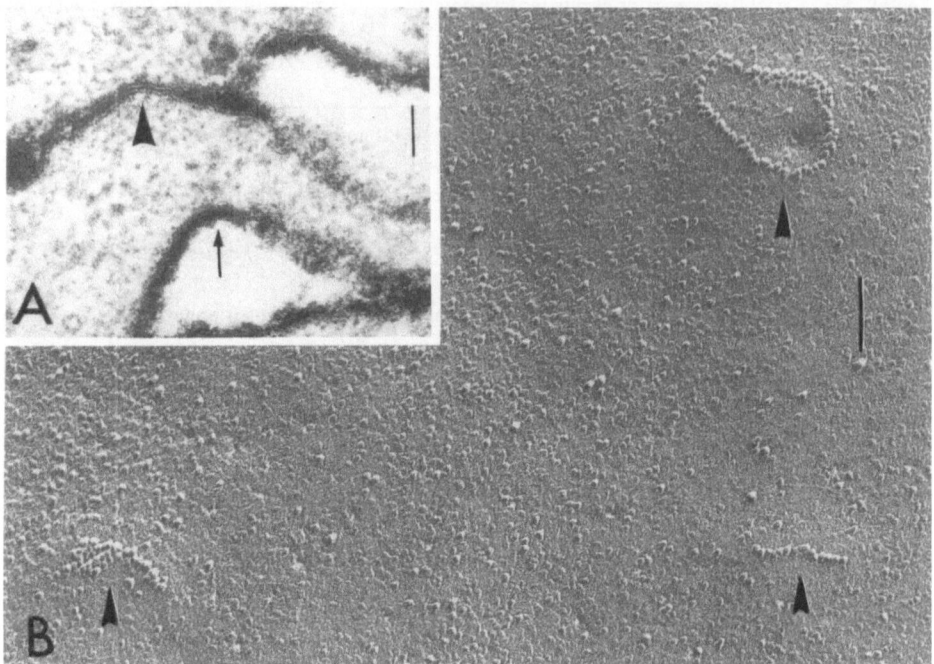

Fig. 88. A Transmission electron micrograph of a gap junction (*arrowhead*) between cells in a heart cell aggregate stained with ruthenium red. The thick surface glycocalyx is indicated by a *small arrow*. × 70 000. **B** Freeze-fracture preparation showing the three typical configurations of junctional intramembranous protein particles (*arrowheads*). × 125 000 (From WILLIAMS and DE HAAN 1981)

loose clusters and linear arrays of junctional intramembrane protein particles (the functional unit of the gap junction) in the 8-day mouse embryo beating heart (GROS et al. 1979a, b). It is not until 10 days post-coitum that a few small gap junctions recognizable by transmission electron microscopy develop (GROS et al. 1979a, b). Gap junctions are also not observed in the contracting myocardium of the 15-day ferret embryonic tubular heart although primitive desmosomes and fasciae adherentes are present (MARINO and SEVERDIA 1982). Myocardial cells cultured as spontaneously beating aggregates from 7-day chick embryos remain electrically coupled after cycloheximide-induced loss of ultrastructurally recognizable gap junctions (WILLIAMS and DE HAAN 1981; DE HAAN et al. 1981). Cycloheximide treatment changes the arrangement of intramembrane protein particles from small arrays characteristic of gap junctions (Fig. 88) to scattered particles over large areas of the intimately-apposed plasma membranes. However, it is not known if electrical couplings without intramembrane protein particles can occur in embryonic heart. Ephaptic impulse transmission (i.e. non-junctional) can occur between ventricular myocardial cells *in vitro* (SUENSON 1984).

3. Timing of First Appearance and Position of the SA Node

Although the pacemaker is localized in the left sinus horn when the tubular heart begins to beat (VAN MIEROP 1967; KAMINO et al. 1981), it is the SA node which becomes the definitive pacemaker of the adult heart. In sheep embryos the SA nodal primordium is first detectable at a crown-rump length of 10.6 mm (MUIR 1951). In human embryos the nodal primordium has been detected at a crown-rump length of 6 mm (WALLS 1947), 10 mm (YAMAUCHI 1965), and 12 mm (ANDERSON et al. 1978). The first histologically-distinct SA node in the embryonic mouse heart appears at 10.5 to 11 days post-coitum (VAN MIEROP and GESSNER 1970; HEINTZBERGER 1974) and coincides with the development of the rest of the conduction system (VIRÁGH and CHALLICE 1980). In rat embryos the SA nodal primordium appears at 13.5 days post-coitum (MUIR 1955).

In mouse and rat heart the first sign of the SA nodal primordium is in the anteromedial wall of the right common cardinal vein, from which the superior vena cava is derived (MUIR 1955; VAN MIEROP and GESSNER 1970; HEINTZBERGER 1974; VIRÁGH and CHALLICE 1980). Initially, the developing nodal cells form a small oval group at this location (Fig. 89), but subsequently a cauda extends towards the lateral side during the embryonic period in mice (VAN MIEROP and GESSNER 1970), but post-natally in the rat (MUIR 1955). Ultimately the SA node becomes a horse-shoe shape, with the head of the node applied to the medial wall of the superior vena cava, the body lying anteriorly and the tail of the node extending laterally and down to the right atrium. The myocardium of the tail of the node blends with the atrial myocardium.

In contrast, the SA node in human embryos develops primarily as an antero-lateral structure (YAMAUCHI 1965; ANDERSON et al. 1978) but is similar in shape, forming a crescent around the anterior quadrants of the caval-atrial junction (Fig. 90, ANDERSON et al. 1978).

4. Cytological Differentiation

The cytological development of SA nodal cells appears to be consistent in different species. In light microscopy few distinguishing features are observed in the early SA node (ANDERSON et al. 1978). Presumptive SA nodal cells can be identified by their distinctive arrangement rather than by cytological characteristics (WALLS 1947). These cells have prominent nuclei and are densely packed. As development proceeds the originally compact structure of the node gradually changes into an interconnected labyrinth of cellular strands or sheets separated by connective tissues (WALLS 1947). Shortly after the appearance of an aggregation of presumptive SA nodal cells, their cytological differentiation from the surrounding myocardium begins (VIRÁGH and CHALLICE 1980; HEINTZBERGER 1974; MUIR 1955).

In comparison to the atrial musculature or surrounding sinus myocardium, the early SA nodal cells stain poorly in histological sections, are smaller, and possess fewer and less well organised myofibrils (YAMAUCHI 1965). The nuclei

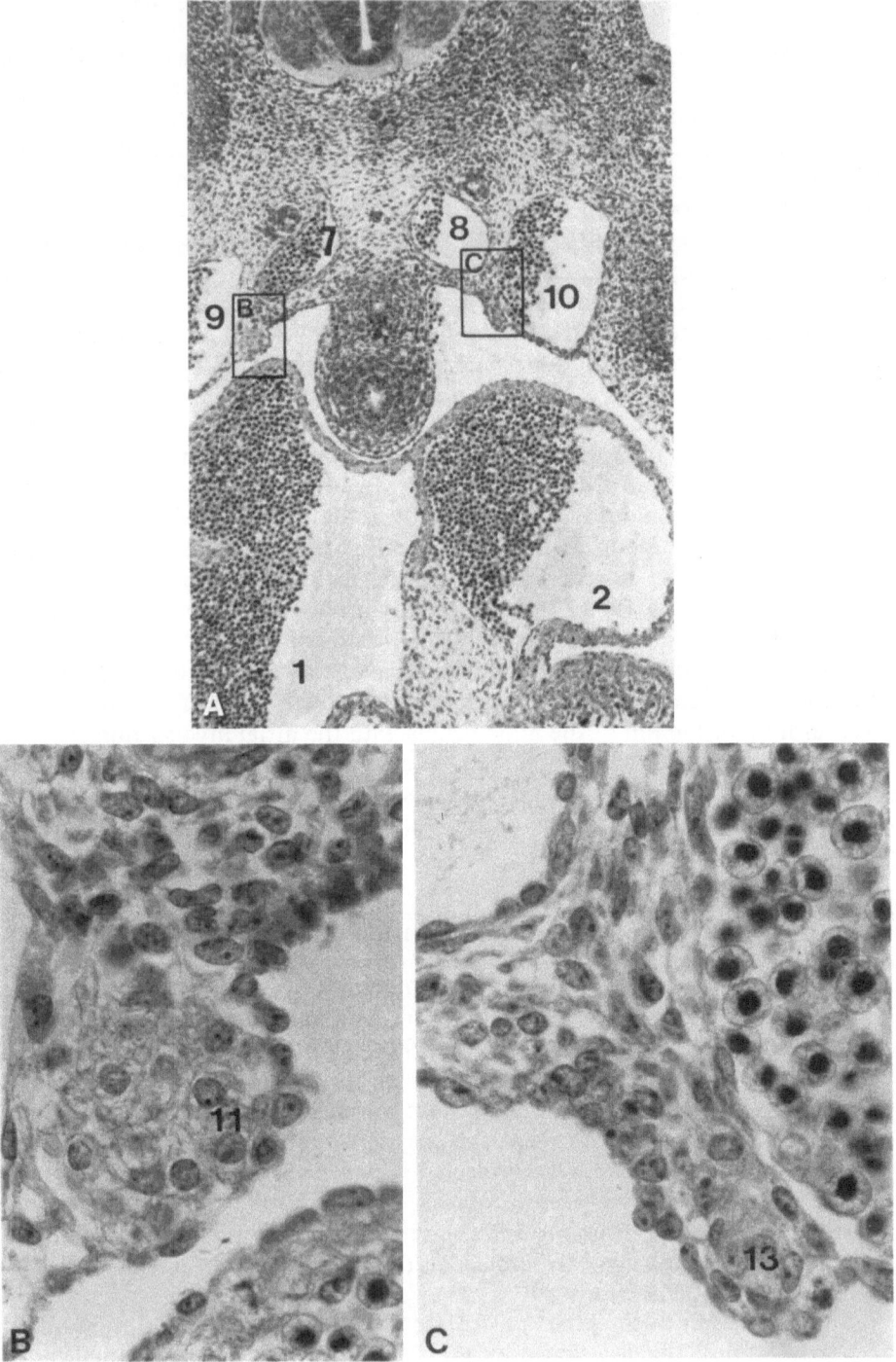

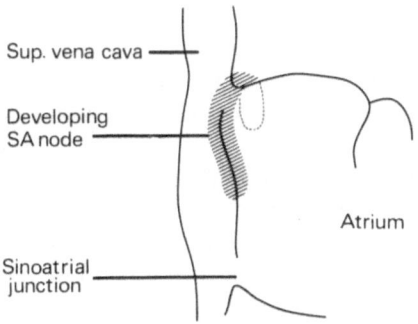

Sup. vena cava

Developing
SA node

Sinoatrial
junction

Atrium

Fig. 90. The early developing SA node extends over the anterior quadrants of the junction of the sinus venosus and the atrium. (Adapted from DAVIES et al. 1983)

of SA nodal cells may have distinctive features although this is not a consistent finding. They have been reported to be more spherical (HEINTZBERGER 1974; MUIR 1955), more darkly stained (MUIR 1955), less intensely stained (HEINTZBERGER 1974), or no different in size, shape or chromatin content (VAN MIEROP and GESSNER 1970) than those of the surrounding myocardium.

With transmission electron microscopy the nodal cells appear oval, 4 to 6 µm in diameter and contain glycogen-filled areas, as well as free ribosomes in all regions (YAMAUCHI 1965). Fasciae adherentes and desmosomes connect the nodal cells but gap junctions, which are already present between ventricular myocytes, develop later (YAMAUCHI 1965; VIRÁGH and CHALLICE 1980). The rate of myofibrillogenesis within nodal cells varies (VIRÁGH and CHALLICE 1980) but at all stages myofibrillar disorganization is greater than in working myocardium (YAMAUCHI 1965; DOMENECH-MATEU and BOYA-VEGUE 1975; VIRÁGH and CHALLICE 1980). With continued development the cells become more interdigitated and irregular and there is an increase in the number of disorganized myofibrils and the amount of sarcoplasmic reticulum (VIRÁGH and CHALLICE 1980). The nodal cells undergo the same process of Z disc loss and restoration during mitosis as the working myocardium but the mitotic index is 0.4% – 10 times less than ventricular working myocardium (EROKHINA and RUMYANYSEV 1983).

The shape and distribution of the SA node primordium is consistent with a derivation from a ring of specialized tissue at the SA junction (ANDERSON et al. 1976; WENINK 1976; ANDERSON et al. 1978; BECKER 1978; DAVIES et al. 1983), even though this ring has never been demonstrated ultrastructurally. Further discussion of this concept and its relation to the development of the conduction system is offered in Chaps. G.III.3 and G.IV.3c.

Fig. 89. **A** Transverse section of the mediastinal region of a mouse embryo, 11 days post-coitum. Right atrium (*1*); left atrium (*2*); right dorsal aorta (*7*), left dorsal aorta (*8*); right sinus horn (*9*); left sinus horn (*10*). Two structures can be distinguished (*boxes B, C*) in close proximity to the sinus horns. × 80. **B** Higher magnification of the box marked B in Fig. 89 A, showing the developing SA node (*11*). × 550. **C** Higher magnification of the box marked C in **A**, showing a node-like structure (*13*) which is smaller than the SA node and related to the left sinus horn near its junction with the atrial wall. × 550. (From HEINTZBERGER 1974, reproduced by permission of Acta Morphol Neerl Scand 12:317–330, copyright 1974, Swets and Zeitlinger, Lisse)

5. Innervation

Ganglion cells become associated with the SA node soon after its appearance (WALLS 1947). Shortly afterwards, nerve processes appear in the SA node of the human embryo between STREETER stages 20 to 23 (YAMAUCHI 1965), and at 12 days post-coitum in the mouse heart (VIRÁGH and CHALLICE 1980). Nerve processes, which frequently contain small clear and large granular vesicles, become more numerous within the SA node than in the adjacent myocardium. Innervation is densest in the junctional region between the SA node and the atrial myocardium, near the sulcus terminalis. A cholinesterase-positive reaction is present in nerves of this region in human embryos of crown-rump length 28 to 30 mm and in nodal cells by 90 mm crown-rump length (10 to 12 weeks gestation) (ANDERSON et al. 1976). In human (YAMAUCHI 1965) and sheep embryos (MUIR 1951) the SA node forms an intimate relationship with nerves early in development, and nerve-nodal cell contacts of less than 30 nm separation are frequent in human embryos of 30 mm crown-rump length (YAMAUCHI 1965).

In contrast, nerve fibres appear no more common in the SA node of newborn rats than elsewhere in the atria and no specialised endings are present although ganglia occur medial to the node (MUIR 1955). In mouse embryos nerves are small and infrequent (VAN MIEROP and GESSNER 1970) and, although increasing in frequency with development, do not reach the level of association with nodal cells observed in human embryos.

6. Bilateral Nodal Primordia

Controversy has surrounded the existence of a left sided equivalent of the SA node. Such a left sided nodal structure has not been observed in developing sheep and rat (MUIR 1951, 1955) or human hearts (ANDERSON et al. 1976; YAMAUCHI 1965). However, a transient and smaller node-like structure is present in an anteromedial position in the wall of the left sinus venosus of the embryonic mouse heart between 11 and 14 days post-coitum (HEINTZBERGER 1974; VIRÁGH and CHALLICE 1980) (Fig. 89). The constituent cells possess a few, small myofibrils and are associated with nerve fibres (VIRÁGH and CHALLICE 1980). As development proceeds this node-like structure becomes incorporated into the wall of the left atrium (HEINTZBERGER 1974; VIRÁGH and CHALLICE 1980).

PATTEN (1956) concluded that during embryonic life paired SA nodes exist and that the left one became the AV node. Bilateral SA nodes do develop in "asplenia syndrome" (VAN MIEROP et al. 1964), which is characterised by a right and a left superior caval vein entering the right and left atria respectively. It has been suggested that the development of the second SA node may be related to the development of the left veno-atrial junction in these cases (BECKER 1978).

The concept of bilateral nodal primordia is also supported by studies of the development and distribution of pacemaking regions in the early tubular heart (DE HAAN 1959; KAMINO et al. 1981; SAKAI et al. 1983). These studies have demonstrated left-sided pacemaker dominance in early chick hearts. At the 7- to 9-somite stage the flexibility of pacemaker location (SAKAI et al. 1983) is consistent with the existence of two sinus node primordia which are poorly defined and independent. The "focusing" of the pacemaker to the left sinus horn (Fig. 87) appears to be due to the emergence of its transient dominance over the right-sided counterpart until about the 19- or 20-somite stage (DE HAAN 1959). Presumably, at a later stage of development the right-sided pacemaker tissue becomes the SA node. Evidence for participation of the left-sided pacemaker in the formation of the AV node is lacking.

III. Development of the AV Node

1. Formation of the AV Node

Most early morphologists considered that the AV node and bundle were in continuity when first distinguishable (MALL 1912; SHANER 1929; WALLS 1947; MUIR 1954). Although this appears to be the case in the mouse (VIRÁGH and CHALLICE 1977a, b, 1982, 1983), more recent studies conclude that in most species two or more separate primordia fuse to form the AV node and bundle (JAMES 1970; ANDERSON and TAYLOR 1972; ANDERSON et al. 1976; TRUEX et al. 1978; MARINO et al. 1979; MARINO and SEVERDIA 1983). DAVIES et al. (1983) suggest that because it resides in the base of the inferior part of the interatrial septum, the definitive AV node is formed after completion of septation. Therefore it is likely that early studies (MALL 1912; SHANER 1929 and WALLS 1947) which reported the existence of the AV node before completion of atrial septation, referred to only one component of this structure.

Studies of the pathology of the AV conduction system support the idea that the AV node is formed from two or more originally separate parts (JAMES 1970; JAMES et al. 1975, 1976; JAMES and MARSHALL 1976). When present, discontinuity of the conduction system between atria and ventricles occurs either at the nodal-bundle junction, or between the node and atrial muscle, and connective tissue invariably 'fills in' the separation (LEV 1972; DAVIES et al. 1983). Maldevelopment of the AV junctional tissues resulting in congenital AV block has been extensively reviewed recently by DAVIES et al. (1983).

A number of structures have been described as AV-node primordia including one of AV canal origin which is continuous with the AV bundle (ANDERSON and TAYLOR 1972; ANDERSON et al. 1976), a well-innervated primordium related to the coronary sinus (ANDERSON and TAYLOR 1972; ANDERSON et al. 1976; GARDNER and O'RAHILLY 1976; TRUEX et al. 1978; MARINO et al. 1979; MARINO and SEVERDIA 1983), and one originating from the base of the interatrial septum below the right venous valve attachment (TRUEX et al. 1978; MARINO et al. 1979). The definitive AV node is formed by the fusion of the primordia in 46 mm to 50 mm human embryos (ANDERSON and TAYLOR 1972; TRUEX et al. 1978), but it has not been determined if the primordium related to the left sinus horn becomes continuous with the AV bundle at this time (ANDERSON and TAYLOR 1972) or earlier (GARDNER and O'RAHILLY 1976; MARINO et al. 1979). There is also controversy over whether there is a primordium of the AV node derived from the AV canal (ANDERSON and TAYLOR 1972; ANDERSON et al. 1976), or if it contributes only to the AV bundle (TRUEX et al. 1978; MARINO et al. 1979; MARINO and SEVERDIA 1983), or if the majority of the AV node develops from the AV canal (VIRÁGH and CHALLICE 1982, 1983). There is also some disagreement about whether the left sinus horn primordium becomes the superficial part of the AV node (ANDERSON and TAYLOR 1972; ANDERSON et al. 1976) or the compact part (TRUEX et al. 1978; MARINO et al. 1979; MARINO and SEVERDIA 1983). The diverging results suggest that there are a number of structures or segments of myocardium in the AV nodal area which are consistently observed to be involved in the development of the AV node and bundle but these have been credited with different roles by different observers.

2. Evidence for Species Differences

In the mouse heart, the AV node and bundle develop solely from large, glycogen-rich cells in the dorsal wall of the AV canal (VIRÁGH and CHALLICE 1977a, b, 1982, 1983). These cells show a strong PAS reaction and can be identified at 9 and 10 days post coitum just prior to the beginning of septation (Fig. 86). Similar cells develop along the ridge of the growing interventricular septum in continuity with the inner layer of the AV canal musculature, and differentiate into the AV bundle and branches (Fig. 91). The AV node primordium is evident in 11-day embryos and enlarges rapidly until the end of septation (VIRÁGH and CHALLICE 1982). The dorsal atrial myocardium and interatrial septum establish contact on both sides of the node (Fig. 92). The development of atrionodal connections leads to the formation of an AV nodal overlay, which probably corresponds to the transitional cell zone of the node described in the human (ANDERSON and TAYLOR 1972; TRUEX et al. 1978).

In contrast to humans, the AV node in mice does not receive a contribution from a left sinus horn primordium. A node-like structure occurs in the left sinus horn of mouse heart but it is incorporated into the atrial musculature (VIRÁGH and CHALLICE 1980, 1982). In ferret heart the presumptive AV nodal cells are spherical or polyhedral, with few cytoplasmic extensions. Most of the cytoplasm consists of free ribosomes and polyribosomes and few myofilaments. A PAS reaction is not present (MARINO and SEVERDIA 1983) implying an absence of glycogen. Whether this is a reflection of a species difference (MARINO and SEVERDIA 1983) or of glycogen loss during tissue preparation is not clear.

3. The Specialized Ring Tissue Theory

The concept that the heart's conduction system is either partly or entirely derived from a series of rings of specialized tissue is based on three observations.

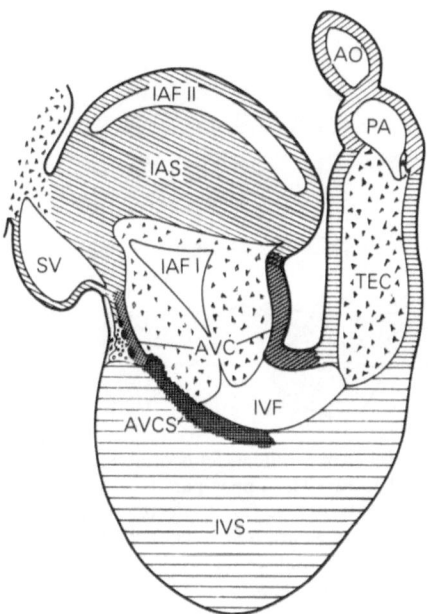

Fig. 91. A sagittal section of a 12-day-old mouse embryo heart. The interventricular septum (*IVS*), primitive atrioventricular conducting system (*AVCS*), interventricular foramen (*IVF*), atrioventricular canal walls (*AVC*), interatrial foramen primum and secundum (*IAF I* and *IAF II*), interatrial septum (*IAS*), sinus venosus (*SV*), truncus endocardial cushion (*TEC*), pulmonary artery (*PA*), and aorta (*AO*) are labelled. Note the continuity between the dorsal wall of the atrioventricular canal and the atrioventricular conducting system primordia. The ventral wall of the atrioventricular canal is continuous with that of the truncus arteriosus. (From VIRÁGH and CHALLICE 1977b)

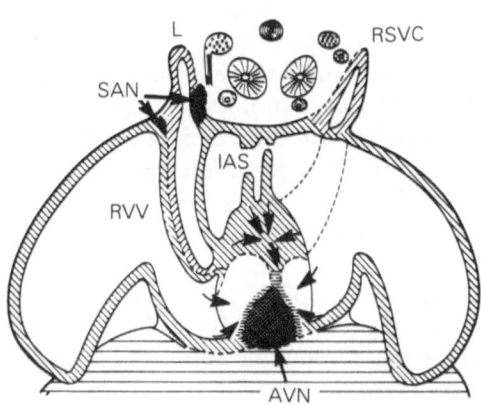

Fig. 92. In a frontal plane, the dorsal wall of the atria near the sinoatrial junction of a 13-dpc mouse embryo. The *heavily cross-hatched area* indicates the cross-sectioned AV node (*AVN*) in contact with the infolded dorsal atrial wall (*lower arrows*). Through the upper atrionodal interconnection (*upper heavy arrow*) the input reaches the node from different directions, i.e. IA septum primum and secundum (*IAS*), and left and right superior venae cavae (*L, RSVC*). In the right venous valve (*RVV*) the apposed layers of atrial and sinus muscle are indicated. The SA node (*SAN*) is a clearly-defined structure at this stage. (From VIRÁGH and CHALLICE 1982)

Firstly, four rings of tissue exist in early developing tubular heart, situated between adjacent chambers. These have been named the Sino-Atrial (SA), Atrio-Ventricular (AV), Bulbo-Ventricular (BV), and Bulbo-Truncal (BT) rings (ANDERSON et al. 1976). No ultrastructural evidence for specialization in the cells

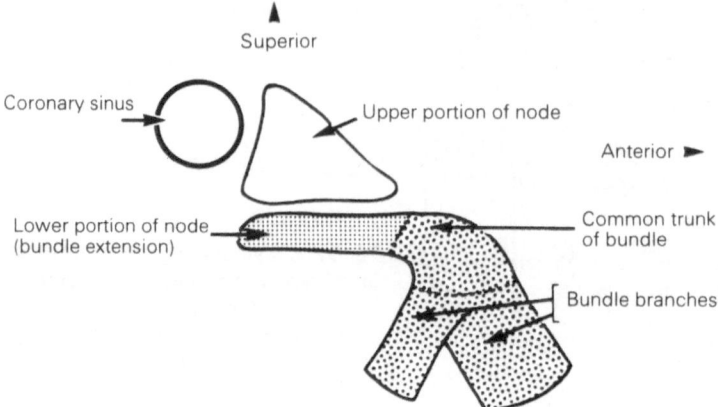

Fig. 93. Diagram of atrioventricular node, bundle and branches in a human fetus of about 190 mm crown-rump length. The node is in two parts: upper and lower. The upper part is derived from sinus horn muscle and does not exhibit cholinesterase activity except for its nerves which stain for acetylcholinesterase. The lower part of the node and the bundle are both pseudocholinesterase-positive and develop from atrioventricular canal musculature. The bundle branches also show pseudo-cholinesterase activity. (From TAYLOR and ANDERSON 1973)

of the AV and BV rings has been found in ferret embryonic heart (MARINO and SEVERDIA 1982). Secondly, these rings resemble rings of histologically specialized tissue in adult hearts of lower vertebrates such as frog and eel (ROBB and PETRIE 1961; ROBB 1965; WENINK 1976) suggesting that ontogeny follows the phylogenetic development of the conduction system (see D.V.1, D.V.3). Thirdly, the location and distribution of adult nodal tissues, and in some mammals extranodal AV specialized tissue, is consistent with origins from an earlier tissue in the form of a ring in fetal life (ANDERSON 1972a, b, c; ANDERSON et al. 1974, 1976).

The SA node primordium appears to be entirely derived from the SA ring tissue, and it has been proposed (see also G.IV.3c) that the compact part of the AV node as well as the AV bundle primordium are derived from AV ring tissue (ANDERSON et al. 1976).

4. Innervation

The AV node forms at the site where the first nerves reach the heart, and in proximity to cardiac ganglion cells (NAVARATNAM 1965). In fetal-lamb heart, nerve bundles form intimate contacts with nodal cells as early as 45 days of gestation (CANALE, unpublished observations). The nerves in that part of the node related to the left sinus horn are AChE-positive but the nodal cells associated with these nerves are AChE-negative (Fig. 93). In contrast, the non-innervated AV bundle cells are pseudocholinesterase-positive at early stages but are AChE-positive in the adult. TAYLOR and ANDERSON (1973) suggest that this change in enzyme pattern may be related to innervation of the AV bundle and branches.

IV. The Development of the Ventricular Purkinje System

1. Differentiation

Since the turn of the century, a number of conflicting views have arisen about the development and origins of the Purkinje system. An early hypothesis suggested that the Purkinje system was an embryonic remnant of developing working myocardium stopped or retarded in its development (RANVIER 1882; KEITH and FLACK 1907; FIELD 1951; PATTEN 1956). This was based on a number of biochemical (see DE HAAN 1961) and morphological similarities between adult Purkinje cells and embryonic myocardial cells. Both cell types are relatively resistant to anoxia, and contain sparse and poorly-orientated myofibrils as well as large glycogen-filled regions (MOWRY and BANGLE 1951; MCCALLION and WONG 1956). However, in recent times there has been growing evidence of qualitative differences between Purkinje tissue and working myocardium (ROBB and PETRI 1961; DE HAAN 1961; THORNELL 1972, 1973a, b, 1975; THORNELL et al. 1978; THORNELL and ERIKSSON 1981), and it is now clear that the development of specialized Purkinje fibres occurs early in fetal life and is distinct from that of working myocardium (ANDERSON and TAYLOR 1972; BOGUSCH 1979; FORSGREN et al. 1980, 1981; VIRÁGH and CHALLICE 1982; VASSALL-ADAMS 1982a, b; MARINO and SEVERDIA 1983; FORSGREN 1985) (see D.III). Purkinje fibres in fetal bovine heart produce an intense fluorescence after staining with antisera against Purkinje fibre desmin (an intermediate-filament protein), whereas the working myocardium stains weakly (FORSGREN et al. 1980). In addition, myofibrillar M bands develop earlier in Purkinje cells than in working myocardium (Fig. 94) (FORSGREN and THORNELL 1981). This has been confirmed by staining with specific antibodies to the principal component of M bands, MM-creatine kinase (FORSGREN et al. 1982). Interestingly, M-band proteins develop earlier in the AV bundle and branches than in AV node (FORSGREN et al. 1983). In culture the morphology and behaviour of Purkinje cells is also distinct from that of working myocardial cells (CANALE et al. 1983a; see J.IV.2).

2. Formation of the Purkinje Network

The development of a network arrangement within the Purkinje system is thought to be related to the breaking up of the myocardium during trabeculation. It appears that the formation of anastomosing trabeculae, reaching almost to the epicardium (Figs. 67, 68) provide a template for the Purkinje network and may also be related to the formation of false tendons (FIELD 1951; PATTEN 1956; VASSALL-ADAMS 1982a).

3. Theories of Purkinje Fibre Development

One of the early theories on the differentiation of Purkinje tissue was that the ventricular Purkinje system arose by extension and active growth of the

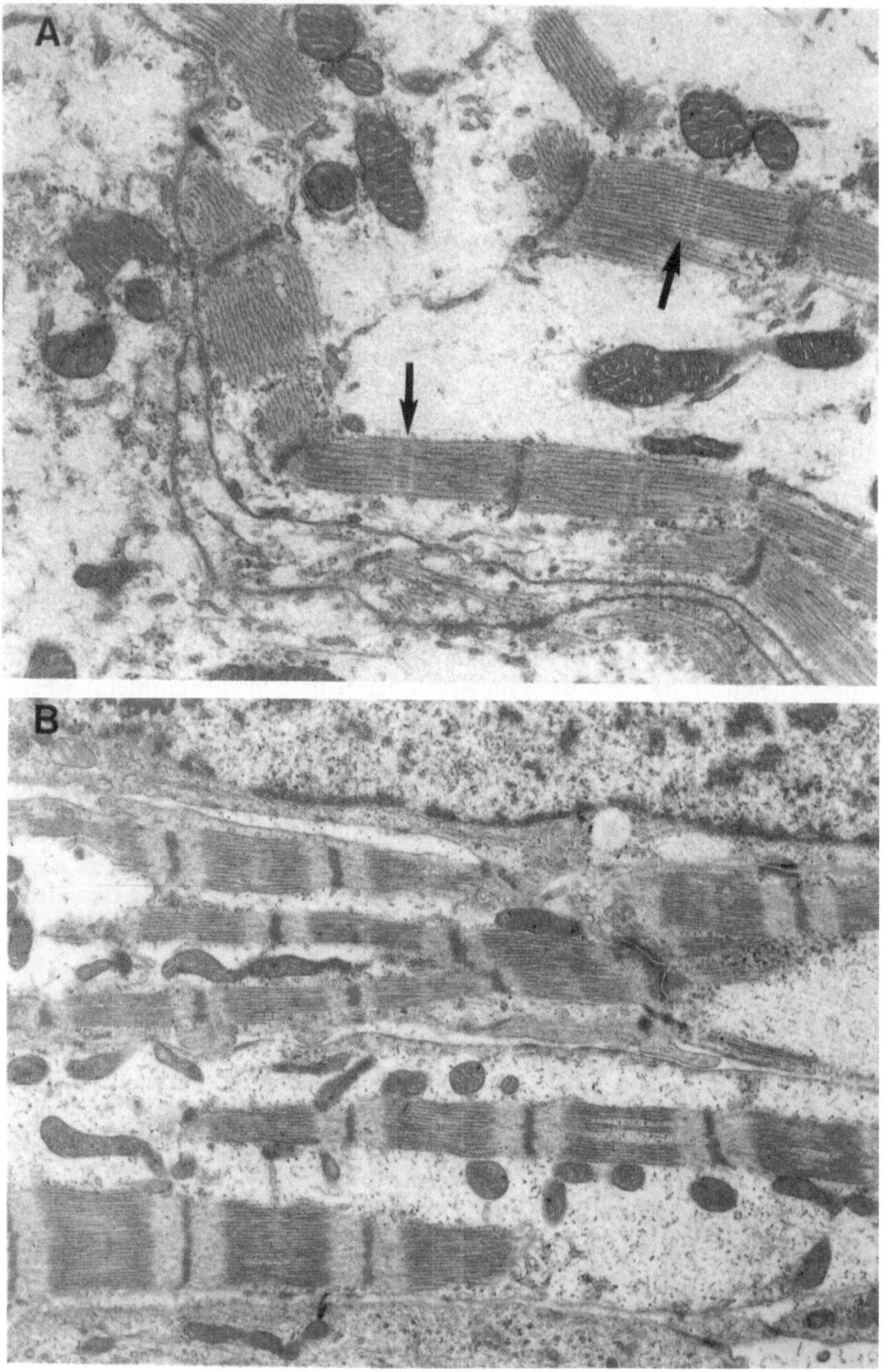

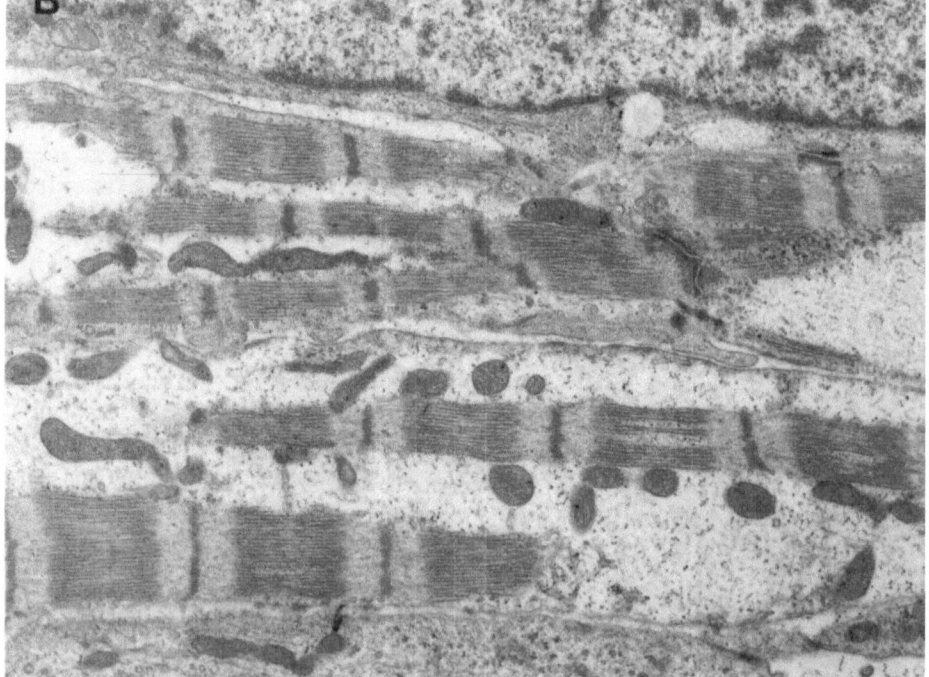

AV bundle, ramifying as it grew (RETZER 1909, 1920; TANDLER 1912; SHANER 1929; WALLS 1947). This postulate was based on the observation that the AV bundle appears first, followed by the bundle branches, and so on to the more distal parts of the Purkinje system (ROBB 1965). This theory has now been discarded because very few mitotic figures are ever found in the developing conduction system (ROBB 1965), and a lack of evidence for active cell migration (FIELD 1951).

Current opinion on the development of the Purkinje system is divided. One theory states that Purkinje tissue differentiates from the early myocardium, – the ventriculomyocardium theory (MALL 1912; MUIR 1954; JAMES 1970; VIR-ÁGH and CHALLICE 1977b; VAN MIEROP 1979; FORSGREN et al. 1980; THORNELL and ERIKSSON 1981; FORSGREN 1985). Another advocates that precursor tissue for the Purkinje system and working myocardium originate separately in the cardiogenic crescent, – the specialised-precursor-tissue theory (DE HAAN 1961; VASSALL-ADAMS 1982a, b). A third suggests that the Purkinje system constitutes the distal portion of the conduction system derived from a series of rings of specialised tissue, – the specialised-ring-tissue theory (ANDERSON and TAYLOR 1972; WENINK 1976; ANDERSON et al. 1976; KIM and YASUDA 1980). However, there is general agreement that Purkinje-fibre development occurs *in situ* and does not involve cell migration.

a) Ventriculomyocardium Theory

MALL (1912) concluded that the ventricular working myocardium is distinct from the Purkinje system at all stages of development, and that Purkinje fibres are derived from working myocardium but follow their own path of differentiation. The distal end of the Purkinje system is continous with the working myocardium at all stages of development (MUIR 1954). At the terminations of the Purkinje system in the adult a "transitional Purkinje" form occurs (see D.III.4a) before the working myocardium is reached (MARTINEZ-PALOMO et al. 1970). The morphology of these cells (Purkinje type III – VIRÁGH and CHALLICE 1973a) is consistent with the concept that the Purkinje system differentiates from primitive working myocardium. However a major weakness of the ventriculomyocardium theory is the absence of an obvious mechanism and of known factors that influence only certain cells of the early myocardium to develop into Purkinje cells.

b) Specialized Precursor Tissue Theory

This theory proposes that there are two populations of myocardial cells, one of which is a source of Purkinje cells and the other a source of working myocardium (ANDERSON and TAYLOR 1972; VASSALL-ADAMS 1982a, b). The

Fig. 94. A Purkinje cells from 76-day fetal lamb. A few myofibrils have developed M bands (*arrowed*). × 20000. **B** At the same stage M bands are not present in myofibrils of working myocardium. Note that working myocardium has more and better organised myofibrils. × 12000

Purkinje precursor tissue theory, by proposing intrinsic properties of the precursor cells better explains the specificity of Purkinje fibre development and also the observation that only a small proportion of the myocardium involved in trabeculation differentiates into Purkinje tissue (VASSALL-ADAMS 1982a).

c) The Specialized Ring Tissue Theory

According to the specialized-ring theory, the entire ventricular network must develop from one or two rings of tissue. DAVIES et al. (1983) modified the hypothesis and suggested the Purkinje system developed from specialised tissue in the form of a broad drape over the developing interventricular septum. This version more easily accounts for the extensiveness of the Purkinje system, but seems to be a significant departure from the central theme of the concept of specialized rings (ANDERSON et al. 1976; WENINK 1976). The main failing of this theory is that it does not adequately explain the extensive and highly-interconnected network of Purkinje fibres in the ventricles (see VASSALL-ADAMS 1982a).

It is difficult to establish whether there is a population of specialized precursor cells in the ventricles. The early heart consists of a morphologically homogeneous population of myocardial cells (MANASEK 1970). Although electrophysiological properties have been suggested as a means of distinguishing two myocardial cell populations prior to morphological differentiation (DE HAAN 1961), the existence of distinctive transmembrane action potentials does not necessarily indicate special intrinsic membrane properties (SPACH and KOOTSEY 1983). Pacemaker capability (spontaneous depolarisation) is not a specific property of potential conduction cells because for example, young working myocardial cells are spontaneously contractile in culture (GOSHIMA 1975; JONGSMA et al. 1975; see J.II.6).

H. Cardiac Hypertrophy

I. General Features

Hypertrophy of the whole heart or either of the ventricles constitutes an adaptive response to an overload by the myocytes, the connective tissue, and vascular elements (ANVERSA et al. 1983). Hypertrophy may also be a response to metabolic factors such as iron-copper deficiency (LIN et al. 1977) or elevated thyroxine levels (PAGE and McCALLISTER 1973b). Genetic factors may also be important, as in hypertrophic cardiomyopathy in man (MARON and FERRANS 1978), the spontaneously hypertensive rat (TOMANEK et al. 1979), and biventricular hypertrophy in normotensive rats (PFEFFER et al. 1979).

The overload may be of either the pressure or volume type. Pressure overload on the heart results from a variety of conditions including aortic and pulmonary valvular stenosis, systemic or pulmonary hypertension, coarctation of the aorta, or in experimental studies, constriction of the aorta or pulmonary trunk. The aetiological factors in volume overload include mitral and tricuspid valvular incompetence, and in experimental studies, complete atrio-ventricular block (WINKLER et al. 1976) and aorto-caval fistula (HATT et al. 1979; RAKUSAN et al. 1980). Isometric exercise results in a pressure-overload (MUNTZ et al. 1981) whilst dynamic exercise produces a volume-overload (ANVERSA et al. 1982).

Fibre geometry and orientation are not changed in the pressure-, volume- and exercise-overloaded heart (CAREW and COVELL 1979) or in pressure-overload hypertrophy with failure (PEARLMAN et al. 1982).

The appearance of the heart in hypertrophy can be classified into three main categories: concentric, eccentric, and asymmetric. Concentric hypertrophy is the result of pressure-overload, and is characterized by an increase in wall thickness and the wall thickness-to-cavity radius ratio, without a change in ventricular-cavity dimensions (LIN et al. 1977). The myocytes increase in length and width, so that the length-to-width ratio remains within the normal range (see BISHOP et al. 1979a, b, 1980; KAWAMURA 1982; KORECKY and RAKUSAN 1978; KAWAMURA et al. 1984).

Eccentric hypertrophy is caused by a volume-overload, and is associated with an enlarged ventricular cavity, an increase in wall thickness in proportion to the increase in the cavity radius, and consequently an unchanged wall thickness-to-cavity radius ratio (LIN et al. 1977). The myocytes increase predominantly in length, and have longer branches than normal (KAWAMURA 1982, 1984).

Asymmetric hypertrophy is observed in hypertrophic cardiomyopathy with preferential thickening of the septum due to an irregular hypertrophy of myocytes. Septal hypertrophy with normal myocyte morphology also occurs as a normal and transient feature of cardiac development (MARON et al. 1978).

The hypertrophic process can be divided into three phases (MEERSON 1969). The first is the stage of damage associated with an increased level of myocardial function in response to a pressure- or volume-overload. The second is a prolonged stage of stable hyperfunction, characterized by ongoing hypertrophic processes which maintain a normal level of cardiac function but with a diminished reserve. The third stage is that of exhaustion with deterioration, and histologically, degeneration of the myocardium.

Cardiac hypertrophy has been divided into physiological and pathological forms, with physiological hypertrophy constituting the response to a stimulus falling within normal physiological limits (for example, exercise or normal growth), and pathological hypertrophy the response to disease states (LINZBACH 1960). However, the difference can be defined more precisely in terms of biochemical and contractility parameters (see WIKMAN-COFFELT et al. 1979), with physiological hypertrophy demonstrating a normal or increased contractile state, maximal velocity of muscle shortening and maximal rate of myosin ATPase activity, and pathological hypertrophy showing a reduced contractile state, with or without failure, accompanied by a decrease in the myosin ATPase activity and velocity of muscle shortening.

The isozymes of myosin (see B.II.6d) may show a characteristic shift in cardiac hypertrophy. In rats subjected to a sudden pressure or volume overload, the quantity of ventricular V_3 isozyme is increased (LOMPRÉ et al. 1979b; MARTIN et al. 1983) in proportion to the degree of hypertrophy (MERCADIER et al. 1981) with a decrease in myosin ATPase activity (see SCHWARTZ et al. 1982). Normal human ventricular myocardium mainly contains the V_3 isozyme so that a similar shift is not observed with hypertrophy of this chamber, although it is present in the atria (MERCADIER et al. 1983; GORZA et al. 1984; TSUCHIMOCHI et al. 1984). The ventricular isozymic pattern and myosin ATPase activity is also unaltered in growing pigs with a gradual onset of a pressure overload (WISENBAUGH et al. 1983).

LINZBACH (1960) suggested that in hypertrophy of the left ventricle of the adult human heart to less than 200 g (or of the whole heart to less than 500 g), only myocyte hypertrophy was present, but that above these "critical weights" hyperplasia also occurred, by longitudinal cleavage of existing myocytes between points of anastomosis. ASTORRI et al. (1971) appeared to corroborate this view by finding increased numbers of myocytes within markedly hypertrophied human right ventricles greater than 100 g in weight. However, WEARN (1941) observed that myocyte diameter increases in proportion to heart weight up to 1100 g, and MOORE et al. (1980) found that in humans, myocyte nuclear density decreased in proportion to cardiac enlargement, even in hearts over 500 g. In adult experimental animals, cardiac hyertrophy is associated with hypertrophy but not hyperplasia (ANVERSA et al. 1973, 1978, 1979, 1980). However, in the postnatal period when mitosis of myocytes is a normal feature of growth, hyperplasia of myocytes, as well as hypertrophy, is seen in response to a pressure-

(BISHOP 1973; DOWELL and McMANUS 1978) or volume-overload (NEFFGEN and KORECKY 1972).

Hypertrophy of the heart is also associated with hyperplasia and hypertrophy of the connective tissue and vascular components (see ANVERSA et al. 1983), and it is hyperplasia of these elements which is responsible for DNA synthesis within the hypertrophic adult heart (see BISHOP and MELSEN 1976; ZAK 1974; ZAK et al. 1979).

The morphologic and morphometric changes observed within the myocytes in cardiac hypertrophy are dependent on the type of stimulus, its onset (that is whether sudden or gradual), duration and severity.

Morphological differences are present within myocytes from spontaneously hypertensive and aortic-constricted rats (LUND and TOMANEK 1978), and in hypertrophy of differing aetiologies in man (see MARON and FERRANS 1978). In addition, sudden constriction of the pulmonary artery is associated with degenernation and fibrotic changes within the right ventricular myocardium (BISHOP and MELSEN 1976).

II. Morphological Changes

The initial morphological changes observed in myocytes exposed to an overload usually occur within the myofibrils, but with progression of the hypertrophic process all organelles and membrane specializations may be affected. These morphological features have been described in experimental animals (see BISHOP and COLE 1969; IMAMURA 1978; LUND and TOMANEK 1978; TOMANEK 1979; TOMANEK et al. 1979; TOMANEK and HOVANEC 1981) and man (MARON and FERRANS 1978; FERRANS 1984), and will not be dealt with in detail here.

The main changes in the myofibrils comprise distortion of the Z discs, disorganization of the myofilaments, and later, loss of thick filaments. The Z disc shows an irregular widening and expansion towards adjacent sarcomeres, and it appears that some of this may be degenerative in nature. ANVERSA et al. (1973) found no preferential localization of newly synthesized protein at areas of accumulation of Z substance in myofibrils, under the sarcolemma or near the intercalated disc. Furthermore, VITALI-MAZZA et al. (1972) noted fragmentation of Z discs with disorganization of myofilaments in the left ventricle 45 min after the constriction of the aorta. However Z discs, together with intercalated discs have been implicated in sarcomerogenesis in other studies (see BISHOP and COLE 1969; BISHOP 1971; KAWAMURA et al. 1984).

The intercalated disc demonstrates enhanced folding, which BISHOP and COLE (1969) and BISHOP (1971) suggest is due to a "push-pull" effect. Other organelles reflect increased synthesis: the Golgi region is enlarged and increased numbers of free and bound ribosomes are evident. In advanced hypertrophy changes include intranuclear and intramitochondrial inclusions, proliferation of sarcoplasmic reticulum as well as degeneration and necrosis of myocytes.

III. Morphometric Changes

Myocyte density is decreased (ANVERSA et al. 1978, 1979; RAKUSAN et al. 1980) in association with an increase in diameter and cross-sectional area. The fibre-to-capillary ratio is not changed significantly (ANVERSA et al. 1978; AS-TORRI et al. 1971; RAKUSAN et al. 1980; SHIPLEY et al. 1937) because of a concomitant decrease in capillary density, but intercapillary and diffusion distances are increased.

The change in myocyte size can occur very rapidly. ANVERSA et al. (1976) found a 20% increase in the cross-sectional area of left ventricular myocytes 20 h after aortic constriction, and 90% of this was due to an increase in the volume fractions of mitochondria, sarcoplasmic reticulum and matrix.

The cell volume of myocytes is increased in hypertrophy, but with an overload of short duration (HATT et al. 1978) or of mild degree (ANVERSA et al. 1978) there appears to be no significant change in myocyte length. However, in sustained hypertrophy of adequate degree, the length-to-width ratio of the myocytes is similar to that of normal myocytes (BISHOP et al. 1979a, b, 1980; KORECKY and RAKUSAN 1978).

Although ANVERSA et al. (1971) and LAKS et al. (1974) found that sarcomere length is reduced in hypertrophy, most studies have not found any significant change (see ANVERSA et al. 1979, 1980; JULIAN et al. 1981; WENDT-GALLITELLI et al. 1979).

In the earliest phase after a pressure overload, the volume fraction of mitochondria increases, whilst that of myofibrils decreases (ANVERSA et al. 1976; MEERSON et al. 1964). The increase in the mitochondrial volume fraction is due to an increase in the absolute volume of mitochondria without an increase in number, whilst the decrease in the myofibrillar volume fraction is related to the increase in myocyte size (ANVERSA et al. 1976). Subsequently, the volume fraction of mitochondria decreases but that of myofibrils increases due to an enhanced growth of myofibrillar components (ANVERSA et al. 1971). These changes produce a characteristic pattern in the mitochondrial-to-myofibrillar ratio: an initial increase is followed by a decrease to below normal levels and further progressive reduction.

In volume overload the mitochondrial-to-myofibrillar ratio remains unchanged (WINKLER et al. 1976), whilst in hypertrophy due to thyroxine administration, a marked increase in the mitochondrial volume fraction results in an elevated ratio (PAGE and MCCALLISTER 1973b).

The volume fractions of matrix, sarcoplasmic and the T system may be increased or unchanged (see ANVERSA et al. 1976, 1978). The surface area of the T system plus the sarcolemma increases in proportion to the increase in cell volume, so that the surface area to volume ratio of the myocyte is unchanged (ANVERSA et al. 1978).

The absolute volumes of all myocyte components (ANVERSA et al. 1979), and the surface area of the sarcoplasmic reticulum, T system and sarcolemma, all increase in hypertrophy (ANVERSA et al. 1979).

IV. Regional Patterns

Myocyte hypertrophy displays regional differences. ANVERSA et al. (1978) found that the volume fraction of myocytes within the myocardium decreased in the subepicardium but not subendocardium, while the mitochondrial-to-myofibrillar ratio decreased in the subendocardium but not subepicardium. Cell dimensions increased to a greater extent in the subepicardium, compared to the subendocardium (ANVERSA et al. 1978; GERDES et al. 1979), although in pressure-overload of short duration the opposite may be the case (HATT et al. 1978). In volume-overload a greater increase in cell volume occurs in the subendocardium compared to the midwall of the left ventricle (HATT et al. 1979). In papillary muscle, myocyte hypertrophy is accomplished by an increase in transverse dimensions with no change in the average cell length (ANVERSA et al. 1980), while both parameters are increased in the ventricular wall. The response of myocytes within a given region may not be uniform because both myocyte diameter and the variability of the measurements increase in response to isometric exercise (MUNTZ et al. 1981).

I. Regeneration

The adult mammalian ventricular myocyte is considered incapable of repairing local injuries, and a connective tissue scar forms at the site of injury (McMINN 1969; FANBURG 1970; BING 1971; RUMYANTSEV 1977, 1981). This loss of regenerative ability is due to an irreversible withdrawal of the cardiac myocyte from the cell cycle (INGWALL 1980) and the fact there are no satellite cells present as in skeletal muscle (RUMYANTSEV 1979; CAMPION 1985). Cardiac muscle of developing animals can however divide but appears to require some loss of Z-disc material and myofibrillar disintegration (OBERPRILLER and OBERPRILLER 1971; RUMYANTSEV 1972; EROKHINA and RUMYANTSEV 1983). Whether or not cells are capable of division appears related to their degree of differentiation or development of specialized characteristics. Maximum incorporation of ^{3}H-thymidine into DNA after injury to the ventricle occurs in the one-week animal (8%), with incorporation of 3.2, 2.2 and 0.2% in the 2-, 3- and 4-week-old animals respectively (NAG et al. 1983a). Mammalian atrial muscle (in which there are a number of less specialized cells) does show some DNA synthesis and mitosis after infarction of the left ventricle (RUMYANTSEV 1974), however direct damage again fills with scar tissue as a result of overgrowth of proliferating nonmuscle cells (RUMYANTSEV 1982). In contrast to the mammalian heart, the myocytes of lower vertebrates, such as amphibians and reptiles, possess a higher capacity for regeneration (OBERPRILLER and OBERPRILLER 1974; RUMYANTSEV 1977; BADER and OBERPRILLER 1979; McDONNELL and OBERPRILLER 1983) but even here direct wounding leads to scar formation (McDONNELL and OBERPRILLER 1984).

J. Cardiac Muscle Cells in Culture

I. Introduction

Cardiac muscle has been grown in culture since 1910 when the independent spontaneous contraction of small clumps of tissue was observed (BURROWS 1910, 1912). Until the 1950's, most of the studies involved explant culture and the modification of nutrient medium to maintain the cells in a differentiated, contractile state (see MURRAY 1965). In 1955 CAVANAUGH enzymatically dispersed cardiac muscle into single cells and systematically observed their behaviour.

Cell culturists were among the first to pronounce that cardiac muscle is not a syncytium but a reticulum of joined cells (LEWIS 1926; HOGUE 1937), with HOGUE (1947) observing the formation of intercalated discs in long-term culture.

Cardiac pacemaker activity was clearly demonstrated in culture when PAFF (1935) showed that tissue fragments from the sino-atrial level contract at a higher frequency than those from more anterior levels, and that cardiac fragments from different levels allowed to fuse contract at the higher rate. Using enzyme-dispersed cells, CAVANAUGH and CAVANAUGH (1957) and HARARY and FARLEY (1960a, b, 1963a, b) showed that beat synchrony follows the establishment of connections between adjacent cells, and that each cell retains its own inherent beat rate since this is reasserted when contact is broken.

It is well established that enzyme-dispersed heart cells from neonatal animals will remain contractile in culture for many weeks (see HALLE and WOLLENBERGER 1970). In contrast, while it is possible to isolate beating cardiac cells from adult animals, most remain functionally viable for only a few hours (HEWETT et al. 1983), although there are some reports of long-term survival (MOSES and CLAYCOMB 1982, 1984; CLAYCOMB and LANSEN 1984; SCHWARZFELD and JACOBSON 1981; NAG et al. 1983b), and even of spontaneous contractions for up to 60 days (JACOBSON 1977).

Since the early 1960's cell culture has been used extensively to study the development and normal biology of cardiac muscle, its myofibrillogenesis, general metabolism, protein biosynthesis and degradation, electrophysiology, ion transport, inotropic and chronotropic responses to drugs and hormones, effect of hypoxia, receptor modulation and innervation. The use of cultured myocytes as models for research has been recently reviewed (WOLLENBERG 1984; ROBINSON 1985). In this chapter we will give an overview of some of the more important and interesting aspects of these studies.

II. Methods of Culture

1. Methodology

a) Monolayer

Disaggregation of cardiac tissue into single cells can be achieved with a variety of proteolytic and mucolytic enzymes such as trypsin, collagenase, pronase and hyaluronidase, by chelation of calcium ions with EDTA, or by mechanical procedures. Mechanical methods involve forcing the tissue through stainless-steel mesh followed by vigorous agitation which produces a low cell yield (see BAROFSKY et al. 1973). By far the most widely-used method utilizes trypsin to dissociate cells from 2 mm³ pieces of neonatal ventricle (HARARY and FARLEY 1960; MARK and STRASSER 1966; KASTEN 1972). The method used by our laboratory is described in detail:

Whole hearts from 4- to 15-day-old chick embryos or 1- to 4-day-old rats are removed sterilely to a Petri dish containing Hanks' balanced salt solution (HBSS), the atria and large vessels removed and the ventricles stripped of epicardium. The ventricles are cut gently into fragments about 2 mm³ using razor blade knives then rinsed in HBSS to remove blood and cellular debris. The tissue is then rinsed in trypsin (1 mg/ml Difco) in HBSS minus Ca^{2+} and Mg^{2+} ions and incubated at 37° C in successive changes of trypsin for 10 min each. With hearts from slightly older animals, the tissue is incubated in 1 mg/ml collagenase (4196 Worthington) in HBSS with Ca^{2+} and Mg^{2+} ions at 37° C for 30 min prior to the trypsin digests. At the beginning and end of each trypsin incubation the tissue is dispersed by gently aspirating up and down in a wide-bore pipette. The first two digests are discarded as these contain a high proportion of non-cardiac muscle cells (endothelium, smooth muscle, fibroblasts), but the third and subsequent digests containing single cells are placed into cold fetal-calf serum (10% final volume) to inactivate the remaining trypsin and centrifuged at 900 rpm. The cell pellets are resuspended in Eagles Minimal Essential Medium with Hanks' salts supplemented with 10% fetal calf serum and 2% chick embryo extract. To obtain pure cultures of myocardial cells, the cell suspension is preplated into 90-mm plastic culture dishes and incubated at 37° C for $1^1/_4$ h. Within this time the non-cardiac muscle cells attach and flatten to the substratum, but the more rigid cardiac muscle cells are still in suspension or only weakly attached. The supernatant is removed and the cardiac muscle cells re-seeded into fresh culture chambers (BLONDEL et al. 1971). These cells attach and flatten within 4 to 6 h. This method of culture is used for morphological, biochemical and functional studies.

b) Synthetic Strand

In order to obtain accurate values for specific electrical constants and to apply the technique of voltage clamp for measurement of non-linear membrane electrical properties, the synthetic strand of cultured cardiac muscle was developed (LIEBERMAN et al. 1972). This consists of enzyme-dissociated heart cells plated onto channels cut in agar gel such that the bundles of cells are directed, forming a strand. Once the strands contract they detach from the channel walls and can be used.

c) Cluster

Also for electrophysiologic studies, clusters of cardiac muscle cells can be obtained by modifying the culture techniques used to prepare synthetic strands. Small openings (20 µm) are formed on the surface of an agar-coated culture dish using a 27 gauge needle. The exposed plastic surface provides points of attachment for preformed small spherical aggregates 50–200 µm diameter (see LIEBERMAN et al. 1981). Similar cardiac muscle cell clusters have been obtained by gyratory reaggregation (FISCHMAN and MOSCONA 1971; ZACCHEI and CARAVITA 1972; SACHS and DE HAAN 1973) or by spontaneous reaggregation on a cellophane or polystyrene sheet (HALBERT et al. 1971; McLEAN and SPERELAKIS 1976).

d) Polystrand

The polystrand was developed for ion-transport studies. This is a nylon-supported growth-orientated preparation of cultured-heart cells consisting of 70 parallel segments of 5 mm-long nylon monofilament supported by a U-shaped silver clip (HORRES et al. 1977; WHEELER et al. 1982).

2. Culture Medium

While it is possible to maintain cardiac muscle cells in a chemically-defined medium (HALLE and WOLLENBERGER 1968; CLAYCOMB 1981; MOHAMED et al. 1983; LIBBY 1984), most laboratories use a serum and/or tissue supplement such as 10% fetal calf serum and 2% chicken embryo extract. The buffer mainly used in cardiac cell culture medium is the $CO_2 - NaHCO_3$, system, as in Eagles Minimal Essential Medium with Hanks' Salts. However, this buffer is not suitable for studies where the culture medium is exposed to room air for long periods as in electrophysiological studies. Alternate buffers are then employed such as phosphate (BLOOM 1970), 2-(N-2 hydroxyethylpiperazin-N^1-yl) ethanesulphonic acid (HEPES) (MASSON-PÉVET 1979) or Tris-HCl (GOSHIMA 1971). Penicillin is usually added to the medium to discourage bacterial contamination. Unlike streptomycin, which interferes with the beating rate and force of contraction (PAYET et al. 1980), penicillin at the recommended concentration has no adverse effects on functional parameters of the cells.

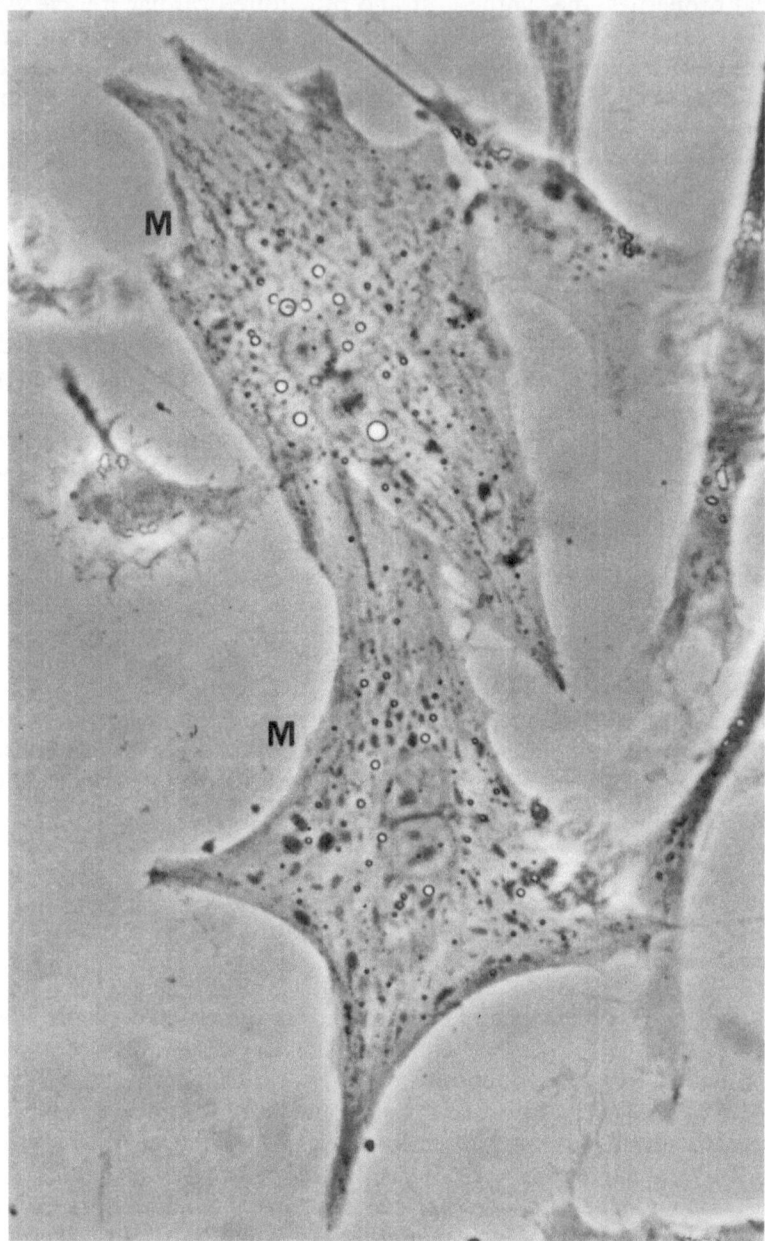

Fig. 95. Two cardiac muscle cells (*M*), 2 days in culture. Phase-contrast microscopy. × 900

III. Working Myocardium in Culture

1. Morphology

As soon as the myocardial cells flatten on a substrate they assume a characteristic morphology, having an elongated star-shape, phase-dense cytoplasm, prominent mitochondria, glycogen granules and lipid droplets and one or two relatively small and round nuclei (Fig. 95). Cells from the sinoatrial node are smaller and elongated or spindle in shape (MARVIN et al. 1984b). Cardiac muscle cells are easily distinguished from fibroblast/smooth muscle/endothelial contaminants which are phase-lucent and considerably less granular and do not stain with FITC-labelled antibodies to cardiac myosin (Fig. 96).

Irrespective of the developmental age of the donor animal, enzyme-isolated cardiac muscle cells from 4- to 15-day-old chick embryos and 1- to 4-day-old rats suffer a reversible disruption of their myofibrils upon culture (see KULI-KOWSKI 1981). This can be seen when the cells 6 to 12 h in culture are reacted with FITC-labelled antibodies to cardiac actin, tropomyosin or myosin. Before the appearance of definitive myofibrils the staining initially appears in clumps and uninterrupted fibrils. Gradually the myofilaments become aligned and sarcomeric cross-banding develops throughout the cell (Fig. 97 a) (GRÖSCHEL-STEWART et al. 1975; CHAMLEY et al. 1977; CHAMLEY-CAMPBELL et al. 1977; GLACY 1983; DLUGOSZ et al. 1984). The myofibrils grow laterally as well as terminally such that intricate patterns of cross-banding develop (Fig. 97B). This is especially evident in well-spread cells where there is no preferred orientation of myofibrils, whereas in elongated cells most of the myofibrils are aligned parallel to the long axis. Spontaneous contractions begin in most cells after 12–18 h at rates of 30–90 per min.

Following the initial reorganization of myofibrils, cultured cardiac muscle, even as isolated cells, maintains its differentiated structure and contractility for many months under optimal conditions of temperature, pH, osmolarity and nutrients.

Intermediate filaments can also be demonstrated in cultured myocardial cells with FITC-labelled antibodies to the 55K dalton protein (LAZARIDES 1978; CAMPBELL et al. 1979; FUSELER and SHAY 1982; DLUGOSZ et al. 1984). In well-differentiated cells the 55K protein is clearly localized in the Z discs of sarcomere units of the myofibril (Fig. 98A). Filaments also extend parallel to the myofibril and others connect Z discs of closely apposed myofibrils giving a ladder-like appearance. In some cells the Z discs of adjacent sarcomeres have come together in close register across the cell giving a striped appearance and bundles of filaments accumulate in the region of intercalated discs (Fig. 98A). In less well-differentiated cells numerous intermediate filaments extend throughout the cytoplasm in a random network (Fig. 98B). Between these two extremes is the stage where the 55K protein is localized in short filament bundles and foci parallel to the long axis of the cell. The sarcomeres that are present and contractile during this period do not contain 55K protein in the Z discs even though

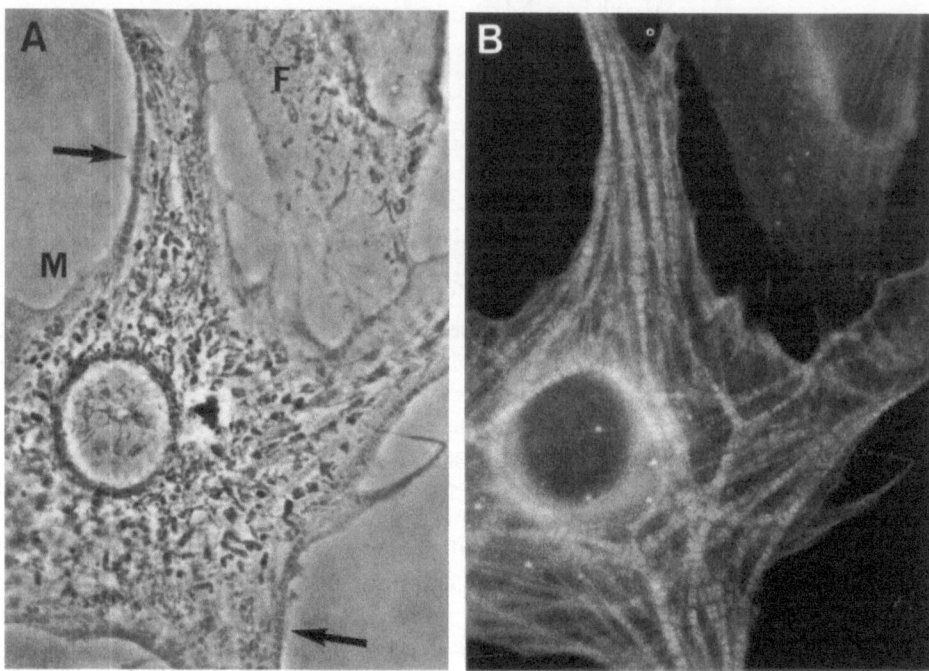

Fig. 96. A Cardiac muscle cell (*M*) and a fibroblast (*F*), 4 days in culture. Phase-contrast microscopy. *Arrows*, sarcomeric cross banding. **B** Same cells as **A**, stained with FITC-labelled antibodies to cardiac myosin. × 1200. Antibody the generous gift of Professor UTE GRÖSCHEL-STEWART, Darmstadt, Federal Republic of Germany

these early sarcomeres are completely formed as manifest by their birefringence and the staining of their Z and H regions by creatine phosphokinase antibody (FUSELER and SHAY 1982).

Ultrastructurally the cells in culture demonstrate the same myofibrillar organization as observed with immunofluorescence (see CEDERGREN and HARARY 1964a, b; LEGATO 1972; KELLY and CHACKO 1977). The sarcomeric length varies between 1.2 and 1.7 μm. In many cells the sarcomeres are aligned with mitochondria located between myofibrils similar to their position in the intact heart. The mitochondria are large (Fig. 99). LE FURGEY et al. (1983) propose that many mitochondria of cultured heart cells are interconnected and that this may represent a parallel functional unit which maintains the energy state of the cell. Cristae are tightly packed as is typical of cells with high rates of oxidative metabolism. Large aggregates of dense osmiophilic material is present between cristae. These are deposits of phosphate salts of calcium, magnesium and other divalent ions. Glycogen is present in large aggregates devoid of any limiting membrane structure. The Golgi apparatus is generally associated with the outer nuclear envelope. It is small and rather undeveloped compared to the extensive apparatus of secretory cells. Microtubules are present and lysosomes are usually few in number. Ribosomes and polyribosomes of 4–6 ribosome units are present, as are both rough and smooth endoplasmic reticulum.

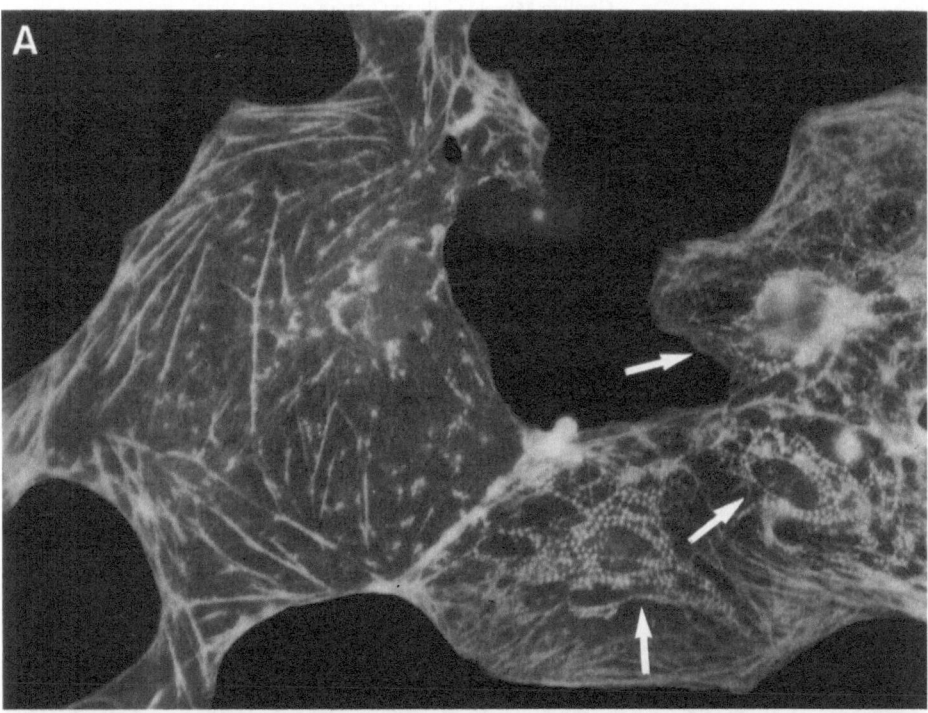

Fig. 97. A Cardiac muscle cells, 1 day in culture. Stained with FITC-labelled antibodies to cardiac actin. Most staining is in clumps and uninterrupted fibrils. Cross banding starts to appear first in the perinuclear region (*arrows*). × 800. **B** Cardiac muscle cells, 3 days in culture. Stained with FITC-labelled antibodies to cardiac myosin. Note intricate patterns of cross banding. × 1500. Antibodies the generous gift of Professor UTE GRÖSCHEL-STEWART, Darmstadt, Federal Republic of Germany

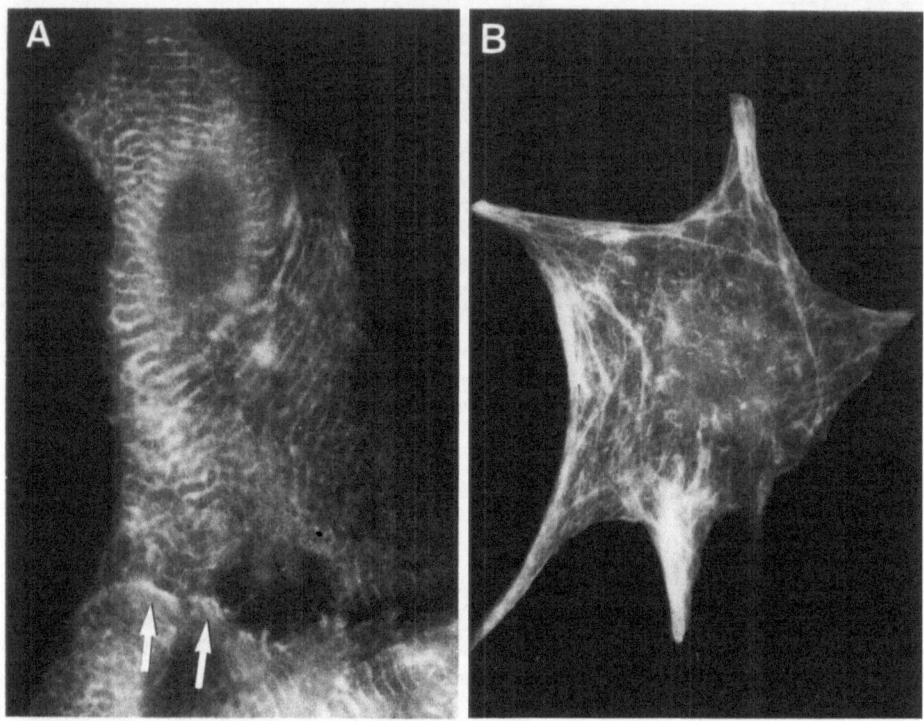

Fig. 98. A Cardiac muscle cells, 4 days in culture. Stained with FITC-labelled antibody to the 55 K dalton protein. Staining is localized in the region of the Z disc and the intercalated disc (*arrows*). × 1100. **B** Cardiac muscle cell, 2 days in culture. Stained with FITC-labelled antibody to the 55 K dalton protein. In this less well-differentiated cell the intermediate filaments form a random network. × 900. Antibodies the generous gift of Professor UTE GRÖSCHEL-STEWART, Darmstadt, Federal Republic of Germany

The smooth endoplasmic reticulum or sarcoplasmic reticulum forms a lace-like sleeve around the myofibrils. It is thought to be the subcellular system which affects myofibrillar relaxation by removal of calcium (B.II.5). The transverse tubular system (T system) is a modification of the sarcolemma invaginating at the level of the Z disc of the myofibrils. It is thought to be involved in excitation-contraction coupling (see KELLY and CHACKO 1977) (B.II.3).

2. Intercellular Junctions

Both *in vivo* and in culture, the junctions which form between cardiac muscle cells are of three types: the nexus, the desmosome and the fascia adherens (see B.II.4). When cells are enzymatically dispersed and allowed to settle onto a coverslip in culture, cells that were not previously coupled become so after the cells adhere to each other. The latent period between adhesion and electrical

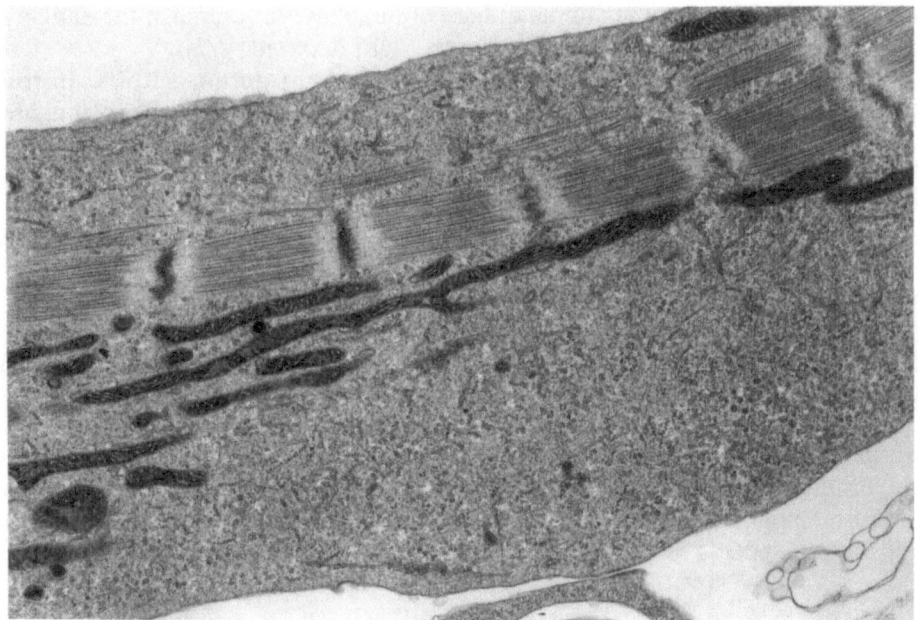

Fig. 99. Cardiac muscle cell from the ten-day chick embryo, 2 days in culture. Myofibrils stretch the length of the cell. Note the long branched mitochondrion. × 9000

coupling is variable, but can be 4 min or less (see PAGE and SHIBATA 1981). The formation of nascent channels in the junctions is detectable as quantal increments in cell-cell conductance up to a steady state value (LOWENSTEIN et al. 1978). Nexus formation in aggregates of chick embryo cardiac muscle cells is prevented by inhibitors of protein synthesis (GRIEPP and BERNFIELD 1978). Furthermore, chick and rat cardiac muscle cells of comparable stages of development will coaggregate in culture and establish intercellular junctions with synchronous beating within 24 h, indicating a high degree of conservation of nexus proteins between different species (NAG and CHENG 1980).

3. Mitosis

In vivo adult cardiac muscle is unable to regenerate by forming new muscle cells either by division of healthy cells in areas adjacent to the damage or by the proliferation and differentiation of a reserve population of myogenic stem cells. However, in very young animals apparently fully-differentiated cardiac muscle cells are capable of division. This occurs *in vivo* and in culture. In culture, cross-banded cardiac muscle cells from young animals often beat spontaneously during all stages of mitosis though there is a decline in the number of beating cells at metaphase and anaphase. There is no consistent pattern of activity, with some cells ceasing contraction throughout the entire mitotic process and others stopping in metaphase or anaphase to resume in the daughter cells (KELLY and CHACKO 1976; MANASEK 1968; KASTEN 1972).

Cardiac muscle cells in culture do not round up during mitosis. Instead the cell is pinched in two, not necessarily in the middle, such that daughter cells of different size frequently occur (CHACKO 1979). There is variable disruption of myofibrils during mitosis, compatible with the presence or absence of contraction in different cells. During mitosis myofibrils are observed at different stages of organization, ranging from mature sarcomeres to small aggregates of partially-aligned myofilaments in which Z discs are difficult to detect (RU-MYANTSEV 1972). KASTEN (1972) has suggested that the partial, transient disorganisation of myofibrils in mid-mitosis is related to a competition for energy at the time of spindle changes and chromosome movements so that priority is given to the mitotic processes rather than the myofibrillar contractions.

A small proportion (about 10%) of muscle cells entering mitosis do not complete the cell division, but show a cleavage furrow that fails to progress and eventually disappears without cytokinesis, resulting in binucleate cells. Most of the remaining mononucleated cells are tetraploid or octaploid (BOGENMANN and EPPENBERGER 1980).

4. Metabolism

a) ATP as an Energy Source

Under conditions of adequate oxygen supply, cardiac muscle cells are almost completely dependent for energy production on aerobic, mitochondria-involved metabolism, with ATP as the immediate energy source for contraction (SOBEL 1972). Many metabolic pathways are used to generate ATP and the exact source depends on availability of substrates. FUJIMOTO and HARARY (1964) and HARARY et al. (1966) reported that the major source of energy in beating cells is from lipids, particularly fatty acids. Glucose is also a major source. Phosphocreatine serves as a source of high-energy phosphate for synthesis of ATP at the site of utilization (the myofibrils) (SERAYDARIAN and ARTAZA 1976).

b) Lipid Metabolism

Serum lipids and fatty acids are essential to maintain the contractility of heart cells in culture (HARARY et al. 1966) and to prevent isozyme shifts (FUJI-MOTO and HARARY 1964).

Triglycerides cannot be taken up by the heart cells until they are hydrolysed to fatty acids by the membrane-bound enzyme lipoprotein lipase (see PINSON et al. 1973). Thus there is a correlation between lipoprotein lipase activity and the ability of the heart to assimilate triglyceride. In culture the development of lipoprotein lipase activity marks the myoblast maturation to a myocyte capable of mechanical work for which the energy is supplied mainly through lipid metabolism. Positive-charged low-molecular-weight molecules present in serum are required for the stabilization and synthesis of lipoprotein lipase in heart cell cultures (CHAJEK et al. 1977, 1978a, b; FRIEDMAN et al. 1980).

Chick-embryo heart cells in culture incorporate radio-labelled fatty acids into cellular triglycerides and phospholipids 15 s after addition to the medium (SAMUEL et al. 1976). The esterification occurs in the sarcoplasmic reticulum (STEIN and STEIN 1968; CRASS 1977; GLOSTER et al. 1978). The fatty acids traverse the membrane via a high-affinity transport mechanism ($K_m = 4.5 - 16$ µM, $V_{max} = 1.8$ nmole/h/10^6 cells), with diffusion-like uptake becoming important at concentrations above 0.02 mM. The rate of entry into cells is not inhibited by glucose, lactate or amino acids and is not affected by the omission of Na^+, K^+, Ca^{2+} or Mg^{2+} from the medium (SAMUEL et al. 1976; PARIS et al. 1979).

In contrast to the rapidity of esterification, radio-labelled fatty acids are oxidized 15 min after they have entered the cultured heart cells (SAMUEL et al. 1976). Long-chain fatty acids are oxidized completely to carbon dioxide in accordance with the mechanism of β-oxidation control in which the intermediates of fatty acid degradation never leave the multienzyme system of the tricarboxylic acid (TCA) cycle.

c) Carbohydrate Metabolism

Heart muscle, well supplied with oxygen, converts glucose to energy via acetyl CoA and the aerobic TCA cycle. In culture, the uptake of glucose by heart cells is 10–20 times higher than *in vivo* (PADIEU et al. 1978; FRELIN et al. 1979). While some of this can be stored as glycogen deposits, most is converted to lactate via anaerobic glycolysis and excreted into the medium (MCCARL et al. 1980). Once glucose has been exhausted, the excreted lactate is taken up and further metabolised (FRELIN et al. 1974). Heart tissue *in vivo* does not excrete lactate unless it is hypoxic, and since in culture the cells have adequate oxygen supply, it has been suggested that alterations in glucose transport have occurred (ANASTASIA and MCCARL 1973; FRELIN et al. 1974, 1979).

d) Amino-Acid Metabolism

Heart cells from neonatal rats grown in monolayer culture take up the essential amino acids provided by the culture medium (FRELIN 1980). Excessive utilization of glutamine, glutamic acid and arginine suggests that these amino acids are involved in intermediary metabolism. Moderate amounts of glycine and proline and high amounts of alanine are released into the culture medium. Release of these amino acids represents *de novo* synthesis and not protein breakdown. Alanine synthesis is stimulated by extracellular glucose, pyruvate, glutamic acid and serum, and cannot be completely accounted for by the high glycolytic rates that occur in culture due to impaired regulation at the hexokinase level. The release of alanine is not due to hypoxia since the pO_2 in the culture medium is 95 mm Hg and the heart cells do not release cytoplasmic enzymes (see ALTONA et al. 1984). It is suggested that a major role for alanine synthesis and release by the myocardium is to feed the citric acid cycle with α-ketoglutarate. As such it should contribute to the energy supply of heart muscle during oxygen deprivation (TAEGTMEYER et al. 1977). The persistence of these metabolic routes in well-oxygenated heart cultures may be a remnant of the metabolic adaptation of the fetal heart to its poorly-oxygenated environment (FRELIN 1980).

5. Synthesis of Prostaglandins

Ischemia, trauma and hormonal stimulation elicit the release of prostaglandins from the heart (see BERGER et al. 1976). It is well known that the coronary arteries synthesize PGI_2 (RAZ et al. 1977), but the source of other prostaglandins is not clear. To determine the capacities for prostaglandin synthesis by individual types of cells comprising the heart, cultures of rat cardiac muscle cells and mesenchymal cells were evaluated by radiochromatography for products obtained by incubating cells with $(1-{}^{14}C)$ arachidonate. Cultured mesenchymal cells synthesized PGE_2, $F_{2\alpha}$ and 6-keto-$F_{1\alpha}$ (a stable metabolite of PGI_2). Cardiac muscle cells synthesized PGE_2 and $PGF_{2\alpha}$ but no PGI_2 (AHUMADA et al. 1980).

6. Contractility

a) Use of Automatic Monitors

Rate and strength of beating of individual cardiac-muscle cells can be measured by the use of automatic monitors (BODER et al. 1971; THOMPSON et al. 1973; PURVES et al. 1974; FAYET et al. 1974; LIEBERMAN et al. 1975; HARARY et al. 1983). This allows observation of the direct physiological response of cardiac muscle cells to a variety of stimuli in a controlled environment. While some of these devices involve embedding electrodes into the cells to measure transmembrane potential and thus individual depolarizations, other procedures are indirect and non-invasive, and consist essentially of measuring changes in light intensity due to contractions of the cells by a light-sensitive diode mounted in a microscope. The signal from the diode is amplified and recorded giving a measure of beats per minute and strength of beat. Cells isolated from the sinoatrial node beat at about 185/min (MARVIN et al. 1984). The reported rates of contraction of atrial and ventricular cells vary greatly depending on species, age and culture conditions.

b) Effect of Anaesthetics

Using this type of monitor the influence of the anaesthetics methoxyflurane, enflurane, halothane and nitrous oxide on beating of individual cardiac muscle cells has been determined. It is well known that there is myocardial depression during anaesthesia, and the concentration of anaesthetic necessary to give 50% inhibition of contractility in culture correlates well with the minimum alveolar concentration values for specific anaesthetics. Thus it is possible to use cultured heart cells to provide a potency value for anaesthetics which is reliable, repeatable and readily obtained (see MCCARL et al. 1980).

c) Effect of Autonomic Agonists and Antagonists

Isolated single cardiac muscle cells in culture are highly sensitive to autonomic agonists with an ED_{50} for noradrenaline of 800 pM and 370 pM for acetylcholine (HERMSMEYER and ROBINSON 1977). Thus under certain conditions of culture, the cells are two orders of magnitude more sensitive than from freshly isolated heart, undoubtedly due to the absence of nerve terminals on the cultured cells.

The rate of beating of cultured cardiac muscle is accelerated in the presence of catecholamines due to stimulation of both α and β receptors (KARSTEN et al. 1977). In the continous presence of L-noradrenaline for 48 h, neonatal rat-ventricle cells cultured in serum-free medium hypertrophy to a maximum of 150% of control (SIMPSON et al. 1982; SIMPSON and McGRATH 1983); half maximal effect is obtained at a concentration of 0.2 μm. This increase in cell size (cell volume, cell surface area and cell protein) is inhibited by the nonselective α-adrenergic antagonist phentolamine and by the α_1-adrenergic antagonists prazosin and terazosin. It is not inhibited by the β-adrenergic antagonist propranolol or by the α_2-adrenergic antagonist yohimbine. The β-adrenergic agonist isoproterenol does not increase cell size. Thus it is concluded that noradrenaline-stimulated hypertrophy of cultured rat myocardial cells is an α_1-adrenergic response.

The β-adrenergic blocking isomer (L) and the non β-adrenergic blocking isomer (D) of propranolol are equally effective in suppressing the spontaneous beating of cultured myocytes (HIGGINS et al. 1979). This suggests that supression of spontaneous beating is not dependent upon β-adrenergic blockade, but on some other effect of the molecule such as membrane stabilization.

d) Effect of Colchicine

Colchicine, a microtubule disrupting agent, increases the rate of contraction of spontaneously beating neonatal rat-heart cells in culture by as much as 98% compared with control (KLEIN 1983). The increase in heart rate is dose dependent and does not occur with the inactive stereoisomer lumicolchicine. The mechanism by which colchicine exerts its positive chronotropic effect is unknown, but it is not blocked by propranolol and does not affect the activity of adenylate cyclase.

e) Effect of Dibutyryl Cyclic AMP

The spontaneous contractions of cultured heart cells are arrested upon the addition of 0.2–2.0 mM N^6 O^2-dibutyryladenosine $3':5'$ cyclic monophosphate (Bt_2 cAMP) (NATH and BOLLON 1978). The contractions are restored by adding colchicine at 1.0 μM (NATH et al. 1978). In Bt_2 cAMP-treated myocytes the intracellular distribution of microtubules is altered such that they appear to form parallel arrays. Treatment with colchicine disaggregates the microtubules. Thus the orientation of microtubules can affect heart-cell contractions.

Dibutyryl cyclic AMP also supresses the mobility of chick heart cells in aggregates (GERSHMAN et al. 1979). There are a number of possibilities as to the mechanism of this effect, however it may be related to the altered intracellular distribution of microtubules.

7. Hypoxia

The quantitative relationship between a given decrease in oxygen tension (pO_2) and the degree of depression of contractile function is difficult to define in the intact myocardium because of uncertainty in estimation of diffusion limitations. Cell culture of spontaneously-contracting monolayers of myocardial cells offers a system with minimal diffusion barriers and fine control of pO_2 at the surface of individual cells (see SPERELAKIS and LEHMKUHL 1967; MARTINEZ and WALKER 1977; BARRY et al. 1980).

When perfusate pO_2 is less than 12 mm Hg, reversible decreases in amplitude and velocity of contraction occur but the cells continue to beat for some time indicating that sufficient energy is stored or can be generated (BARRY et al. 1980; SMITH et al. 1982). On reoxygenation, recovery of amplitude and velocity of contraction occurs in less than 15 s. During hypoxia, very little energy is obtained from oxidative phosphorylation, as evidenced by the low rate of flux through the TCA cycle, and high-energy phosphates derived from anaerobic glycolysis supply much of the needed energy. There is a quantitative relationship between the degree of depression of contractility and adenosine triphosphate and phosphocreatine levels during mild to moderate degrees of inhibition of oxidative phosphorylation (DOOREY and BARRY 1983).

A combination of anoxia and substrate deprivation causes release of several intracellular marker enzymes from cultured neonatal rat heart cells (ACOSTA et al. 1978; VAN DER LAARSE et al. 1978, 1979a, b; ALTONA and VAN DER LAARSE 1982; ALTONA et al. 1984). The degree of depletion is high (>80%) for cytoplasmic enzymes, moderate (50%) for sarcolemmal and lysosomal enzymes and enzymes associated with the mitochondrial outer membrane, and small (<10%) for enzymes associated with the mitochondrial inner membrane. Changes in sarcolemmal cholesterol content precede sarcolemmal enzyme release, showing that energy deprivation in cardiac cells destroys sarcolemmal function secondary to cholesterol loss. Mild acidosis (pH 6.7) protects the cells against hypoxic injury as reflected by a decrease in lactate dehydrogenase leakage, while a pH of 6.5 does not protect the cells but exacerbates the cell injury (ACOSTA and LI 1980). Pretreatment of cells with methylprednisolone for 24 h provides protection to cells challenged by hypoxia and glucose deprivation (ACOSTA et al. 1980). The survival of anoxic cultures is closely associated with glycolytic activity (HIGGINS et al. 1981). Glycolysis rate falls and enzyme release increases as the glucose concentration of the medium is reduced. If glycolysis is inhibited, enzyme release under anoxic conditions is enhanced. Elevation of the calcium content of the culture medium exacerbates the damage caused to cardiac myocytes by anoxic insult.

A role for free radical-mediated processes in myocardial damage has been suggested, especially with respect to the reoxygenation of hypoxic tissue (SCHLAFER et al. 1982). SCOTT et al. (1985) examined this problem in myocardial cultures utilizing fluorescence-activated sorting of myosin antibody-labelled cells to quantify cell damage. They found that oxygen-derived free radicals cause significant loss of membrane integrity, and that the species of radical formed is dependent both on pH and the concentration of iron salts. The injury is, at least in part, preventable by the administration of exogenous radical scavenging agents.

Under anoxic incubation conditions, heart cell cultures show enhanced uptake of radio-labelled fatty acids into neutral lipids, mainly triglycerides (BAILEY and HIGGINS 1983). When anoxic cultures are reoxygenated, fatty acid, endogenous lipid and glucose oxidation is depressed and mitochondrial function is readily uncoupled by 2,4-dinitrophenol. This suggests that anoxia-stimulated lipid accumulation may prove injurious to subsequent mitochondrial function and be contributory to the pathological processes associated with hypoxic injury of cardiac muscle.

8. Electrical Properties

a) Monolayers and Reaggregates

As pointed out by SPERELAKIS and MCLEAN (1978), monolayers of cardiac muscle cells in culture have advantages over intact myocardium for studying certain questions in electrophysiology. For example, the simultaneous recording of transmembrane potentials and contractions on single cells using photoelectric techniques; studies of electrotonic spread of current in a two-dimensional system; electrogenesis of various components of the action potential or pacemaker potential; voltage clamp experiments with fine voltage control over the entire cell membrane; direct effect of agents without interference of neural input; no diffusion lag thus facilitating ion-flux studies; and direct observation of electrical and mechanical effects following microelectrophoretic injection.

However, it has been suggested that monolayers of cardiac muscle cells dedifferentiate electrophysiologically to some extent in culture. That is, fast Na^+ channels are lost and slow Na^+ channels are gained (see MCLEAN and SPERELAKIS 1974). The cells also have a lowered (Na^+, K^+)-ATPase specific activity (SPERELAKIS and LEE 1971) and an elevated cyclic AMP level (MCLEAN et al. 1975). These changes are probably due to suboptimal culturing conditions as more recent reports show that fast Na^+ channels as well as maximum diastolic potential can be maintained (LOMPRE et al. 1979a; ATHIAS et al. 1979; ROBINSON and LEGATO 1980; ROBINSON 1982, 1985; ROBINSON et al. 1984). Also, BERNARD and COURAND (1979) have shown that latent fast Na^+ channels in cultured heart cells can be activated by binding scorpion toxin to the membrane.

Early in cell culture, ventricular cells beat spontaneously at independent rhythms, indicating that these nonpacemaker cells have regained automaticity. The resting potential of these cells is low and many cells exhibit pacemaker potentials, consistent with low K^+ permeability (SPERELAKIS and MCLEAN 1978). Culturing the cells in media containing elevated K^+ (12–60 mM) helps the cells to retain more highly differentiated electrical properties (MCLEAN and SPERE-LAKIS 1974). The addition of 5 mM ATP to the medium, particularly in combination with elevated K^+, also helps maintain differentiated membrane properties.

Most cells in clusters or cylindrical strands retain fully-differentiated electro-physiological properties identical to those of the myocardium *in situ*. The cells have high resting potentials of about -80 mV, and pacemaker potentials are absent indicating that K^+ permeability remains high. In addition, the cells retain their full complement of fast Na^+ channels (SPERELAKIS and MCLEAN 1978; NATHAN and DE HAAN 1978; BERNARD and COURAND 1979; JOURDON and SPERELAKIS 1980; NATHAN 1981).

b) Changes in Membrane Composition

Cholesterol depletion of 33% in cultured rat heart cells by exposure to high-density apolipoprotein-sphingomyelin mixtures causes a significantly faster rate of depolarization than in control cells which is presumed to be related to the facilitation of Ca^{2+} and Na^+ influx during depolarization (HASIN et al. 1980). Increase in cholesterol content of cultured rat heart myocytes alters their action potential by an increased activity of membrane ATPase (ALIVISATOS et al. 1977). An increase in the linoleic/oleic acid ratio in the membrane lipids augments the initial phase of repolarization with no effect on the depolarization phase. The enhanced repolarization may be due to an increase in total membrane conductance by the higher linoleic acid content (HASIN et al. 1982).

9. Transmembrane Ion Fluxes

a) Calcium Homeostasis

Calcium homeostasis in the myocardium occurs in three related processes: calcium influx across the sarcolemma, uptake by sarcoplasmic reticulum and mitochondria, and efflux. Transmembrane calcium fluxes play a central role in cardiac muscle electromechanical coupling (WINEGRAD 1979). Calcium enters the cell during the plateau of the action potential via voltage-sensitive ion-specific channels (REUTER 1974), and may trigger release of a larger amount of calcium from stores within the sarcoplasmic reticulum resulting in a rise in Ca^{2+} to levels adequate to produce contraction of the myofilaments (FABIATO 1981). Calcium may also enter the cardiac muscle cell via a Na-Ca-exchange carrier, since studies on cultured chick-embyro ventricular cells have shown that 80% of Ca influx occurs by Na-Ca exchange (BARRY and SMITH 1982). The stoichiometry of the Na-Ca exchange is probably 3 Na^+:1 Ca^{2+} (WAKA-BAYASHI and GOSHIMA 1981; BURT 1982; BRIDGE and BASSINGTHWAIGHTE 1983), and thus this process is electrogenic and calcium entry is stimulated during cell depolarization.

The total amount of calcium entering the cell with each depolarization must be extruded before the next cycle. Thus, mechanisms for calcium efflux must respond to Ca^{2+} increases and extrude calcium against a large calcium-concentration gradient. REUTER and SEITZ (1967) found that efflux of ^{45}Ca was sensitive to the Na_e and suggested that energy from the Na gradient could be utilized to extrude calcium from the myocardial cell via Na^+-Ca^{2+} exchange (see also GOSHIMA et al. 1980; CLUSIN 1981). However BARRY and SMITH (1984) found that ^{45}Ca efflux from cultured monolayers of chick embryo ventricular cells is not affected by removal of sodium from extracellular fluid but is slowed by metabolic blockade of ATP, which is consistent with the alternative or additional hypothesis that an ATP-dependent calcium pump in the sarcolemma contributes significantly to calcium efflux.

b) Calcium Regulation of Cyclic AMP and Beating

Catecholamines affect rate and strength of beating of cardiac muscle through increasing the cellular concentration of cyclic AMP (KRAUSE et al. 1970). Experiments utilizing beating rat cardiac muscle cells in culture have shown that both noradrenaline and Ca^{2+} control the concentration of cyclic AMP in the cells, which in turn regulates the utilization of Ca^{2+} (HARARY et al. 1976). Addition of noradrenaline or depletion of Ca^{2+} in the external medium results in cells which can beat in a lower concentration of external Ca^{2+} compared with controls. Ca^{2+} affects the concentration of cyclic AMP by an effect on cyclic AMP phosphodiesterase (HARARY and WALLACE 1978). This enzyme is stimulated by an "activator" protein on the sarcoplasmic reticulum, which is activated by Ca^{2+} (KAKIUCHI and YAMAZAKI 1970). Thus a decrease in Ca^{2+} causes a decrease in cyclic AMP phosphodiesterase activity and a resultant increase in cyclic AMP. An increase in sarcoplasmic reticulum Ca^{2+} leads to decreased cyclic AMP. It has been proposed that the levels vary cyclically, one step out of synchrony with each other. Thus, depolarization of the sarcolemma and entry of Ca^{2+} into the cells cause contraction and a series of events leading to the restoration of the state necessary for the repetition of the cycle (HARARY and WALLACE 1978).

c) Calcium in Myocardial Cell Injury

Calcium has been hypothesized to play a major role in myocardial ischemia and cell injury by affecting mitochondrial function and activating endogenous phospholipases (NAYLER et al. 1979; FARBER 1981). Inhibitors of the Na^+-K^+ pump in cultured heart cells promote the elevation of intracellular Ca^{2+} and the formation of mitochondrial densities similar to that seen in mitochondria after reperfusion of the ischemic myocardium (SHEN and JENNINGS 1972). Using cell cultures of embryonic-chick cardiac muscle, MURPHY et al. (1983) found that Ca^{2+} loading of the cells and their mitochondria caused by Na^+-K^+ pump inhibition does not result in the release of cytoplasmic enzymes or an irreversible decline in the level of cellular ATP, thus showing that significant increases in total cell Ca^{2+} can occur without irreversible myocardial cell injury.

d) Other Ion Fluxes

Using cultures of cardiac muscle it has been shown that there is a rapid Na^+-H^+-exchange mechanism in the muscle cell membrane (PIWNICA-WORMS and LIEBERMAN 1983). The greater part of the steady-state Na^+ efflux may be due to mechanisms other than the Na^+-K^+ pump (WHEELER et al. 1982), and Cl^- transport across the cell membrane is more rapid than K^+ transport and is largely electrically silent (PIWNICA-WORMS et al. 1983). Studies on the Na^+-K^+-exchange characteristics, ouabain-binding kinetics, and Na^+ pump turnover rates of contracting heart cells in culture, show that the behaviour of the Na^+ pump in these cells is very similar to that in intact tissues (McCALL 1979). MEAD and CLUSIN (1984) studied the effects of abrupt Na^+ removal on membrane current and conductance in voltage-clamped chick embryonic myocardial cell aggregates in the presence of various sodium flux inhibitors. They found that removal of extracellular Na^+ interrupts a cesium-sensitive background current that may be related to the time-dependent pacemaker current. Na^+ removal also causes gradual activation of a nonspecific conductance which can ultimately depolarize the cells and which may be gated by cytoplasmic Ca^{2+}. KIM et al. (1984) showed that prolonged exposure of spontaneously-beating monolayers of cultured chick heart cells to low extracellular K^+ or ouabain causes induction of additional functional sarcolemmal Na^+ pump sites. This long-term effect on Na^+, K^+-ATPase appears to be mediated by intracellular Na^+.

10. Effect of Hormones and Drugs

a) Parathyroid Hormone

Patients with advanced renal failure commonly suffer from myocardial disease (see DEROW 1954). Several consequences of uremia have been implicated including anaemia, electrolyte disturbances, acidosis, and hypertension. Since these patients often have secondary hyperparathyroidism and markedly elevated blood levels of parathyroid hormone (PTH), a role for PTH in the cardiac disease of uremia has been suggested (LEHR 1966). Using rat-heart cells grown in culture, BOGIN et al. (1981) examined the direct effect of PTH. Both the amino-terminal (1–34) PTH and intact (1–84) PTH, but not the carboxy-terminal (53–84) PTH, produced immediate and sustained significant rise in beats per minute which was dose-dependent and the cells died earlier than control cultures. The effect was reversed if PTH was removed from the medium, and was abolished by inactivation of the hormone. The effect of PTH required calcium, was mimicked by calcium ionophore, was prevented by verapamil and was not abolished by α- or β-adrenergic blockers. Sera from uremic parathyroidectomized rats did not affect heart beat, but sera from uremic rats with intact parathyroid glands or from uremic-parathyroidectomized rats treated with PTH had effects similar to PTH. From these studies in culture it was concluded that the cardiac muscle cell is a target for PTH and may have receptors for the hormone; PTH increases beating rate of heart cells and causes early death of cells; the PTH effect appears to be due to calcium entry; the locus of action through which PTH induces calcium entry is different from that for catecholamines (see also LARNO et al. 1980).

b) Diazepam

Studies on the action of diazepam (Valium) on the heart are often contradictory, some reporting an increase in heart rate and others citing a depression of heart rate or no effect (see ACOSTA and CHAPPELL 1977). The consensus of most investigators is that diazepam exerts its myocardial actions mainly via centrally-mediated mechanisms, however, there are some suggestions that it may have a direct effect on the myocardium. To examine this question, cultures of rat myocardial cells have been exposed to different concentrations of diazepam for 1, 4 or 24 h. After 1-h exposure, the cultures exhibit dose-dependent tachycardia. The longer exposures produce either arrhythmias or cessation of beating. Moderate amounts of lactic dehydrogenase leak from the cells into the media after a 4-h exposure to diazepam indicating cell injury, while after 24-h exposure, cell viability is greatly reduced. While the levels of diazepam used in this study are considerably higher than those encountered in the blood with prescribed usage, it suggests that direct myocardial damage may occur with severe overdosing (ACOSTA and CHAPPELL 1977).

11. Receptor Modulation

a) Adrenergic

The properties and responses of adrenergic membrane receptors to pharmacological agents and simulated pathological stimuli have been examined using purified populations of chick embryo cardiac muscle cells in culture (BOBIK et al. 1981). Activation of β-receptors by 50 µM isoprenaline causes a rapid, but transient, increase in intracellular cyclic AMP. The transient nature of the increase in cyclic AMP is primarily due to the inability of β-receptors to maximally activate the enzyme adenylate cyclase for prolonged periods. Loss of β-adrenergic binding sites on continued exposure to isoprenaline is relatively slow and dependent upon the concentration of isoprenaline to which the cells are exposed. From experiments in which the rate of receptor recovery has been measured following prolonged exposure to isoprenaline, the turnover rate for cardiac β-receptors has been estimated to be approximately 12 h.

The effects of partial β-receptor agonists on the β-receptor-linked adenylate cyclase system have also been examined in culture. In heart failure cardiac noradrenaline stores in sympathetic terminals are greatly reduced and hence sympathetic stimulation of the heart impaired. In theory, partial agonists should be able to at least partially compensate for the impaired sympathetic activity by directly stimulating cardiac β-receptors. This is dependent upon whether during chronic therapy these drugs induce significant impairment of receptor function which results in the development of drug tolerance. The partial β-

receptor agonist prenalterol, in contrast to isoprenaline, does not induce any significantly-demonstrable rise in intracellular cyclic AMP in cultured cardiac cells (BOBIK et al. 1981). Its partial agonist activity on cardiac adenylate cyclase, about 12% of that due to isoprenaline, can be demonstrated *in vitro* in cell homogenate preparations where phosphodiesterase activity has been impaired by isobutylmethylxanthine. Despite these very small effects on adenylate cyclase activity, overnight exposure of cardiac cells to prenalterol concentrations approximating those achieved in plasma during therapy, reduces the ability of adrenoceptors to activate adenylate cyclase. Maximum reduction in receptor responsiveness is about 60%. Prolongation of the exposure time to 48 h or increasing the concentration of prenalterol have no further influence on this response. Thus chronic exposure to partial β-receptor agonists also impairs receptor function. It is not possible to predict the degree of impairment of receptor function (desensitization) from the partial agonist activity of the drug. This is probably dependent upon the turnover of the various catalytic units which constitutes the β-receptor-adenylate cyclase system.

Many mechanisms have been proposed to account for the desensitization effects of β-adrenoceptor agonist drugs. These include loss of β-receptors (MUK-HERJEE et al. 1975; CHUANG et al. 1980), impairment of adenylate cyclase (KEBA-BIAN et al. 1975), or activitation of phosphodiesterase activities (BROWNING et al. 1976), "uncoupling" of the β-receptor-linked adenylate cyclase (FISHMAN et al. 1981), the synthesis of inhibitory proteins (DE VELLIS and BROOKER 1974) and activation of phospholipase A_2 activity (MALLORGA et al. 1980). Studies on the mechanism of desensitization of cardiac β-receptors in culture suggest that the prime event responsible for the desensitization is the initial uncoupling of the β-receptor from adenylate cyclase (BOBIK et al. 1981). Analysis of phospholipase A_2 activity in intact cells during exposure to isoprenaline suggests that this mechanism cannot account for the loss in receptor responsiveness. Furthermore, mepacrine and chlorpromazine, two drugs possessing phospholipase A_2 inhibitory activity, do not influence the apparent rate of desensitization by isoprenaline. A two-fold elevation of intracellular cyclic AMP in cardiac cells by phosphodiesterase inhibition with 0.5 mM isobutylmethylxanthine, reduces both β-receptor responsiveness as well as β-receptor concentrations (BOBIK and LITTLE 1984). This suggests that intracellular cyclic AMP concentration in the vicinity of the sarcolemma is the prime determinant of β-receptor responsiveness. The desensitization phenomenon of β-receptors may be dependent on phosphorylation of one or more components of the β-receptor-linked adenylate-cyclase system. This phosphorylation would appear to be cyclic AMP dependent and use ATP as the substrate. Recently it has been shown that the acetylcholine receptor (DAVIS et al. 1982) and insulin receptor undergo phosphorylation (KA-SUGA et al. 1982).

The effects of endogenous phospholipase A_2 activation on adrenergic receptor function has also been examined in cultured cells. This enzyme is activated during myocardial ischemia and is at least partially responsible for impairing sarcolemmal enzyme function in ischemic hearts (KATZ and MESSINEO 1981). Activation of phospholipase A_2 by melittin impairs the ability of cardiac cells to increase intracellular cyclic AMP content during exposure to isoprenaline

(BOBIK et al. 1983). Studies in which membrane phospholipids have been isotopically labelled in cultured cells, suggest that the mechanism by which the function of the β-adrenoceptor-linked adenylate-cyclase system is impaired is primarily due to partial loss of membrane integrity. This appears to be due to loss of hydrolysed phospholipids rather than their accumulation within the cell membrane.

b) Cholinergic

Exposure of cultured heart cells to muscarinic agonists causes a biphasic decrease in the number of muscarinic receptors as measured by binding of the potent muscarinic antagonist ^3H quinuclidinyl benzilate ^3H QNB to homogenates of these cultures (GALPER and SMITH 1980). A rapid loss of 26% of receptors occurs during the first minute of exposure to agonist, followed by a gradual loss of another 44% of receptors by 3 h. Studies of the rapid phase of agonist-induced receptor loss indicate that 1) this phase is accompanied by a shift in the affinity of the remaining receptors for agonist; 2) exposure of cell homogenates to guanosine $5'$-(β,γ-imino) triphosphate causes a similar shift in affinity unaccompanied by alteration in receptor number; 3) incubation with guanine nucleotides causes the reappearance of binding sites lost during the first minute of agonist exposure; and 4) incubation with colchicine has no effect on the rapid phase of receptor loss. Studies of the slow phase of agonist-induced receptor loss indicates that: 1) inhibition of microtubule function by colchicine and vinblastine inhibits up to two-thirds of agonist-induced receptor loss; 2) recovery of receptors following washout of agonist is preceded by a 3-h lag period after which receptor number returns to 95% of control levels over 9 h; the protein-synthesis inhibitor cycloheximide (2 μg/ml) prevents recovery. These data indicate a complex set of interactions of muscarinic agonists, guanine nucleotides, any cytokinetic events in the modulation of muscarinic receptor activity.

Exposure of cultured heart cells to the muscarinic agonist carbamylcholine causes a 15% decrease in beating rate and a 33% decrease in the rate of K^+ efflux from the cells (GALPER et al. 1982). A pre-exposure of the cells to muscarinic agonists causes a decrease in the ability of carbamylcholine to stimulate K^+ efflux and to decrease beating rate. A close correlation is found between loss of the subclass of muscarinic receptors subject to agonist control and the loss of physiologic responsiveness after agonist exposure.

IV. Cellular Interactions

1. Nerve – Muscle

a) Tropic Interactions

In ovo, sympathetic cardioacceleratory function can be detected by 5 days of embryonic chick development, coinciding with the first morphological detection of nerve fibre entry into the heart (FISCHMAN and CULVER 1981). In the rat *in vivo*, nerve fibre appearance and function are not detected until close to birth (OWMAN et al. 1971).

In culture, sympathetic nerve fibre growth from chick, mouse and rat ganglion explants is stimulated towards explants of cardiac muscle (LEVI-MONTALCINI et al. 1954; CHAMLEY et al. 1973; CHAMLEY and DOWEL 1975; EBENDAL and JACOBSON 1977; EBENDAL 1979). The "attraction" effect is evident as soon as the nerve fibres emerge from the sympathetic ganglion explant and occurs over a distance of 2 mm. With the chick, the maximum neurite-stimulating activity occurs with explants from 16- to 18-day-old embryo hearts, reflecting the phase of maximum sympathetic ingrowth *in ovo* (EBENDAL 1979). The reason for this is not clear, however, recently FURUKAWA et al. (1984) have shown that mouse heart cells in culture synthesize and secrete a Nerve Growth Factor (NGF) which is immunologically, biologically, and physicochemically indistinguishable from mouse submaxillary gland β-NGF. In addition, a factor which stimulates a dense overgrowth of neurites from spinal, sympathetic and ciliary ganglia has been partially purified by gel filtration of extract from 18-day embryonic chick hearts (EBENDAL et al. 1979). The active factor has an apparent molecular weight of about 40000 daltons and is distinct from NGF by stimulating parasympathetic ciliary neurons and by lack of cross-reactivity with antibodies to NGF (see also COUGHLIN et al. 1981).

The initial interactions between sympathetic nerve fibres and isolated cardiac muscle cells in culture have been studied using time-lapse microcinematography (MARK et al. 1973). There is no obvious "attraction" from a distance of the nerve fibres to the single cells, with chance contacts occurring through random growth. Once a nerve growth cone encounters a cell, it halts in its growth and palpates the cell for about 50 min. The nerve fibre then appears to be able to distinguish whether or not the cell is a suitable target to innervate. That is, contact with a fibroblast or very immature cardiac muscle cell leads to a transitory relationship lasting no longer than a few hours, whereas contact with a mature cardiac muscle cell leads to an association that lasts for days or weeks. Receptor sites for noradrenaline or acetylcholine on the muscle cells do not appear to be involved in the process of nerve-muscle recognition since the presence of propranolol or hyoscine does not inhibit the formation of nerve-cardiac muscle associations in culture (CAMPBELL et al. 1978).

b) Functional Junctions

Sympathetic and spinal nerve fibres form functional junctions in culture with cardiac muscle cells (PURVES et al. 1974; FURSHPAN et al. 1976; MARVIN et al. 1984a). Electrophysiological studies performed on microcultures (300–500 μm in diameter) of solitary sympathetic neurons plated on cardiac muscle from the newborn rat have shown that some neurons inhibit, some excite, and others first inhibit then excite the muscle. Application of drugs has provided evidence for secretion of acetylcholine by the first group, catecholamines by the second, and both acetylcholine and catecholamines by the third. Ultrastructurally, the cholinergic endings contain small agranular vesicles, the adrenergic endings small granular vesicles, and the dual function endings mainly agranular vesicles with an occasional granular vesicle (LANDIS 1976; JOHNSON et al. 1976, 1980a, b). Enzyme-dispersed hearts of embryonic mice 9 days *in utero* grown in culture contain a number of endogenous neurons. These neurons are electrically excitable, have ultrastructural characteristics of cholinergic embryonic neurons and functionally innervate the heart cells in culture via a muscarinic cholinergic mechanism (LANE et al. 1976).

With reference to the ability of sympathetic neurons to synthesize both adrenaline and acetylcholine, it is interesting that in cell culture sympathetic neurons dissociated from the superior cervical ganglia of newborn rats and grown in the absence of non-neuronal cells are predominantly adrenergic, and synthesize, store, release and take up noradrenaline while acetylcholine synthesis is negligible (MAINS and PATTERSON 1973). However, if the same neurons are grown in the presence of non-neuronal cells from ganglia or the heart or in medium conditioned by them, the cultures synthesize significant amounts of acetylcholine as well as catecholamines. If sufficiently high concentrations of conditioned medium from heart cultures are used, then the neurons become predominantly cholinergic (PATTERSON and CHUN 1974, 1977; JOHNSON et al. 1976; LANDIS 1980). Differential survival and selection of subclasses of the neuronal population have been ruled out indicating that the heart-conditioned medium acts by altering the differentiated fate of individual neurons and that many of the neurons isolated from ganglia of newborn rats are plastic with respect to transmitter functions. However, the development of cholinergic characteristics in adrenergic neurons is age dependent, with a decrease in the ability of sympathetic ganglion explants taken from rats older than 2–5 days to develop choline acetyltransferase activity in culture (HILL and HENDRY 1977; ROSS et al. 1977; JOHNSON et al. 1980a). This suggests that most sympathetic neurons may become committed to a certain neurotransmitter at a critical stage and lose their ability to respond to alternative instructions after that time.

2. Muscle – Non Muscle

Most cultures of cardiac muscle contain some non-muscle cells such as fibroblasts from the connective tissue and smooth muscle and endothelial cells from blood vessels. While the initial proportion of these contaminating cells can be reduced by differential attachment technique (BLONDEL et al. 1971), their high proliferative rate compared with the generally non-proliferative myocardial cells often results in non-muscle overgrowth of the culture. The non-muscle cells contribute properties which may be confused with those of the muscle cells. For example, an increase in non-myocardial creatine kinase with non-muscle cell proliferation with time in culture, may be misinterpreted as dedifferentiation of the myocardial cells (VAN DER LAARSE et al. 1979). Also, myocardial cells and non-muscle cells possess a similar number of β-adrenergic receptors per cell and similar affinities for the radioligand ^{125}I-iodohydroxybenzylpindolol. The receptors can, however, be distinguished on the basis of subclassification according to the relative binding potency of different β-agonists (LAU et al. 1980).

There is also a direct effect of non-muscle cells on myocardial cell differentiation. In myocardial cultures containing proliferating non-muscle cells there is a lower maximum response to L-isoproterenol compared with bromodeoxyuridine (BrdU)-treated cultures (which have no proliferating non-muscle cells), even though baseline beating rates are the same. The proportion of living myocardial cells with cross-striations is greater in BrdU cultures than with proliferating non-muscle cells on day 8 (SIMPSON and SAVION 1982). It has also been reported that in cultures with more non-muscle cells, the myocardial cells have slower beating rates, less negative membrane potentials, and slower action potential upstroke velocities (HYDE et al. 1969; LOMPRÉ et al. 1979; JOURDAN 1980; JOURDAN and SPERELAKIS 1980), effects that could be mediated by electrical coupling between the two types of cells. However, ROBINSON (1982) disputes these latter findings, and shows that non-muscle overgrowth to the extent of 50–66% does not cause a reduction in either the muscle cell maximum diastolic potential or in the rapid and tetrodotoxin-sensitive action potential upstroke.

GROSS (1982) maintains that fibroblasts play an important role in bringing about muscle-muscle interconnections in culture through guiding the rather immobile cardiac muscle cells to nearby like cells. Metabolic functional synergy between cardiac muscle and non-muscle is suggested by the response of mixed cultures to insulin and serum deprivation compared with cultures enriched in either muscle cells or fibroblasts (SCHROEDL and HARTZELL 1983). The mixed cultures also exhibit greater increases in cellular protein with time in culture and greater resistance to the stress of nutrient deprivation.

V. Other Cardiac Muscle

1. Cardiac Muscle from Blood Vessels

When the azygos vein from rats is dispersed into single cells and placed in culture, two types of muscle cells are observed (CULLINAN et al. 1986). One type is from 50 to 90 µm long, spindle-shaped and contracts spontaneously at a rate of 30 to 90 times per minute with a contraction cycle of 160–240 msec (Fig. 100A). With FITC-labelled antibody to cardiac myosin, these cells stain in a striated pattern typical for cardiac muscle (Fig. 100B). The second type of muscle, which is in the minority, is typical of vascular smooth muscle being ribbon or irregular ribbon in shape with few spontaneous contractions (Fig. 100A). When contractions are present their rate is 1–7 per minute with a contraction cycle of several seconds (see CHAMLEY-CAMPBELL et al. 1979). They do not stain with FITC-labelled antibody to cardiac muscle but stain throughout the cytoplasm with labelled antibody to smooth muscle myosin (GRÖSCHEL-STEWART et al. 1975).

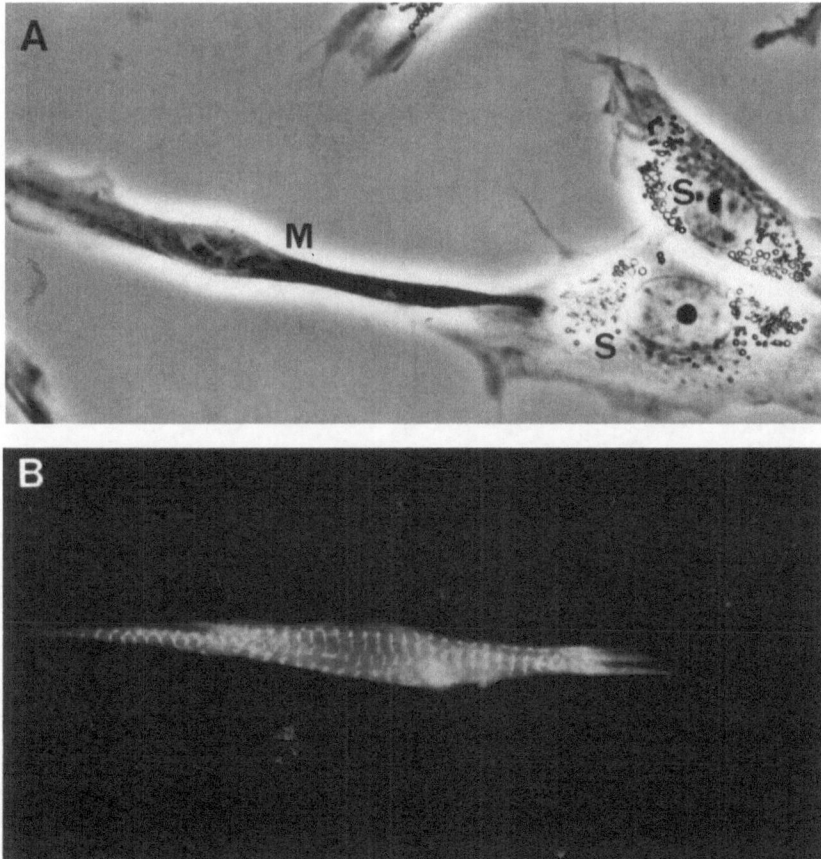

Fig. 100. A Cells from the azygos vein, 4 days in culture. Phase-contrast microscopy. Smooth muscle cells (*S*). Cardiac muscle (*M*). × 1000. **B** Cardiac muscle cell from the azygos vein, 4 days in culture. Stained with FITC-labelled antibody to cardiac myosin. × 1000. Antibody the generous gift of Professor UTE GRÖSCHEL-STEWART, Darmstadt, Federal Republic of Germany

Extension of myocardium over the pulmonary and caval veins occurs in many species (see KARRER 1959; NATHAN and GLOOBE 1970). In Section B.IV. of this volume it is shown that cardiac muscle also constitutes the wall of the rat azygos vein. Because this vessel runs alongside the aorta embedded in the same fascia, the azygos vein may be accidentally included in cultures of rat aorta, giving rise to reports of high shortening velocity vascular cells in cultures of rat aorta (HERMSMEYER 1979).

2. Conducting System

To date there is only one report of the successful culture of Purkinje cells of the conducting system of the heart (CANALE et al. 1983a). The main difficulty

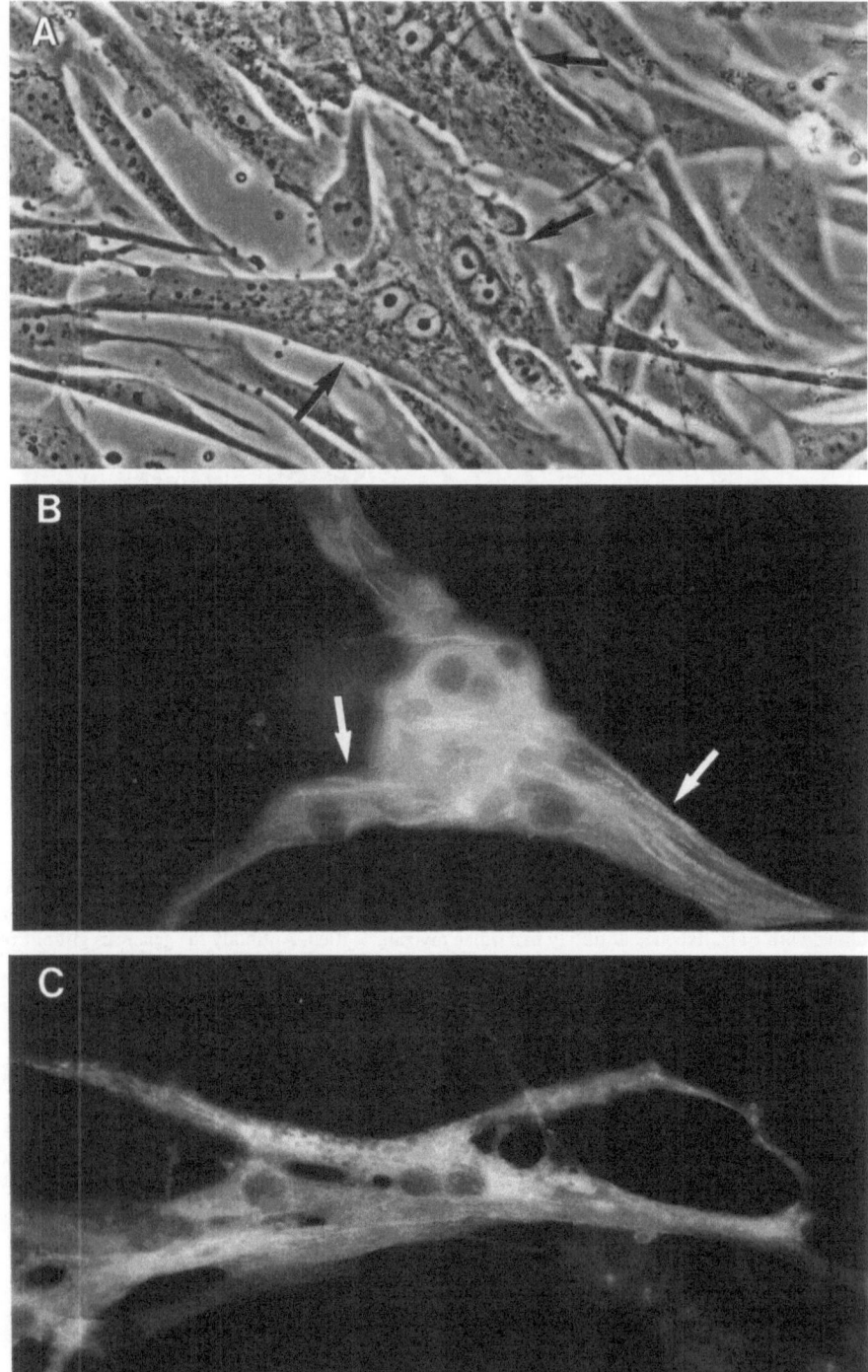

encountered is removal of the dense connective tissue sheath surrounding the fibres, which is overcome by use of the enzymes collagenase and elastase. Contamination by working myocardium is avoided by using only false tendons in which Purkinje cells are usually the only myocardial cell type. Most non-muscle cells are removed by the preplating technique of BLONDEL et al. (1971).

Purkinje cells from false tendons of young rabbits, pigs and fetal lambs remain viable in culture for at least two weeks. After 3–4 days they are larger than non-muscle cells, often binucleate, and contract spontaneously at 15 to 30 beats per minute (Fig. 101 A). With FITC-labelled antibodies to cardiac myosin or tropomyosin most cells exhibit a cross-banding pattern due to the sarcomeric arrangement of myofibrils, while others have no banding and a diffuse staining (Fig. 101 B). After 7 days in culture, contraction in most cells has ceased. Cells exhibit more diffuse staining and non-striated fibrils than sarcomeric cross-banding (Fig. 101 C), and disorganization of the myofilament system is evident ultrastructurally with Z disc material as well as actin and myosin filaments dispersed in the cytoplasm. Phenotypic modulation of isolated smooth muscle from a contractile to sythetic state occurs after 5–8 days in culture (see CHAMLEY-CAMPBELL et al. 1979), while working myocardial cells generally retain their contractile phenotype throughout long periods in culture (see CLAYCOMB 1983). Purkinje cells are thus different from working myocardial cells in this respect and show a similar response to isolation in culture as smooth muscle.

Fig. 101. A Purkinje cells (*arrows*) from false tendons of a young rabbit, 4 days in culture. Phase-contrast microscopy. Cells in the background are fibroblasts of the connective sheath. × 500. **B** Purkinje cells, 4 days in culture. Stained with FITC-labelled antibodies to cardiac myosin. Note sarcomeric cross banding (*arrows*). × 500. **C** Purkinje cells, 7 days in culture. Stained with FITC-labelled antibodies to cardiac myosin. Note lack of sarcomeric staining, with stain diffuse and in non-striated fibrils. × 500. Antibodies the generous gift of Professor UTE GRÖSCHEL-STEWART, Darmstadt, Federal Republic of Germany

References

Abe H, Ito Y, Tada M, Opie LH (1984) Regulation of cardiac function: molecular, cellular and pathophysiological aspects. Japan Scientific Societies Press, Tokyo/VNU Science Press BV, Utrecht, The Netherlands

Abraham A (1983) Ultrastructure investigations into the impulse – generating center in the heart of the european pond turtle (*Emys orbicularis*). Z Mikrosk Anat Forsch (Leipz) 97:254–268

Abrahamson T, Holmgren S, Nilsson S, Pettersson K (1979) On the chromaffin system of the African lungfish, *Protopterus aethiopicus*. Acta Physiol Scand 107:135–139

Ackermann U, Irizawa TG (1984) Synthesis and atrial activity of rat atrial granules depend on extracellular volume. Am J Physiol 247:R750–R752

Acosta D, Chappell R (1977) Cardiotoxicity of diazepam in cultured heart cells. Toxicology 8:311–317

Acosta D, Li CP (1980) Actions of extracellular acidosis on primary cultures of rat myocardial cells deprived of oxygen and glucose. J Mol Cell Cardiol 12:1459–1463

Acosta D, Puckett M, McMillan R (1978) Ischemic myocardial injury in cultured heart cells: Leakage of cytoplasmic enzymes from injured cells. In Vitro 14:728–732

Acosta D, Puckett M, Li CP (1980) Reduction of cell injury in hypoxic cultures of rat myocardial cells by methylprednisolone. In Vitro 16:93 – 96

Adelstein RS (1983) Regulation of contractile proteins by phosphorylation. J Clin Invest 72:1863–1866

Aherne WA, Dunnill MS (1982) Morphometry. Edward Arnold, London

Ahumada GG, Sobel BE, Needleman P (1980) Synthesis of prostaglandins by cultured rat heart myocytes and cardiac mesenchymal cells. J Mol Cell Cardiol 12:685–700

Akester AR (1981) Intercalated discs, nexuses, sarcoplasmic reticulum and transitional cells in the heart of the adult domestic fowl (*Gallus gallus domesticus*). J Anat 133:161–179

Akester AR, Akester BV (1971) Some ultrastructural characteristics of conducting cells in the avian heart. J Anat 108:616–617

Alanis J, Benitez D, Lopez E, Martinez-Palomo A (1973) Impulse propagation through the cardiac junctional regions of the axolotl and the turtle. Jap J Physiol 23:149–164

Alivisatos SGA, Papastavrou C, Drouka-Liapati E, Molyodas AP, Nikitopoulou G (1977) Enzymatic and electrophysiological changes of the function of membrane proteins by cholesterol. Biochem Biophys Res Commun 79:677–683

Allen JR, Carstens CA (1967) Ultrastructural modifications in the fetal rhesus monkey heart. J Cell Biol 35:159A

Alpert NR, Gordon MS (1962) Myofibrillar adenosine triphosphatase activity in congestive heart failure. Am J Physiol 202:902–946

Altona JC, Van der Laarse A (1982) Anoxia-induced changes in composition and permeability of sarcolemmal membranes in rat heart cell cultures. Cardiovasc Res 16:138–143

Altona JC, Van der Laarse A, Bloys Van Treslong CHF (1984) Release of compartment-specific enzymes from neonatal rat heart cell cultures during anoxia and reoxygenation. Cardiovasc Res 18:99–106

Anastasia JV, McCarl RL (1973) Effects of cortisol on cultured rat heart cells, lipase activity, fatty acid oxidation, glycogen metabolism, and ATP levels as related to the beating phenomenon. J Cell Biol 57:109–116

Anderson BG, Anderson WD (1980) Microvasculature of the canine heart demonstrated by scanning electron microscopy. Am J Anat 158:217–227

Anderson DF, Bissonnette J, Faber JJ, Thornborg KL (1981) Central short flows and pressures in the mature fetal lamb. Am J Physiol 241:460–466

Anderson KR, Ho SY, Anderson RH (1979) The location and vascular supply of the sinus node in the human heart. Br Heart J 41:28–32

Anderson RH (1972a) Histologic and histochemical evidence concerning the presence of morphologically distinct cellular zones within the rabbit atrioventricular node. Anat Rec 173:7–23

Anderson RH (1972b) The disposition, morphology and innervation of cardiac specialised tissue in the guinea-pig. J Anat 111(3):453–468

Anderson RH (1972c) The disposition and innervation of atrio-ventricular ring specialised tissue in rats and rabbits. J Anat 113(2):197–211

Anderson RH, Taylor IM (1972) Development of atrioventricular specialised tissue in human heart. Br Heart J 34:1205–1214

Anderson RH, Janse MJ, Van Capelle FJL, Billette J, Becker AE, Durrer D (1974) A combined morphological and electrophysiological study of the atrioventricular node of the rabbit heart. Circ Res 35:909–922

Anderson RH, Becker AE, Brechenmacher C, Davies MJ, Rossi L (1975a) The human atrioventricular junctional area. A morphological study of the AV node and bundle. Eur J Cardiol 3:11–25

Anderson RH, Becker AE, Brechenmacher C, Davies MJ, Rossi L (1975b) Ventricular preexcitation. A proposed nomenclature for its substrates. Eur J Cardiol 3:27–36

Anderson RH, Becker AE, Wenink ACG, Janse MJ (1976) The development of the cardiac specialised tissue. In: Wellens HJ, Lie KI, Janse MJ (eds) The conduction system of the heart. Lea and Febiger, Philadelphia, pp 3–28

Anderson RH, Ho SY, Becker AE, Gosling JA (1978) The development of the sinoatrial node. In: Bonke FIM (ed) The sinus node. Martinus Nijhoff, The Hague, pp 166–182

Anderson RH, Ho SY, Smith A, Becker AE (1981) The internodal atrial myocardium. Anat Rec 201(1):75–82

Angelakos ET, Bernadini P, Barrett WC Jr (1964) Myocardial fibre size and capillary fiber ratio in the right and left ventricles of the rat. Anat Rec 149:671–676

Angelakos ET, Glassman PM, Millard RW, King M (1965) Regional distribution and subcellular localisation of catecholamines in the frog heart. Comp Biochem Physiol 15:313–324

Angelakos ET, King MP (1967) Demonstration of nerve terminals containing adrenaline by a new histochemical technique. Nature 213:391–392

Angelakos ET, Fuxe K, Torchiana ML (1963) Chemical and histochemical evaluation of the distribution of catecholamines in the rabbit and guinea pig hearts. Acta Physiol Scand 59:184–192

Anthonioz P, Mohsen T, Jadoun G (1978) Preuves histochimiques et ultrastructurales de l'innervation du cœur de *Protopterus annectens* (poisson dipneuste). CR Soc Biol Paris 172:208–211

Anversa P, Vitali-Mazza L, Visioli O, Marchetti G (1971) Experimental cardiac hypertrophy: A quantitative ultrastructural study in the compensatory state. J Mol Cell Cardiol 31:213–227

Anversa P, Hagopian M, Loud AV (1973) Quantitative radioautographic localization of protein synthesis in experimental cardiac hypertrophy. Lab Invest 29:282–292

Anversa P, Vitali-Mazza L, Loud AV (1975a) Morphometric and autoradiographic study of developing ventricular and atrial myocardium in fetal rats. Lab Invest 33:696–705

Anversa P, Vitali-Mazza L, Gandolfi A, Loud AV (1975b) Morphometry and autoradiography of early hypertrophic changes in the ventricular myocardium of adult rat. A light microscopic study. Lab Invest 33:125–129

Anversa P, Loud AV, Vitali-Mazza L (1976) Morphometry and autoradiography of early hypertrophic changes in the ventricular myocardium of adult rat. An electron microscopic study. Lab Invest 35:475–483

Anversa P, Loud AV, Giacomelli F, Wiener J (1978) Absolute morphometric study of myocardial hypertrophy in experimental hypertension. II. Ultrastructure of myocytes and interstitium. Lab Invest 38:597–609

Anversa P, Olivetti G, Melissari M, Loud AV (1979) Morphometric study of myocardial hypertrophy induced by abdominal aortic stenosis. Lab Invest 40:341–349

Anversa P, Olivetti G, Melissari M, Loud AV (1980) Stereological measurement of cellular and subcellular hypertrophy and hyperplasia in the papillary muscle of adult rat. J Mol Cell Cardiol 12:781–795

Anversa P, Olivetti G, Bracchi P, Loud AV (1981) Postnatal development of the M band in rat cardiac myofibrils. Circ. Res 48:561–568

Anversa P, Melissari M, Beghi C, Olivetti G (1984) Structural compensatory mechanisms in rat heart in early spontaneous hypertension. Am J Physiol 246:H739–H746

Arbel ER, Liberthson R, Langendorf R, Pick A, Lev M, Fishman AP (1977) Electrophysiological and anatomical observations of the heart of the african lungfish. Am J Physiol 232:H24–H34

Arluk DJ, Rhodin JAG (1974) The ultrastructure of calf heart conducting fibres with special reference to nexuses and their distribution. J Ultrastruct Res 49:11–23

Armiger LC, Urthaler F, James TN (1979) Morphological changes in the right ventricular septomarginal trabeculae (false tendon) during maturation and ageing in the dog heart. J Anat 129(4):805–817

Aschoff L (1910) Referat über die Herzstörungen in ihren Beziehungen zu dem spezifischen Muskelsystem des Herzens. Verh Dtsch Path Ges 14:3–35

Ashley LM (1945) A determination of the diameter of ventricular fibres in man and other mammals. Am J Anat 77:325–363

Ashraf M, Halverson C (1977) Structural changes in the freeze-fractured sarcolemma of the ischemic myocardium. Am J Pathol 88:583–587

Ashraf M, Halverson C (1978) Ultrastructural modifications of nexuses (gap junctions) during early myocardial ischemia. J Mol Cell Cardiol 10:263–270

Ask JA (1983) Comparative aspects of adrenergic receptors in the hearts of lower vertebrates. Comp Biochem Physiol 76A:543–552

Assali NS, Brinkman CR III, Nuwaymid B (1974) Comparison of maternal and foetal cardiovascular functions in acute and chronic experiments in the sheep. Am J Obstet Gynec 120:411–425

Astorri E, Chizzola A, Visioli O, Anversa P, Olivetti G, Vitali-Mazza L (1971) Right ventricular hypertrophy – a cytometric study on 55 human hearts. J Mol Cell Cardiol 2:99–110

Athias P, Frelin C, Groz B, Dumas JP, Klepping J, Padieu P (1979) Myocardial electrophysiology: Intracellular studies on heart cell cultures from newborn rats. Pathol Biol 27:13–19

Atlas SA, Kleinert HD, Camargo MJ, Januszewicz A, Sealey JE, Laragh JH, Schilling JW, Lewicki JA, Johnson LK, Maack T (1984) Purification, sequencing and synthesis of natriuretic and vasoactive rat atrial peptide. Nature 309:717–719

Augustinsson KB, Fange R, Johnels A, Ostlund E (1956) Histological, physiological and biochemical studies on the heart of two cyclostomes, hagfish (*Myxine*) and lamprey (*Lampetra*). J Physiol 131:257–276

Ayettey AS, Navaratnam V (1978) The T-tubule system in the specialised and general myocardium of the rat. J Anat 127(1):125–140

Ayettey AS, Navaratnam V (1980) The fine structure of myocardial cells in the grey seal. J Anat 131:748

Ayettey AS, Navaratnam V (1981) The ultrastructure of myocardial cells in the golden hamster *Cricetus auratus*. J Anat 132:519–524

Azuma T, Binia A, Visscher MB (1965) Adrenergic mechanisms in the bullfrog and turtle. Am J Physiol 209(6):1287–1294

Bader D, Oberpriller J (1979) Autoradiographic and electron microscopic studies of minced cardiac muscle regeneration in the adult newt, *Notophthalmus viridescens*. J Exp Zool 208:177–194

Bailey PH, Higgins TJC (1983) Metabolism of palmitate by anoxic and reoxygenated heart cell cultures. J Cell Science 60:204–216

Bakeeva LE, Chentsov Yu S, Skulachev VP (1983) Intermitochondrial contacts in myocardiocytes. J Mol Cell Cardiol 15:413–420

Baldwin KM (1979) Cardiac gap junction configuration after an uncoupling treatment as a function of time. J Cell Biol 82:66–75

Balogh I, Rubinyi G, Schulze W, Kovach AG, Sotonyi P, Somogyi E (1981) Electron cytochemical studies on changes of the adenyl cyclase activity following experimental hypoxia and ischaemia in perfused rat hearts. Exp Pathol 20:121–127

Balogh I, Rubinyi G, Kovagh AG, Sotonyi P, Somogyi E (1981) Electron cytochemical detection of nickel ions under pathological conditions in the rat heart. Acta Morphol Acad Sci Hungary 29:87–90

Baluk P, Fujiwara T (1984) Direct visualization by scanning electron microscopy of the preganglionic innervation and synapses on the true surfaces of neurons in the frog heart. Neuroscience Lett 51:265–270

Banerjee SK (1983) Comparative studies of atrial and ventricular myosin from normal, thyrotoxic, and thyroidectomized rabbits. Circ Res 52:131–136

Banister J, Mann SP (1965) An investigation of the adrenergic innervation of the heart and major blood vessels of the frog by Falck's method of fluorescence microscopy. J Physiol 181:13p–15p

Barany M (1967) ATPase activity of myosin correlated with speed of muscle shortening. J Gen Physiol 50(Suppl):197–218

Barany M, Barany K (1981) Protein phosphorylation in cardiac and vascular smooth muscle. Am J Physiol 241:H117–H128

Barofsky AL, Feinstein M, Halkerston IDK (1973) Enzymatic and mechanical requirements for the dissociation of cortical cells from rat adrenal glands. Exp Cell Res 79:263–274

Barry WH, Smith TW (1982) Mechanisms of transmembrane calcium movement in cultured chick embryo ventricular cells. J Physiol (Lond) 325:243–260

Barry WH, Smith TW (1984) Movement of Ca across the sarcolemma: Effects of abrupt exposure to zero external Na concentration. J Mol Cell Cardiol 16:155–164

Barry WH, Pober J, Marsh JD, Frankl SR, Smith TW (1980) Effects of graded hypoxia on contraction of cultured chick embryo ventricular cells. Am J Physiol 239:H651–657

Battig CG, Low FN (1961) The ultrastructure of human cardiac muscle and its associated tissue space. Am J Anat 108:199–252

Beasley D, Malvin RL (1985) Atrial extracts increase glomerular filtration rate in vivo. Am J Physiol 248:F24–F30

Becker AE (1978) General comments. In: Bonke FIM (ed) The sinus node. Martinus Nijhoff, The Hague, pp 212–222

Becker AE, Anderson RH (1976) Morphology of the human atrioventricular junctional area. In: Wellens HJJ, Lie KI, Janse MJ (eds) Conduction system of the heart – Structure function and clinical implications. Lea and Febiger, Philadelphia, pp 263–286

Behrendt H (1977) Effect of anabolic steroids on rat heart muscle cells I. Intermediate filaments. Cell Tissue Res 180:303–315

Bencosme SA, Berger JM (1971) Specific granules in mammalian and non-mammalian vertebrate cardiocytes. Methods and Achievements in Exp Pathol 5:173–213

Bencosme SA, Berger JM (1972) Specific granules in human and nonhuman vertebrate cardiocytes. In: Bajusy E, Rona G (eds) Recent advances in studies on cardiac structure and metabolism, vol 1. University Park Press, Baltimore, pp 327–339

Bencosme SA, Trillo A, Alanis J, Benitez D (1969) Correlative ultrastructural and electrophysiological study of the Purkinje system of the heart. J Electrocardiol 2:27–38

Bennett GS, Fellini SA, Croop JM, Otto JJ, Bryant J, Holtzer H (1978) Differences among 100 Å-filament subunits from different cell types. Proc Natl Acad Sci USA, 75:4364–4368

Bennett HS (1936) The development of the blood supply to the heart in the embryo pig. Am J Anat 60:27–53

Bennett MR, Merrillees NCR (1966) An analysis of the transmission of excitation from autonomic nerves to smooth muscle. J Physiol 185:520–535

Bennett MVL, Goodenough DA (1978) Gap junctions, electrotonic coupling, and intercellular communication. Neuroscience Res Prog Bull 16:373–386

Benninghoff A (1923) Über die Beziehungen des Reitzleitungssystems und der Papillarmuskeln zu den Konturfasern des Herzschlauches. Ver Anat Ges 57:125–208

Benninghoff A (1930) Blutgefäße und Herz. In: v. Möllendorff W (ed) Handbuch der mikroskopischen Anatomie des Menschen, vol VI/1. Springer, Berlin, pp 1–232

Berger HJ, Zaret BC, Speroff L, Cohen LS, Wolfson S (1976) Regional cardiac prostaglandin release during myocardial ischemia in anesthetized dogs. Circ Res 38:566–571

Beringer T, Hadek R (1972) Cardiac innervation in the lamprey (Petromyzon marinus). Anat Rec 172:269–270

Beringer T, Hadek R (1973) Ultrastructure of sinus venosus innervation in Petromyzon marinus. J Ultrastruct Res 42:312–323

Bernard P, Courand F (1979) Electrophysiological studies on embryonic heart cells in culture, scorpion toxin as a tool to reveal latent fast sodium channel. Biochim Biophys Acta 553:154–168

Berne RM, Rubio R (1979) Coronary circulation. In: Berne R, Sperelakis N, Geiger S (eds) Handbook of physiology, Sect 2, The cardiovascular system, vol 1, The heart, American Physiological Society, Bethesda, p 873–952

Bhatnagar KP, Spoonamore BA (1979) Ultrastructure of the atrioventricular node of the big brown bat, Eptesicus fuscus. Acta Anat 105:157–180

Bing RJ (1971) Seminar on coronary microcirculation. 1. Introduction. Am J Cardiol 29:591–592

Bishop SP (1971) Ultrastructural alterations in canine myocardial hypertrophy and congestive heart failure. In: Alpert NR (ed) Cardiac hypertrophy. Academic Press, New York, pp 107–124

Bishop SP (1973) Effect of aortic stenosis on myocardial cell growth, hyperplasia and ultrastructure in neonatal dogs. Recent Advances in Studies in Cardiology Structure and Metabolism 3:637–656

Bishop SP, Cole CR (1969) Ultrastructural changes in the canine myocardium with right ventricular hypertrophy and congestive heart failure. Lab Invest 20:219–229

Bishop SP, Drummond JL (1979) Surface morphology and cell size measurement of isolated rat cardiac myocytes. J Mol Cell Cardiol 11:423–433

Bishop SP, Hine P (1975) Cardiac muscle cytoplasmic and nuclear development during canine neonatal growth. Recent Advances in Studies in Cardiology Structure and Metabolism 8:77–98

Bishop SP, Melsen LR (1976) Myocardial necrosis, fibrosis and DNA synthesis in experimental cardiac hypertrophy induced by sudden pressure overload. Circ Res 39:238–245

Bishop SP, Drummond J, Reynolds RH (1979a) Regional cardiac myocyte growth in normotensive and spontaneously hypertensive rats. J Mol Cell Cardiol 11(Suppl 1):8

Bishop SP, Oparil S, Reynolds RH, Drummond JL (1979b) Regional myocyte size in normotensive and spontaneously hypertensive rats. Hypertension 1:378–383

Bishop SP, Dillon D, Naftilan J, Reynolds R (1980) Surface morphology of isolated cardiac myocytes from hypertrophied hearts of aging spontaneously hypertensive rats. Scanning Electron microsc 2:193–199

Black-Schaffer B, Grinstead CE, Braunstein JN (1965) Endocardial fibroelastosis of large mammals. Circ Res 16:383–390

Bleeker WK, Mackaay AJC, Masson-Pévet M, Bouman LN, Becker AE (1980) Functional and morphological organization of the rabbit sinus node. Circ Res 46:11–22

Bleeker WM, Mackaay AJC, Masson-Pévet M, Op't Hof KT, Jongsma HJ, Bouman LN (1982) Asymmetry of the sinoatrial conduction in the rabbit heart. J Mol Cell Cardiol 14:633–644

Bloom GE, Ostlund E, Von Euler US, Ushajko F, Ritzen M, Adams-Ray J (1961) Studies on catecholamine-containing granules of specific cells in cyclostome hearts. Acta Physiol Scand 53(Suppl 185):1–34

Bloom S (1970) Spontaneous rhythmic contraction of separated heart muscle cells. Science 167:1727–1729

Blondel B, Roijen I, Cheneval JP (1971) Heart cells in culture: A simple method for increasing the proportion of myoblasts. Experientia 27:356–358

Bobik A, Little PJ (1984) Role of cyclic AMP in cardiac β-adrenoceptor desensitization: Studies using prenalterol and inhibitors of phosphodiesterase. J Cardiovasc Pharmacol 6:795–801

Bobik A, Campbell JH, Carson V, Campbell GR (1981) Mechanism of isoprenaline-induced refractoriness of the β-adrenoceptor-adenylate cyclase system in chick embryo cardiac cells. J Cardiovasc Pharmacol 3:541–553

Bobik A, Campbell J, Snow P, Little PJ (1983) The effects of endogenous phospholipase A2 activation on beta adrenoceptor function in cardiac cells. J Mol Cell Cardiol 15:759–768

Boder GB, Harley RJ, Johnson IS (1971) Recording system for monitoring automaticity of heart cells in culture. Nature 231:531–532

Bogenmann E, Eppenberger HM (1980) DNA synthesis and polyploidization of chicken heart muscle cells in mass cultures. J Mol Cell Cardiol 12:17–27

Bogin E, Massry SG, Harary I (1981) Effect of parathyroid hormone on rat heart cells. J Clin Invest 67:1215–1227

Bogusch G (1974) Investigations on the fine structure of Purkinje fibres in the atrium of the avian heart. Cell Tissue Res 150:43–56

Bogusch G (1975) Electron microscopic investigations on leptomeric fibrils and leptomeric complexes in the hen and pigeon heart. J Mol Cell Cardiol 7:733–745

Bogusch G (1976) Enzymatic digestion and urea extraction of leptomeric structures and normomeric myofibrils in heart muscle cells. J Ultrastruct Res 55:245–256

Bogusch G (1979) Electron microscopic investigations on the differentiation of Purkinje cells in the ontogenic development of the chicken heart. Anat Embryol 155:259–271

Boineau JP, Schuessler RB, Hackel DB, Miller CB, Brockus CN, Wylds AC (1980) Widespread distribution and rate differentiation of the atrial pacemaker complex. Am J Physiol 239:H406–H415

Bojsen-Møller F, Tranum-Jensen J (1971) On nerves and nerve endings in the conducting system of the moderator band. J Anat (Lond) 108:387–395

Bojsen-Møller F, Tranum-Jensen J (1972) Rabbit heart nodal tissue, sinuatrial ring bundle and atrioventricular connections identified as a neuromuscular system. J Anat 112:367–382

Borelli V (1975) Topographic and histological study of the sinoatrial node in the Jaffarabadi buffalo (*Bubalus bubalis* Linnaeus, 1758). Acta Anat 92:122–128

Borg TK, Caulfield JB (1981) The collagen matrix of the heart. Fed Proc 40:2037–2041

Borg TK, Ronson WF, Mosleby FA, Caulfield JB (1981) Structural basis of ventricular stiffness. Lab Invest 44:49–54

Borgers M, Thone F, Verheyen A, Terkeyrs HEDJ (1984) Localisation of calcium in skeletal and cardiac muscle. J Histochem 16:295–309

Born G (1889) Beiträge zur Entwicklungsgeschichte des Säugetierherzens. Archiv Mikrosk Anat 33:284–377

Bosman FT (1983) Some recent developments in immuno-cytochemistry. J Histochem 15:189–200

Bossen EH, Sommer JR (1984) Comparative stereology of the lizard and frog myocardium. Tissue Cell 16:173–178

Bossen EH, Sommer JR, Waugh RA (1978) Comparative stereology of the mouse and finch left ventricle. Tissue Cell 10:773–784

Bouman LN, Gerlings ED, Biersteker PA (1968) Pacemaker shift within the sinoatrial node during vagal stimulation. Pflüg Archiv 302:255–267

Bouman LN, Mackaay AJC, Bleeker WO, Becker AE (1978) Pacemaker shifts in the sinus node. Effects of vagal stimulation, temperature and reduction of extracellular calcium. In: Bonke FIM (ed) The sinus node. Martinus Nijhoff, The Hague, pp 245–257

Bourne GH (ed) (1980) Hearts and heart-like organs, vol 1. Comparative anatomy and development. Academic Press, New York

Bouvagnet P, Leger J, Pons F, Dechesne C, Leger JJ (1984) Fiber types and myosin types in human atrial and ventricular myocardium. An anatomical description. Circ Res 55:794–804

Brady AJ, Dubkin CH (1964) Coronary circulation in the turtle ventricle. Comp Biochem Physiol 13:119–128

Breisch EA, Bove AA, Phillips SJ (1980) Myocardial morphometrics in pressure overload left ventricular hypertrophy and regression. Cardiovasc Res 14:161–168

Breisch EA, White F, Jones HM, Laurs RM (1983) Ultrastructural morphometry of the myocardium of *Thunnus alalunga*. Cell Tissue Res 233:427–438

Bridge JHB, Bassingthwaighte JB (1983) Uphill sodium transport driven by an inward calcium gradient in heart muscle. Science 219:178–180

Brook WH, Connell S, Cannata J, Maloney JE, Walker AM (1984) Ultrastructure of the myocardium during development from early fetal life to adult life in sheep. J Anat 137:729–741

Brooks C McC, Lu HH (1972) In: The sinu atrial pacemaker of the heart. Thomas, Springfield, Ill, pp 109–110

Brown HF (1982) Electrophysiology of the sino-atrial node. Physiol Rev 62:505–526

Brown OM (1976) Cat heart acetylcholine: Structural proof and distribution. Am J Physiol 231:H781–H785

Browning ET, Brostrom CO, Groppi VE (1976) Altered adenosine cyclic 3′,5′-monophosphate synthesis and degradation by C-6 astrocytoma cells following prolonged exposure to norepinephrine. Mol Pharmacol 12:32–40

Bruns RR, Palade GE (1968) Studies on blood capillaries. I. General organization of blood capillaries in muscle. J Cell Biol 37:244–276

Buja LM, Ferrans VJ, Maron BJ (1974) Intracytoplasmic junctions in cardiac muscle cells. Am J Pathol 74:613–647

Burger JW, Bradley SE (1951) The general form of circulation in the dogfish (*Squalus acanthias*). J Cell Comp Physiol 37:387–402

Burnett JC, Granger JP, Opgenorth TJ (1984) Effects of synthetic atrial natriuretic factor on renal function and renin release. Am J Physiol 247:F863–866

Burnstock G (1969) Evolution of the autonomic innervation of the visceral and cardiovascular systems in vertebrates. Pharmacol Rev 21:247–324

Burrows MT (1910) The cultivation of tissues of the chick-embryo outside the body. JAMA 55:2057–2058

Burrows MT (1912) Rhythmische Kontraktionen der isolierten Herzmuskelzellen ausserhalb des Organismus. Münch Med Wschr 59:1473; Science n.s. 26:90

Burt J (1982) Electrical and contractile consequences of Na^+ or Ca^+ gradient reduction in cultured heart cells. J Mol Cell Cardiol 14:99–110

Burt JM, Frank JS, Berns MW (1982) Permeability and structural studies of heart cell gap junctions under normal and altered ionic conditions. J Membr Biol 68:227–238

Cabrini RL (1981) Practical applications of the microphotometric quantification of histoenzyme reactions. J Histochem 13:241–250

Caesar R, Edwards GA, Ruska H (1958) Electron microscopy of the impulse conducting system of the sheep heart. Z Zellforsch Mikrosk Anat 48:698–719

Cameron JS (1979) Autonomic nervous tone and regulation of heart rate in the goldfish, *Carassius auratus*. Comp Biochem Physiol 63C:341–349

Campbell GR, Uehara Y, Malmfors T, Burnstock G (1971) Degeneration and regeneration of smooth muscle transplants in the anterior eye chamber: an ultrastructural study. Z Zellforsch Mikrosk Anat 177:155–175

Campbell GR, Chamley JH, Burnstock G (1978) Lack of effect of receptor blockers on the formation of long-lasting associations between sympathetic nerves and cardiac muscle cells *in vitro*. Cell Tissue Res 187:551–553

Campbell GR, Chamley-Campbell JH, Gröschel-Stewart U, Small JV, Anderson P (1979) Antibody staining of 10 nm (100 Å) filaments in cultured smooth, cardiac and skeletal muscle cells. J Cell Science 37:303–322

Campbell KP, MacLennan DH, Jorgensen AO, Mintzer MC (1983) Purification and characterization of calsequestrin from canine cardiac sarcoplasmic reticulum and identification of the 53,000 dalton glycoprotein. J Biol Chem 258:1197–1204

Campion DR (1985) The muscle satellite cell: A review. Int Rev Cytol 87:225–251

Canale ED, Campbell JH, Campbell GR (1983a) Cardiac Purkinje cells in culture. J Mol Cell Cardiol 15:197–206

Canale ED, Campbell GR, Uehara Y, Fujiwara T, Smolich JJ (1983b) Sheep cardiac Purkinje fibers: Configurational changes during the cardiac cycle. Cell Tissue Res 232:97–110

Canale ED, Fujiwara T, Campbell GR (1983c) The demonstration of close nerve-Purkinje fiber contacts in false tendons of sheep heart. Cell Tissue Res 230:105–111

Cantin MS, Benchimol S, Castonguay Y, Berlinguet JC, Huet M (1975) Ultrastructural cytochemistry of atrial muscle cells V. Characterization of specific granules in the human left atrium. J Ultrastruct Res 52:179–192

Cantin M, Timm-Kennedy M, El-Khatib E, Huet M, Yunge L (1979) Ultrastructural cytochemistry of atrial muscle cells VI. Comparative study of specific granules in right and left atrium of various animal species. Anat Rec 193:55–70

Cantin M, Gutkowska J, Tibault G, Milne RW, Ledoux S, Minli S, Chapeau C, Garcia R, Hamet P, Genest J (1984) Immunocytochemical localization of atrial natriuretic factor in the heart and salivary glands. J Histochem 80:113–127

Cardell RR Jr (1971) Action of metabolic hormones on the fine structure of rat liver cells. I. Effects of fasting on the ultrastructure of hepatocytes. Am J Anat 131:21–54

Cardwell JC, Abramson DI (1931) The atrioventricular conduction system of the beef heart. Am J Anat 49(2):167–192

Carew TE, Covell JW (1979) Fibre orientation in hypertrophied canine left ventricle. Am J Physiol 236:H487–H493

Carey RA, Bove AA, Coulson RL, Packham DL, Spann JF (1979) Correlation between cardiac muscle myosin ATPase activity and velocity of muscle shortening. Biochem Med 21:235–245

Carlsson E, Kjörell V, Thornell L-E, Lambertsson A, Strehler E (1982) Differentiation of the myofibrils and the intermediate filament system during postnatal development of the rat heart. Eur J Cell Biol 27:62–73

Cartwright JJr, Goldstein MA (1985) Microtubules in the heart muscle of the postnatal and adult rat. J Mol Cell Cardiol 17:1–7

Caspar DLD, Goodenough D, Makowski L, Phillips WC (1977) Gap junction structures. I. Correlated electron microscopy and X-ray diffraction. J Cell Biol 74:605–628

Casta A, Wolff GS, Mehta AV, Tamer D, Garcia OL, Pickoff AS, Ferrer PL, Sung RJ, Gelband H (1980) Dual atrioventricular nodal pathways: A benign finding in arrhythmia-free children with heart disease. Am J Cardiol 46:1013–1018

Casta A, Wolff GS, Tamer D, Flinn CJ, Mehta AV, Smith KG, Gelband H (1983) Multiple atrioventricular nodal pathways – A new electrophysiological phenomenon in children. J Electrocardiol 16(4):331–338

Caulfield JB, Borg TK (1979) The collagen network of the heart. Lab Invest 40:364–372

Cavanaugh MW (1955) Pulsation, migration and division in dissociated chick embryo heart cells *in vitro*. J Exp Zool 128:573

Cavanaugh MW, Cavanaugh DJ (1957) Studies on the pharmacology of tissue cultures. I. The action of quinidine on cultures of dissociated chick embryo heart cells. Arch Int Pharmacodyn 110:43

Cedergren B, Harary I (1964a) *In vitro* studies on single beating rat heart cells: VI Electron microscopic studies of single cells. J Ultrastruct Res 1:428–442

Cedergren B, Harary I (1964b) *In vitro* studies on single beating rat heart cells. VII. Ultrastructure of the beating cell layer. J Ultrastruct Res 11:443–454

Chacko K (1972) Ultrastructural observations in mitosis in myocardial cells of the rat embryo. Am J Anat 135:305–310

Chacko KJ (1976) Observations on the ultrastructure of developing myocardium of rat embryos. J Morphol 150:681–710

Chacko S (1979) Cardiac muscle differentiation and growth in developing chick embryos. In: Mauro A et al. (eds) Muscle regeneration. Raven Press, New York, p 363

Chajek T, Stein O, Stein Y (1977) Rat heart in culture as a tool to elucidate the cellular origin of lipoprotein lipase. Biochim Biophys Acta 488:140–144

Chajek T, Stein O, Stein Y (1978a) Lipoprotein lipase of cultured mesenchymal rat heart cells. Biochim Biophys Acta 528:456–465

Chajek T, Stein O, Stein Y (1978b) Lipoprotein lipase of cultured mesenchymal rat heart cells. 2 Hydrolysis of labelled very low density lipoprotein triacylglycerol by membrane-supported enzyme. Biochim Biophys Acta 528:466–474

Chalcroft JP, Bullivant S (1970) An interpretation of liver membrane and junction structure based on observation of freeze-fracture replicas of both sides of the fracture. J Cell Biol 47:49–60

Challice CE (1966) Studies on the microstructure of the heart. I. The sino-atrial node and the sinoatrial ring bundle. J Roy Microsc Soc 85:1–21

Challice CE, Edwards EA (1961) The micromorphology of the developing ventricular muscle. In: Carvalho AP, Mello WC, Hoffman B (eds) The specialized tissues of the heart. Elsevier, Amsterdam, pp 44–73

Challice CE, Virágh S (eds) (1973) Ultrastructure of the mammalian heart. Academic Press, New York

Challice CE, Virágh S (1973a) The architectural development of the early mammalian heart. Tissue Cell 6:447–462

Challice CE, Virágh S (1973b) The embryologic development of the mammalian heart. In: Challice CE and Virágh S (eds) Ultrastructure of biological systems, vol 6. Ultrastructure of the mammalian heart. Academic Press, New York and London, pp 91–126

Challice CE, Wilkens JL, Chohan KS, White F (1975) Musculature of the pulmonary veins. In: Roy PE, Rona G (eds) Recent advances in studies on cardiac structure and metabolism, vol 10. The metabolism of contraction. University Park Press, Baltimore, pp 571–582

Chambers DJ, Braimbridge MV, Frost GTB, Nahir AM, Chayen J (1982) A quantitative cytochemical method for the measurement of β-hydroxyacyl CoA dehydrogenase activity in rat heart muscle. Histochemistry 75:67–76

Chamley JH, Dowel JJ (1975) Specificity of nerve fibre "attraction" to autonomic effector organs in tissue culture. Exp Cell Res 90:1–7

Chamley JH, Goller I, Burnstock G (1973) Selective growth of sympathetic nerve fibers to explants of normally densely innervated autonomic effector organs in tissue culture. Dev Biol 31:362–379

Chamley JH, Gröschel-Stewart U, Campbell GR, Burnstock G (1977) Distinction between smooth muscle, fibroblasts and endothelial cells by the use of fluoresceinated antibodies against smooth muscle actin. Cell Tissue Res 177:445–457

Chamley-Campbell J, Campbell GR, Gröschel-Stewart U, Burnstock G (1977) FITC-labelled antibody staining of tropomyosin-containing fibrils in smooth, cardiac and skeletal muscle cells and 3T3 cells in culture. Cell Tissue Res 183:153–166

Chamley-Campbell J, Campbell GR, Ross R (1979) The smooth muscle cell in culture. Physiol Rev 59:1–61

Chenard J, Samson M, Beaulieu M (1965) Embryonal sinusoids in the myocardium: Report of a case successfully treated surgically. Can Med Assoc J 92:1356–1359

Cheng YP (1971) The ultrastructure of the rat sino-atrial node. Acta Anat Nippon 46:339–358

Chiba T, Yamauchi A (1970) On the fine structure of the nerve terminals in the human myocardium. Z Zellforsch Mikrosk Anat 108:324–338

Chiba T, Yamauchi A (1973) Fluorescence and electron microscopy of the monoamine-containing cells in the turtle heart. Z Zellforsch Mikroskop Anat 140(1):25–37

Chiesi M, Ho MM, Inesi G, Somlyo AV, Somlyo AP (1981) Primary role of sarcoplasmic reticulum in phasic contractile activation of cardiac myocytes with shunted myolemma. J Cell Biol 92:728–742

Chizzonite RA, Zak R (1984) Regulation of myosin isoenzyme composition in fetal and neonatal rat ventricle by endogenous thyroid hormones. J Biol Chem 259(2):12628–12632

Chizzonite R, Everett A, Clark W, Jakovic S, Rabinowitz M, Zak R (1982) Isolation and characterization of two molecular variants of myosin heavy chain from rabbit ventricle change in their content during normal growth and after treatment with thyroid hormone. J Biol Chem 257:2056–2065

Chowrashi PK, Pepe FA (1982) The Z-band: amorphin and alpha-actinin and their relation to structure. J Cell Biol 94:565–573

Chuang DM, Kinnier WJ, Farber L, Costa E (1980) A biochemical study of receptor internalization during β-adrenergic receptor desensitization in frog erythrocytes. Mol Pharmacol 18:348–355

Clancy MJ, Herlihy PD (1978) Assessment of changes in myofibre size in muscle. Cited in: Aherne WA, Dunnill MS (1982) Morphometry. Edward Arnold, London, pp 113–115

Clark WA, Chizzonite RA, Everett AW, Rabinowitz M, Zak R (1982) Species correlations between cardiac isomyosins and a comparison of electrophoretic and immunological properties. J Biol Chem 257:5449–5454

Claycomb WC (1981) Culture of cardiac muscle cells in serum-free medium. Exp Cell Res 131:231–236

Claycomb WC (1983) Cardiac muscle cell proliferation and cell differentiation in vivo and in vitro. Adv Exp Med Biol 161:249–265

Claycomb WC, Lanson NJr (1984) Isolation and culture of the terminally differentiated adult mammalian ventricular cardiac muscle cell. In Vitro 20:647–651

Clubb FJJr, Bishop SP (1984) Formation of binucleated myocardial cells in the neonatal rat. An index for growth hypertrophy. Lab Invest 50:571–577

Clusin WT (1981) The mechanical activity of chick embryonic myocardial cell aggregates. J Physiol (Lond) 320:149–174

Cobb JLS, Santer RM (1973) Electrophysiology of cardiac function in teleosts: Cholinergically mediated inhibition and rebound excitation. J Physiol (Lond) 230:501–533

Coffman JD, Lewis FB, Gregg DE (1960) Effect of prolonged periods of anoxia on atrioventricular conduction and cardiac muscle. Circ Res 8:649–659

Cohen AS (1963) The fine structure of the visual receptors of the pigeon. Exp Eye Res 2:88–97

Colborn GL, Carsey E Jr (1972) Electron microscopy of the sino-atrial node of the squirrel monkey (Saimiri sciureus). J Mol Cell Cardiol 4:525–536

Coleridge HM, Coleridge JCG, Kidd C (1964) Cardiac receptors in the dog, with particular reference to two types of afferent ending in the ventricular wall. J Physiol 174:323–339

Coleridge JCG, Hemingway A, Holmes RL, Linden RJ (1957) The location of atrial receptors in the dog: A physiological and histological study. J Physiol 136:174–197

Colgan JH, Lazarus ML, Sachs HG (1978) Postnatal development of the normal and cardiomyopathic Syrian hamster heart: A quantitative electron microscopic study. J Mol Cell Cardiol 10:43–54

Conte G, Pelegrini A (1984) On the development of the coronary arteries in human embryos, stages 14–19. Anat Embryol 169:209–218

Cooper CJ, Delalande IS, Tyler MJ (1966) The catecholamines in lizard heart. Aust J Exp Biol Med Sci 44:205–210

Cooper MH, O'Rahilly R (1971) The human heart at seven post ovulatory weeks. Acta Anat (Basel) 79:280–299

Cooper T, Napolitano LM, Fitzgerald MJT, Moore KE, Daggett WM, Willman VL, Sonnenblick EH, Hanlon CR (1966) Structural basis of cardiac valvular function. Arch Surg 93:767–771

Copenhaver W (1981) A re-study of cardiac conduction pathways by techniques for visualisation of cholinesterase reaction. Anat Rec 201:51–54

Cote G, Mohiuddin SM, Roy PE (1970) Occurrence of Z-band widening in human atrial cells. Exp Mol Pathol 13:307–318

Coughlin MD, Bloom EM, Black IB (1981) Characterization of a neuronal growth factor from mouse heart-cell-conditioned medium. Dev Biol 82:56–68

Coupland RE (1972) The chromaffin system. In: Blaschko H, Muscholl E (eds) Handbook of experimental pharmacology, vol 33, Catecholamines. Springer, Berlin Heidelberg New York, pp 16–45

Cowin P, Mattey D, Garrod D (1984) Distribution of desmosomal components in the tissues of vertebrates, studied by fluorescent antibody staining. J Cell Science 66:119–132

Crass MF III (1977) Regulation of triglyceride metabolism in the isotopically prelabelled perfused heart. Fed Proc 36:1995–1999

Cullinan V, Campbell JH, Mosse PRL, Campbell GR (1986) The morphology and cell culture of the striated musculature of the rat azygos vein. Cell Tiss Res 243:185–191

Cummins P, Price KM, Littler WA (1980) Fetal myosin light chain in human ventricle. J Muscle Res Cell Motility 1(3):357–366

Currie MG, Geller DM, Cole BR, Siegel NR, Fok KF, Adams SP, Eubanks SR, Galluppi GR, Needleman P (1984) Purification and sequence analysis of bioactive atrial peptides (atriopeptins). Science 223:67–69

Dahl E, Ehinger B, Falck BV, Mecklenbourg C, Myhrberg H, Rosengren E (1971) On the mono-amine-storing cells in the heart of *Lampetra fluviatilis* and *L. planeri*. Gen Comp Endocr 17:241–246

Dahlström A, Fuxe K, Mya-tu M, Zetterström BEM (1965) Observations on adrenergic innervation of dog heart. Am J Physiol 209:689–692

Dalla Libera L, Sartore S, Schiaffino S (1980) Foetal myosin light chain in human ventricle. J Muscle Res Cell Motility 1:357–366

Danillo PJr, Reder RF, Binah O, Legato MJ (1984) Fetal canine purkinje fibers: electrophysiology and ultrastructure. Am J Physiol 246:H250–H260

Danto SI, Fischman DA (1984) Immunocytochemical analysis of intermediate filaments in embryonic heart cells with monoclonal antibodies to desmin. J Cell Biol 98:2179–2191

Dashow L, Epple A, Nibbio B (1982) Catecholomines in adult lampreys: Baseline levels and stress induced changes, with a note on cardiac cannulation. Gen Comp Endocr 46:500–504

David H, Meyer R, Marx I, Guski H, Wenzelides K (1979) Morphometric characterization of left ventricular myocardial cells of male rats during postnatal development. J Mol Cell Cardiol 11:631–638

Davies F (1930a) The conduction system of the bird's heart. J Anat 64:129–146

Davies F (1930b) Further studies of the conducting system of the bird's heart. J Anat 64:319–323

Davies F, Francis ETB (1940) The heart of the salamander (*Salamandra salamandra* L.) with special reference to the conducting (connecting) system and its bearing on the phylogeny of the conducting systems of mammalian and avian hearts. Phil. Trans Roy Soc Lond, B 231:99–130

Davies F, Francis ETB (1946) The conducting system of the vertebrate heart. Biological Review. Cambridge Philadelphia Society 21:173–188

Davies F, Francis ETB (1952) The conduction of the impulse for cardiac contraction. J Anat 86:302–309

Davies F, Francis ETB, King TS (1952) The conducting (connecting) system of the crocodilian heart. J Anat 86(2):152–161

Davies MJ, Anderson RH, Becker AE (eds) (1983) The conduction system of the heart. Butterworths, Sevenoaks, England

Davis CG, Gordon AS, Diamond I (1982) Specificity and localization of the acetylcholine receptor kinase. Proc Natl Acad Science USA 79:3666–3670

Davis CL (1924) The cardiac jelly of the chick embryo. Anat Rec 27:201–202

Davis CL (1927) Development of the human heart from its first appearance to the stage found in embryos of 20 paired somites. Contrib Embryol Carnegie Inst 19:245–284

Dbaly J, Ostadal B, Rychter Z (1968) Development of the coronary arteries in rat embryos. Acta Anat (Basel) 71:209–222

De Bold AJ (1979) Heart atria granularity effects changes in water-electrolyte balance. Proc Soc Exp Biol Med 161:508–511

Decker RS, Decker ML, Pool AR (1980) The distribution of lysosomal Cathepsin D in cardiac myocytes. J Histochem Cytochem 28:231–237

De Felice LJ, Challice CE (1969) Anatomical and ultrastructural study of the electrophysiological
   AV node of the rabbit. Circ Res 24:457–474
De Haan RL (1959) Cardia bifida and the development of pacemaker function in the early chick
   heart. Dev Biol 1:586–602
De Haan RL (1961) Differentiation of the atrioventricular conducting system of the heart. Circ
   24:458–470
De Haan RL (1963a) Organisation of the cardiogenic plate in the chick embryo. Acta Embryol
   Morphol Exper 6:26–38
De Haan RL (1963b) Migration patterns of the pre-cardiac mesoderm in the early chick embryo.
   Exp Cell Res 29:544–560
De Haan RL (1965) Development of pacemaker tissue in the embryonic heart. Ann NY Acad
   Sci 127:7–18
De Haan RL, Williams EH, Ypey DL, Calpham DE (1981) Intercellular coupling of embryonic
   heart cells. In: T Pexieder (ed) Perspectives in cardiovascular research, vol 5. Mechanisms of
   cardiac morphogenesis and teratogenesis. Raven Press, New York, pp 299–316
De La Lande IS, Tyler MJ, Pridmore BR (1962) Pharmacology of the heart of Tiliqua (Trachysaurus)
   rugosa (the sleepy lizard). Aust J Exp Biol Med Sci 40:129–138
Délèze J (1970) The recovery of resting potential and input resistance in sheep heart injured by
   knife or laser. J Physiol (Lond) 208:547–564
De Mello WC (1972) The healing process in cardiac and other muscle fibers. In: De Mello WC
   (ed) Electrical phenomena in the heart. Academic Press, New York, pp 323–351
De Mello WC (1975) Effect of intracellular injection of calcium and strontium on cell communication
   in heart. J Physiol (Lond) 250:231–245
De Mello WC (1979a) Effect of 2–4-dinitrophenol on intercellular communication in mammalian
   cardiac fibres, Pflüg Arch 380:267–276
De Mello WC (ed) (1979b) Intercellular communication. Plenum Press, New York, pp 87–125
De Mello WC (1982a) Cell-to-cell communication in heart and other tissues. Prog Biophys Mol
   Biol 39:147–182
De Mello WC (1982b) Intercellular communication in cardiac muscle. Circ Res 51:1–9
De Mello WC, Hoffman BF (1960) Potassium ions and electrical activity of specialised cardiac
   fibers. Am J Physiology 199:1125–1130
De Mello WC, Motta E, Chapeau M (1969) A study of the healing-over of myocardial cells of
   toads. Circ Res 24:475–487
De Mello WC, Castillo MG, Van Loon P (1983) Intercellular diffusion of Lucifer Yellow CH
   in mammalian cardiac fibers. J Mol Cell Cardiol 15:637–643
Dennis MJ, Sargent PB (1978) Multiple innervation of normal and re-innervated parasympathetic
   neurones in the frog cardiac ganglion. J Physiol (Lond) 281:63–76
Derow HA (1954) The heart in renal disease. Circulation 10:114–128
De Vellis J, Brooker G (1974) Reversal of catecholamine refractoriness by inhibitors of RNA and
   protein synthesis. Science 186(4170):1221–1223
De Vries PA, Saunders (1962) Development of the ventricles and spiral outflow tract in the human
   heart. A contribution to the development of the human heart from age group 9 to age group
   15. Contrib Embryol Carnegie Inst, Washington 37:87–114
Dewey MM (1969) The structure and function of the intercalated disc in vertebrate cardiac muscle.
   Experientia Suppl 15:10–28
Dickson DW, Lund DD, Subieta AR, Prall JL, Schmid PG, Roskoski JR R (1981) Regional distribu-
   tion of tyrosine hydroxylase and dopamine-β-hydroxylase activities in guinea pig heart. J Auto-
   nomic Nerv Sys 4:319–326
Dlugosz AA, Antin PB, Nachmias VT, Holtzer H (1984) The relationship between stress fibre-like
   structures and nascent myofibrils in cultured cardiac myocytes. J Cell Biol 99:2268–2278
Dolber PC, Sommer JR (1980) Freeze-fracture appearance of rabbit cardiac sarcoplasmic reticulum.
   In: Bailey G (ed) Thirty-eighth Annual EMSA Meeting; Claitor's Publ Dir, Baton Rouge, pp 630–
   631
Dolber PC, Sommer JR (1984) Corbular sarcoplasmic reticulum of rabbit cardiac muscle. J Ultras-
   truct Res 87:190–196
Doležel S, Gerová M, Gero J, Sládek T, Vašků J (1978) Adrenergic innervation of the coronary
   arteries and the myocardium. Acta Anat 100:306–316

Doležel S, Gerová M, Gero J, Sládek T, Vašků J (1980) Monoaminergic pathways to the coronary arteries and to the myocardium. Acta Anat 108:490–497

Domenech-Mateu JM, Boya-vegue J (1975) An ultrastructural study of sinuatrial node cells in the embryonic rat heart. J Anat 119:77–84

Domenech-Mateu JM, Orts-Llorca F (1976) Arterial vascularisation of the sinuatrial node in the embryonic rat heart. Acta Anat 94:343–355

Donald J, Campbell G (1982) A comparative study of the adrenergic innervation of the teleost heart. J Comp Physiol B 147:85–92

Doorey AJ, Barry WH (1983) The effects of inhibition of oxidative phosphorylation and glycolysis on contractility and high-energy phosphate content in cultured chick heart cells. Circ Res 53:192–201

Dopirak MR, Schaal SF, Leier CV (1980) Triple AV nodal pathways in man. J Electrocardiol 13:185–188

Dowd DA (1969) The coronary vessels and conducting system in the heart of monotremes. Acta Anat 74:547–573

Dowell RT, McManus RE (1978) Pressure-induced cardiac enlargement in neonatal and adult rats. Left ventricular functional characteristics and evidence of cardiac muscle cell proliferation in the neonate. Circ Res 42:303–310

Downing SE (1979) Baroreceptor regulation of the heart. In: Berne RM, Sperelakis N, Geiger SR (eds) Handbook of physiology, sect 2, The cardiovascular system, vol 1, The heart, chap 17. Williams & Wilkins, Baltimore, pp 621–652

Draper MH, Mya-tu M (1959) A comparison of the conduction velocity in cardiac tissue of various mammals. Quart J Exp Physiol 44:91–109

Duffy PE, Markesbery WR (1970) Granulated vesicles in sympathetic nerve endings in the pineal gland: Observations on the effects of pharmacologic agents by electron microscopy. Am J Anat 128:97–116

Dufour JJ, Hunziker U, Postenak J (1956) Effets inotropes et chronotropes de l'acetylcholine et de l'adrenaline sur le cœur de la tortue. J Physiol (Paris) 48:521–524

Dulhunty AF, Franzini-Armstrong C (1975) The relative contributions of the folds and caveolae to the surface membrane of frog skeletal muscle fibres at different sarcomere lengths. J Physiol (Lond) 250:513–539

Dusek H, Ostadal B, Duskova M (1975) Postnatal persistence of spongy myocardium with embryonic blood supply. Arch Pathol 99:312–317

Ebendal T (1979) Stage-dependent stimulation of neurite outgrowth exerted by nerve growth factor and chick heart in cultured embryonic ganglia. Dev Biol 72:276–290

Ebendal T, Jacobson C-O (1977) Tissue explants affecting extension and orientation of axons in cultured chick embryo ganglia. Exp Cell Res 105:379–388

Ebendal T, Belew M, Jacobson C-O, Porath J (1979) Neurite outgrowth elicited by embryonic chick heart: partial purification of the active factor. Neuroscience Lett 14:91–95

Ebrecht GH, Rupp R, Jacob R (1982) Alterations of mechanical parameters in chemically skinned preparations of rat myocardium as a function of isoenzyme pattern of myosin. Basic Res Cardiol 77:220–234

Edge MB, Walker SM (1970) Evidence for a structural relationship between sarcoplasmic reticulum and Z lines in dog papillary muscle. Anat Rec 166:51–65

Ehara T (1974) Effects of adrenaline, ouabain and some ionic conditions on twitch contraction of bullfrog ventricle. Jap J Physiol 24:317–328

Eisenberg BR, Gilai A (1979) Structural changes in single muscle fibres after stimulation at a low frequency. J Gen Physiol 74:1–16

Eisenberg BR, Mobley BA (1975) Size changes in single muscle fibres during fixation and embedding. Tissue Cell 7:383–387

Elias EA, De Vries GP, Elias RA, Tigges AJ, Meijer AEFH (1980) Enzyme histochemical studies on the conducting system of the human heart. Histochem J 12:577–589

Elias EA, Elias RA, De Vries GP, Meijer AEFH (1982) Early and late changes in the metabolic pattern of the working myocardial fibres and Purkinje fibres of the human heart under ischaemic and inflammatory conditions: An enzyme histochemical study. Histochem J 14:445–459

Eliska O (1983) Ligation of the artery of the dog sino-atrial node. Folia Morphol (Prague) 31(2):181–184

Ellison JP, Hibbs RG (1973) The atrioventricular valves of the guinea-pig. I. A light microscopic study. Am J Anat 138:331–345

Emery DG, Foreman RD, Coggeshall RE (1976) Fiber analysis of the feline inferior cardiac sympathetic nerve. J Compar Neurol 116:457–468

Eriksson A, Thornell L-E (1979) Intermediate (skeletin) filaments in heart Purkinje fibers a correlative morphological and biochemical identification with evidence of a cytoskeletal function. J Cell Biol 80:231–247

Eriksson A, Thornell L-E, Stigbrand T (1978) Cytoskeletal filaments of heart conducting system localised by antibody against a 55000 dalton protein. Experientia 34:792–794

Eriksson A, Thornell L-E, Stigbrand T (1979) Skeletin immunoreactivity in heart Purkinje fibres from several species. J Histochem Cytochem 27:1604–1609

Eriksson A, Kjörell U, Thornell L-E, Stigbrand T (1980) Skeletin immunoreativity in peripheral nerves. Experientia 36:594–596

Erlij D, Centrangold R, Valdez R (1965) Adrenotropic receptors in the frog. J Pharmacol Exp Ther 149(1):65–70

Erokhina IL, Rumayantsev PP (1983) The ultrastructure of sinoatrial node myocytes of mouse embryo heart during mitosis. Acad Sci Tsitologiya 25:371–379

Ezerman EB, Ishikawa H (1967) Differentiation of the sarcoplasmic reticulum and T system in developing chick skeletal muscle in vitro. J Cell Biol 35:405–420

Esmond WG, Moulton GA, Cowley RA, Attar S, Blair E (1963) Peripheral ramification of the cardiac conducting system. Circulation 27:732–8

Esperanca PJA, Pereira AT, Dos Santos Ferreira A (1975) Vascularisation arterielle du nœud sinoauriculaire du cœur chez le chien. Acta Cardiol 30:67–78

Euler US von, Fänge R (1961) Catecholamines in nerves and organs of *Myxine glutinosa, Squalus acanthias* and *Gadus callarias*. Gen Comp Endocr 1:191–194

Evans RR, Robson RM, Stroner MH (1984) Properties of smooth muscle vinculin. J Biol Chem 259:3916–3924

Fabiato A (1981) Myoplasmic free calcium concentration reached during the twitch of an intact isolated cardiac cell and during calcium-induced release of calcium from the sarcoplasmic reticulum of a skinned cardiac cell from the adult rat or rabbit ventricle. J Gen Physiol 78:457–497

Fabiato A (1983) Calcium-induced release of calcium from the cardiac sarcoplasmic reticulum. Am J Physiol 245:C1–C14

Falck B, Häggendal J, Owman C (1963) The localization of adrenaline in adrenergic nerves in the frog. Quart J Exp Physiol 48:253–257

Falck B, Von Mecklenburg C, Myhrberg H, Persson H (1966) Studies on adrenergic and cholinergic receptors in the isolated hearts of *Lampetra fluviatilis* (Cyclostomata) and *Pleuronectes platessa* (Teleostei). Acta Physiol Scand 68:64–71

Fanburg BL (1970) Experimental cardiac hypertrophy. N Engl J Med 282:723–732

Farber JL (1981) Minireview: The role of calcium in cell death. Life Sciences 29:1289–1295

Fawcett DW (1968) The sporadic occurrence in cardiac muscle of anomalous Z bands exhibiting a periodic structure suggestive of tropomyosin. J Cell Biol 36:266–270

Fawcett DW, McNutt NS (1969) The ultrastructure of the cat myocardium. I. Ventricular papillary muscle. J Cell Biol 42:1–45

Fawcett DW, Selby CC (1958a) Observations on the fine structure of the turtle atrium. J Biophys Biochem Cytol 4:63–72

Fawcett DW, Selby CC (1958b) Observations on the fine structure of the turtle atrium. J Biophys Biochem Cytol 10(Suppl 4):89–109

Fayet G, Couraud F, Miranda F, Lissitzky S (1974) Electro-optical system for monitoring activity of heart cells in culture: application to the study of several drugs and scorpion toxins. Eur J Pharmacol 27:165–174

Feldt RH, Rahimtoola SH, Davis GD, Swan HJC, Titus JL (1969) Anomalous ventricular myocardial patterns in a child with complex congenital heart disease. Am J Cardiol 23:732–734

Ferguson DG, Leeson TS (1983) Postnatal development of sarcolemmal invaginations in right atrial myocardium of the rat. Acta Anat (Basel) 117:289–302

Ferrans VJ (1984) Ultrastructure and function of the myocardium: An overview of recent correlative studies. In: Abe H, Ito Y, Tada M, Opie LH (eds) Regulation of cardiac function. Japan Scientific Societies Press, Tokyo/VNU Science Press, Ultrecht, pp 81–105

Ferrans VJ, Roberts WC (1973a) Intermyofibrillar and nuclear – myofibrillar connections in human and canine myocardium. An ultrastructural study. J Mol Cell Cardiol 5:247–257

Ferrans VJ, Roberts WC (1973b) Structural features of cardiac myxomas. Histology, histochemistry and electron microscopy. Human Pathol 4:111–146

Ferrans VJ, Thiedemann K-U (1983) Ultrastructure of the normal heart. In: Silver MD (ed) Cardiovascular pathology, vol 1. Churchill Livingstone, New York, pp 31–86

Feuvray D, De Leiris J (1975) Ultrastructural modifications induced by re-oxygenation in the anoxic isolated rat heart perfused without exogenous substrate. J Mol Cell Cardiol 7:307–314

Field EJ (1951) The development of the conducting system of the heart of sheep. Brit Heart J 13:129–147

Finlay M, Anderson RH (1974) The development of cholinesterase activity in the rat heart. J Anat 117:239–248

Fischman DA, Culver NG (1981) Coaggregation of embryonic chick sympathetic neurons with cardiac myocytes: Evidence for functional synaptic development *in vitro*. In: Pexieder T (ed) Perspectives in cardiovascular research, vol 5, Mechanisms of cardiac morphogenesis and teratogenesis. Raven Press, New York, p 349

Fischman DA, Moscona AA (1971) Reconstitution of heart tissue from suspensions of embryonic myocardial cells: Ultrastructural studies on dispersed and reaggregated cells. In: Alpert NK (ed) Cardiac hypertrophy. Academic Press, New York, London, p 125

Fisher DT, Hexmann MA, Rudolph AM (1982) Regional myocardial blood flow and oxygen delivery in fetal, newborn and adult sheep. Am J Physiol 243:H729–H731

Fishman PH, Mallorga P, Tallman JF (1981) Catecholamine-induced desensitization of adenylate cyclase in rat glioma C6 cells, evidence for a specific uncoupling of beta-adrenergic receptors from a functional regulatory component of adenylate cyclase. Mol Pharmacol 20:310–318

Flink IL, Rader JH, Banerjee SK, Morkin E (1978) Atrial and ventricular cardiac myosin contain different heavy chain species. FEBS Lett 94:125–130

Floyd K (1979) Light microscopy of nerve endings in the atrial endocardium. Hainsworth R, Kidd C, Linden RJ (eds) Cardiac receptors, chapter 1. Cambridge University Press, Oxford, pp 3–26

Forbes MS, Sperelakis N (1971) Ultrastructure of lizard ventricular muscle. J Ultrastruct Res 34:439–451

Forbes MS, Sperelakis N (1973) A labyrinthine structure formed from a transverse tubule of mouse ventricular myocardium. J Cell Biol 56:865–869

Forbes MS, Sperelakis N (1976) The presence of transverse and axial tubules in the ventricular myocardium of embryonic and neonatal guinea pigs. Cell Tissue Res 166:83–90

Forbes MS, Sperelakis N (1977) Myocardial couplings: Their ultrastructural variations in the mouse. J Ultrastruct Res 58:50–65

Forbes MS, Sperelakis N (1980) Structures located at the level of the Z bands in mouse ventricular myocardial cells. Tissue Cell 12:467–489

Forbes MS, Sperelakis N (1982a) Bridging junctional processes in couplings of skeletal, cardiac, and smooth muscle. Muscle Nerve 5:674–681

Forbes MS, Sperelakis N (1982b) Association between mitochondria and gap junctions in mammalian myocardial cells. Tissue Cell 14:25–37

Forbes MS, Sperelakis N (1983) The membrane systems and cytoskeletal elements of mammalian myocardial cells. In: Dowben RM, Shay JW (eds) Cell and muscle motility, vol 3. Plenum Publishing, New York London, pp 89–155

Forbes MS, Sperelakis N (1984) Ultrastructure of mammalian cardiac muscle. In: Sperelakis N (ed) Physiology and pathophysiology of the heart. Martinus Nijhoff, Boston/The Hague, pp 3–42

Forbes MS, Plantholt BA, Sperelakis N (1977) Cytochemical staining procedures selective for sarcotubular systems of muscle: applications and modifications. J Ultrastruct Res 60:306–327

Forbes MS, Hawkey LA, Sperelakis N (1984) The transverse – axial tubular system (TATS) of mouse myocardium: Its morphology in the developing and adult animal. Am J Anat 170:143–162

Forsgren S (1985) The differentiation of the Purkinje fibres in the mammalian heart – comparisons with the ordinary myocytes. In: Legato MJ (ed) The developing heart. Martinus Nijhoff, The Hague, pp 47–67

Forsgren S, Thornell L-E (1981) The development of Purkinje fibres and ordinary myocytes in the bovine fetal heart. Anat Embryol 162:127–136

Forsgren S, Thornell L-E, Eriksson A (1980) The development of the Purkinje fibre system in the bovine fetal heart. Anat Embryol 159:125–135

Forsgren S, Carlsson E, Strehler E, Thornell L-E (1982a) Ultrastructural identification of human fetal Purkinje fibres – A comparative immunocytochemical and electron microscopic study of composition and structure of myofibrillar M-regions. J Mol Cell Cardiol 14:437–449

Forsgren S, Strehler E, Thornell LE (1982b) Differentiation of Purkinje fibres and ordinary ventricular and atrial myocytes in the bovine heart – an immunochemical and enzyme histochemical study. J Histochem 14:929–942

Forsgren S, Strehler E, Thornell L-E (1983) Differentiation of the atrioventricular node, the atrioventricular bundle and the bundle branches in the bovine heart: An immunohistochemical and enzyme histochemical study. J Histochem 15:1099–1111

Forssmann WG, Girardier L (1966) Untersuchungen zur Ultrastruktur des Rattenherzmuskels mit besonderer Berücksichtigung des sarkoplasmatischen Retikulums. Z Zellforsch Mikroskop Anat 72:249–275

Forssmann WG, Girardier L (1970) A study of the T system in rat heart. J Cell Biol 44:1–19

Forssmann WG, Birr C, Carlquist M, Christmann M, Finke R, Henschen A, Hock D, Kirchheim, Kreye V, Lottspeich F, Metz J, Mutt V, Reinecke M (1984) The auricular myocardiocytes of the heart constitute an endocrine organ: Characterisation of a porcine cardiac peptide hormone, Cardiodilatin-126. Cell Tissue Res 238:425–430

Frank JS, Langer GA (1974) The myocardial interstitium: Its structure and its role in ionic exchange. J Cell Biol 60:586–601

Frank JS, Rich TL (1983) Ca depletion and repletion in rat heart: Age-dependent changes in the sarcolemma. Am J Physiol 245:H343–H353

Frank JS, Langer GA, Nudd LM, Seraydarian K (1977) The myocardial cell surface, its histochemistry, and the effect of sialic acid and calcium removal on its structure and cellular ionic exchange. Circ Res 41:702–714

Frank JS, Beydler S, Kreman M, Rau EE (1980) Structure of the freeze-fractured sarcolemma in the normal and anoxic rabbit myocardium. Circ Res 47:131–143

Frank JS, Rich TL, Beydler S, Kreman M (1982) Calcium depletion in rabbit myocardium: Ultrastructure of the sarcolemma and correlation with the Ca paradox. Circ Res 51:117–130

Frank JS, Philipson KD, Beydler S (1984) Ultrastructure of isolated sarcolemma from dog and rabbit myocardium: comparison to intact tissue. Circ Res 54:414–423

Frank MI, Albrecht W, Sleator W, Robinson RB (1975) Stereological measurements of atrial ultrastructures in the guinea pig. Experientia 31:578–579

Franke WW, Weber K, Osborn M, Schmid E, Freudenstein C (1978) Antibody to prekeratin: Decoration of tonofilament like arrays in various cells of epithelial character. Exp Cell Res 116:429–445

Franke WW, Schmid E, Osborn M, Weber K (1979) Intermediate-sized filaments of human endothelial cells. J Cell Biol 81:570–580

Franke WW, Moll R, Schiller DL, Schmid E, Katrenbeck J, Mueller H (1982) Desmoplakins of epithelial and myocardial desmosomes are immunologically and biochemically related. Differentiation 23:115–127

Franzini-Armstrong C (1980) Structure of sarcoplasmic reticulum. Fed Proc 39:2403–2409

Freeman GL, LeWinter MM, Engler RL, Covel JW (1985) Relationship between myocardial fiber direction and segment shortening in the midwall of the canine left ventricle. Circ Res 56:31–39

Frelin C (1980) Amino acid metabolism by newborn rat heart cells in monolayer cultures. J Mol Cell Cardiol 12:479–491

Frelin C, Pinson A, Moalic J-M, Padieu P (1974) Energy metabolism of beating rat heart cell cultures. 2. Glucose metabolism. Biochimie 56:1597–1602

Frelin C, Pinson A, Athias P, Surville JM, Padieu P (1979) Glucose and palmitate metabolism by beating rat heart cells in culture. Pathol Biol (Paris) 47:45–50

Friedman G, Stein O, Stein Y (1980) Lipoprotein lipase activity in F, heart cells cultures. Effect of dialyzable serum factors on enzyme stability and enzyme synthesis. Biochim Biophys Acta 619:650–659

Friedman PL, Stewart JR, Fenoglio JJ, Wit AL (1973) Survival of subendocardial Purkinje fibers after extensive myocardial infarction in dogs: in vitro and in vivo correlations. Circ Res 33:597–611

Fujii S, Hirota A, Kamino K (1980) Optical signals from early embryonic chick heart stained with potential sensitive dyes: Evidence for electrical activity. J Physiol (Lond) 304:503–518

Fujii S, Hirota A, Kamino K (1981a) Optical recording of development of electrical activity in embryonic chick heart during early phases of cardiogenesis. J Physiol (Lond) 311:147–60

Fujii S, Hirota A, Kamino K (1981b) Optical indications of pace-maker potential and rhythm generation in early embryonic chick heart. J Physiol 312:253–263

Fujii S, Hirota A, Kamino K (1981c) Action potential synchrony in embryonic pre-contractile chick heart. Optical monitoring with potentiometric dyes. J Physiol (Lond) 319:529–542

Fujimoto A, Harary I (1964) Studies in vitro on single beating rat-heart cells. IV: The shift from fat to carbohydrate metabolism in culture. Biochim Biophys Acta 86:74–80

Fujita S (1983) The micro-computer-based color image analyzer and its application to histochemistry. J Histochem Cytochem 31:238–240

Fukuda M (1983) Application of the microcomputer in multiparametric fluorescence cytophotometry. J Histochem Cytochem 31:241–243

Fukuyama U (1970) On the origin and termination of the mammalian cardiac nerves. Nippon Iga-kukai Zasshi (in Japanese) 64:1027–1044

Furshpan EJ, MacLeish PR, O'Lague PH, Potter DD (1976) Chemical transmission between rat sympathetic neurons and cardiac myocytes developing in microcultures: Evidence for cholinergic, adrenergic, and dual-function neurons. Proc Natl Acad Sci USA 73:4225–4229

Furukawa Y, Furukawa S, Satoyoshi E, Hayashi K (1984) Nerve growth factor secreted by mouse heart cells in culture. J Biol Chem 259:1259–1264

Fuseler JW, Shay JW (1982) The association of desmin with the developing myofibrils of cultured embryonic rat heart myocytes. Dev Biol 91:448–457

Fuseler JW, Shay JW, Feit H (1981) The role of intermediate (10 nm) filaments in the development and integration of the myofibrillar contractile apparatus in the embryonic mammalian heart. In: Dowben RM, Shay JW (eds) Cell and muscle motility, vol 1. Plenum Press, New York, pp 205–260

Gabella G (1973) Fine structure of smooth muscle. Phil Trans R Soc Ser B (London) 265:7–16

Gabella G (1978) Inpocketings of the cell membrane (caveolae) in the rat myocardium. J Ultrastruct Res 65:135–147

Gadsby DC, Wit AL, Cranefield PF (1978) The effects of acetylcholine on the electrical activity of canine cardiac Purkinje fibre. Circ Res 43:29–35

Gallagher JJ, Sealy UC, Kasell J, Wallace AG (1976) Multiple accessory pathways in patients with the pre-excitation syndrome. Circulation 54:571–590

Gallagher JJ, Smith WM, Kasell JH, Woodrow Benson DW, Sterba R, Grant AO (1981) Role of Mahaim fibres in cardiac arrhythmias in man. Circulation 64:176–189

Galper JB, Smith TW (1980) Agonist and guanine nucleotide modulation of muscarinic cholinergic receptors in cultured heart cells. J Biol Chem 255:9571–9579

Galper JB, Dziekan LC, Miura DS, Smith TW (1982) Agonist-induced changes in the modulation of K permeability and beating rate by muscarinic agonists in cultured heart cells. J Gen Physiol 80:231–256

Gannon BJ (1971) A study of the dual innervation of teleost heart by a field stimulation technique. Comp Gen Pharmacol 2:175–183

Gannon BJ, Burnstock G (1969) Excitatory adrenergic innervation of the fish heart. Comp Biochem Physiol 29:765–773

Gannon BJ, Campbell GD, Satchell GH (1972) Monoamine storage in relation to cardiac regulation in the Port Jackson shark, Heterodontus portus jacksoni. Z Zellforsch Mikrosk Anat 131:437–450

Ganote CE, Seabra-Gomes R, Nayler WG, Jennings RB (1975) Irreversible myocardial injury in anoxic perfused rat hearts. Am J Pathol 80:919–938

Garant PR (1968) Glycogen-membrane complexes within mouse striated muscle cells. J Cell Biol 36:648–652

Garcia R, Thibault G, Cantin M, Genest J (1984) Effect of a purified atrial natriuretic factor on rat and rabbit vascular strips and vascular beds. Am J Physiol 247:R34–R39

Gardner E, O'Rahilly R (1976) The nerve supply and conducting system of the human heart at the end of the embryonic period proper. J Anat 121:571–587

Garrey WE, Bass G (1937) Effects of acetylcholine of the frog's ventricle. Am J Physiol 119:314

Garrey WE, Chastain LC (1937a) Acetylcholine action on the turtle heart. Am J Physiol 119:314–315

Garrey WE, Chastain LL (1937b) Inotropic and chronotropic effects of acetylcholine upon the chelonian heart. Am J Physiol 119:315

Gaskell WH (1882) Observations of the innervation of the heart. J Physiol 4:43–127

Gaskell WH (1883) On the innervation of the heart with special reference to the heart of the tortoise. J Physiol 4:43–127

Gaskell WH (1884a) On the augmentor (accelerator) nerves of the heart of cold blooded animals. J Physiol 5:46–48

Gaskell WH (1884b) On the action of the sympathetic nerves upon the heart of the frog. J Physiol 5:13P–15P

Gaskell WH, Gadow H (1884) On the anatomy of the cardiac nerves in certain cold blooded vertebrates. J Physiol (Lond) 5:362

Geffen LB, Livett BG (1971) Synaptic vesicles in sympathetic neurons. Physiol Rev 51:98–157

Gerdes AM, Callas E, Kasten FH (1979) Differences in regional capillary distribution and myocyte sizes in normal and hypertrophic rat heart. Am J Anat 156:523–531

Gerdes AM, Kasten FH (1980) Morphometric study of endomyocardium and epimyocardium of the left ventricle in adult dogs. Am J Anat 159:389–394

Gerdes AM, Kriseman J, Bishop SP (1982) Morphometric study of the cardiac muscle: The problem of tissue shrinkage. Lab Invest 46:271–274

Gershman H, Weis G, Barstow N (1979) Dibutyryl cyclic AMP suppresses mobility of embryonic chick heart cells in aggregates. J Cell Sci 37:243–255

Gevers W (1984) Protein metabolism of the heart. J Mol Cell Cardiol 16:3–32

Gilson AS (1942) The locus and nature of the AV pause in the spread of cardiac activation. Am J Physiol 138:113–125

Girardier L, Pollet M (1964) Demonstration de la continuité entre l'espace interstitiel et la lumière de canaux intercellulaires dans le myocard de rat. Helvet Physiolo Pharmacolog Acta 22:C72–C73

Glacy SD (1983) Pattern and time course of rhodamine-actin incorporation in cardiac myocytes. J Cell Biol 96:1164–1167

Glick D (1981) Trends in quantification in histochemistry and cytochemistry. J Histochem 13:227–240

Gloster J, Achillea M, Harris P (1978) Subcellular distribution of [1-^{14}C] palmitate and [1-^{14}C] oleate incorporated into lipids in the perfused rat heart: A comparison under isothermal and hypothermic conditions. J Mol Cell Cardiol 10:439–448

Godlewsky E (1902) Die Entwicklung des Skelet- und Herzmuskelgewebes der Säugethiere. Arch Mikrosk Anat 60:111:156

Goldberg JM (1975) Intra-SA nodal pacemaker shifts induces by autonomic nerve stimulation in the dog. Am J Physiol 229:1116–1123

Goldberg JM, Geesbrecht JM, Randall WC, Brynjolfsson G (1973) Sympathetically induced pacemaker shifts following sinus node excision. Am J Physiol 224:1468–1474

Goldsmith JB, Butler HW (1937) Development of the cardiac-coronary circulation system. Am J Anat 60:185–201

Goldstein MA, Cartwright JJr (1982) Microtubules in adult mammalian muscle. In: Dowben RM, Shay JW (eds) Cell and muscle motility, vol 2. Plenum Press, New York, pp 85–92

Goldstein MA, Entman ML (1979) Microtubules in mammalian heart muscle. J Cell Biol 80:183–195

Goldstein MA, Schroeter JP, Sass RL (1977) Optical diffraction of the Z lattice in canine cardiac muscle. J Cell Biol 75:818–836

Goldstein MA, Schroeter JP, Sass RL (1979) The Z lattice in canine cardiac muscle. J Cell Biol 83:187–204

Goldstein MA, Schroeter JP, Sass RL (1982) The Z band lattice in slow skeletal muscle. J Muscle Res Cell Motility 3:333–348

Gonzales-Sanches A, Bader D (1985) Characterisation of a myosin heavy chain in the conductive system of the developing chicken heart. J Cell Biol 100:270–275

Goodenough DA (1976) In vitro formation of gap function vesicles. J Cell Biol 68:220–231

Goodenough DA (1979) Lens gap junctions: a structural hypothesis for non-regulated low resistance intercellular pathways. Invest Ophthal Vis Sci 18:1104–1122

Goodenough DA, Gilula NB (1974) The splitting of hepatocyte gap junctions and zonulae occludentes with hypertonic disaccharides. J Cell Biol 61:575–590

Goodenough DA, Revel JP (1970) A fine structural analysis of intercellular junctions in the mouse liver. J Cell Biol 45:272–290

Goodenough DA, Stoeckenius W (1972) The isolation of mouse hepatocyte gap junctions. Preliminary chemical characterization and X-ray diffraction. J Cell Biol 54:646–656

Goodenough DA, Paul DL, Culbert KE (1978) Correlative gap junction ultrastructure. In: Birth Defects Original Article Series 14:83–97

Goodman D, Van der Steen ABM, Van Dam RT (1971) Endocardial and epicardial activation pathways of the canine right atrium. Am J Physiol 220:1–11

Gorza L, Mercadier JJ, Schwartz K, Thornell L-E, Sartore S, Schiaffino S (1984) Myosin types in the human heart: An immunofluorescence study of normal and hypertrophied atrial and ventricular myocardium. Circ Res 54:694–702

Goshima K (1971) Synchronized beating of myocardial cells mediated by FL cells in monolayer culture and its inhibition by trypsin-treated FL cells. Exp Cell Res 65:161–169

Goshima K (1975) Beating of myocardial cells in culture. In: Lieberman M, Sano T (eds) Developmental and physiological correlates of cardiac muscle. Perspectives in cardiovascular research, vol 1. Raven Press, New York, pp 197–208

Goshima K, Wakabayashi S, Masuda A (1980) Ionic mechanism of morphological changes of cultured myocardial cells on successive incubation in media without and with Ca^{2+}. J Mol Cell Cardiol 12:1135–1157

Gossrau R (1968) Über das Reizleitungssystem der Vögel. Histochemie 13:111–159

Gotoh T (1983) Quantitative studies on the ultrastructural differentiation and growth of mammalian cardiac cells. The atria and ventricles of the cat. Acta Anat (Basel) 115:168–177

Granger BL, Lazarides E (1978) The existence of an insoluble Z disc scaffold in chicken skeletal muscle. Cell 15:1253–1268

Granger BL, Lazarides E (1979) Desmin and vimentin coexist at the periphery of myofibril Z disc. Cell 18:1053–1063

Grant RP (1926) Development of the cardiac coronary vessels in the rabbit. Heart 13:261–271

Grantham JJ, Edwards RM (1984) Natriuretic hormones: At last, bottled in bond? J Lab Clin Med 103:333–336

Gregory MA, Brouckaert CJ, Whitton ID (1983) Characterisation of normal human myocardium by means of morphometric analysis. Cardiovasc Res 17:177–183

Green CR, Severs NJ (1984) Gap junction connexon configuration in rapidly frozen myocardium and isolated intercalated disks. J Cell Biol 99:453–463

Greene CW (1900) Contributions to the physiology of the California hagfish, *Polistotrema stouti*. I. The anatomy and physiology of the caudal heart. Am J Physiol 3:366–382

Greene CW (1902) Contributions to the physiology of the California hagfish, *Polistotrema stouti*. II. The absence of regulative nerves for the systemic heart. Am J Physiol 6:318–324

Griepp EB, Bernfield MR (1978) Acquisition of synchronous beating between embryonic heart cell aggregates and layers. Exp Cell Res 113:263–272

Grimley PM, Edwards GA (1960) The ultrastructure of cardiac desmosomes in the toad and their relationship to the intercalated disc. J Biophys Cytol 8:305–318

Gröschel-Stewart U, Chamley JH, McConnell JD, Burnstock G (1975) Comparison of the reaction of cultured smooth and cardiac muscle cells and fibroblasts to specific antibodies to myosin. Histochemistry 43:215–224

Gros D, Mocquard JP, Challice CE, Schrevel J (1979a) Formation and growth of gap junctions in mouse myocardium during ontogenesis: A freeze cleave study. J Cell Sci 30:45–61

Gros D, Mocquard JP, Challice CE, Schrevel J (1979b) Formation and growth of gap junctions in mouse myocardium during ontogenesis: Quantitative data and their implications on the development of intercellular communication. J Mol Cell Cardiol 11:543–554

Gros D, Potreau D, Mocquard J-P (1980) The myocardial plasma membrane during development: Influence of glutaraldehyde fixation of the density and size of intramembranous particles. J Cell Sci 43:301–317

Gros D, Bruce B, Challice CE, Schrevel J (1982) Ultrastructural localisation of concanavalin A and wheat germ agglutinin binding sites in adult and embryonic mouse myocardium. J Histochem Cytochem 30:193–200

Gross WO (1977) Z-line elementary bodies. Morphological study of sarcomere genesis in chick heart myocytes. J Submicrosc Cytol 9:285–298

Gross WO (1982) Fibroblast-myocyte interactions in *in vitro* cardio myogenesis. Exp Cell Res 142:341–356

Guieysse-Pellisier A (1945) Les appareils respiratoires dans la série animale. Payot, Paris, pp 233–235

Hachenbroch CR (1972) States of activity and structure in mitochondrial membrane. Ann NY Acad Sci 195:492–505

Hadek R, Talso PJ (1967) A study of nonmyelinated nerves in the rat and rabbit heart. J Ultrastruct Res 17:257–265

Hagopian M, Anversa P, Loud AV (1975) Quantitative radioautographic localization of newly synthesized protein in the postnatal rat heart. J Mol Cell Cardiol 7:357–367

Hagopian M, Nunez EA (1972) Sarcolemmal scalloping at short sarcomere lengths with incidental observations on the T tubules. J Cell Biol 53:252–258

Hagopian M, Spiro D (1970) Derivation of the Z line in the embryonic chicken heart. J Cell Biol 44:683–687

Halbert SP, Bruderer R, Lin TM (1971) *In vitro* organization of dissociated rat cardiac cells into beating three-dimensional structures. J Exp Med 133:677–695

Halle W, Wolenberger A (1968) Die Differenzierung isolierter Herzzellen in einem chemisch definierten Nährmedium. Z Zellforsch Mikrosk Anat 87:292–314

Hamburger V, Hamilton HL (1951) A series of normal stages in the development of the chick embryo. J Morphol 88:49–92

Hanak H, Bock P (1971) Die Feinstruktur der Muskel-Sehnenverbindung von Skelett- und Herzmuskel. J Ultrastruct Res 36:68–85

Hansen JT, Yates RD (1975) Light, fluorescence and electron microscopic studies of rabbit subclavian glomera. Am J Anat 144:477–490

Hara T (1967) Morphological and histochemical studies on the cardiac conduction system of the dog. Arch Histol Jap 28(3):227–246

Harary I, Farley B (1960a) *In vitro* studies of single isolated beating heart cells. Science 131:1674–1675

Harary I, Farley B (1960b) *In vitro* organization of single beating rat heart cells into beating fibres. Science 131:1839–1840

Harary I, Farley B (1963a) *In vitro* studies on single beating rat heart cells. I. Growth and organization. Exp Cell Res 29:451–465

Harary I, Farley B (1963b) *In vitro* studies on single beating rat heart cells. II. Intercellular communication. Exp Cell Res 29:466–474

Harary I, Wallace GA (1978) The effect of the reciprocal relationship of Ca and cAMP on the control of beating in cultured rat heart cells. In: Kobayashi T, Ito Y, Rona G (eds) Recent advances in studies on cardiac structure and metabolism, vol 2, Cardiac adaptation, University Park Press, Baltimore, pp 635–643

Harary I, McCarl RL, Farley B (1966) Studies *in vitro* on single beating heart cells. IX. The restoration of beating by serum lipids and fatty acids. Biochim Biophys Acta 115:15–22

Harary I, Renaud J-F, Sato E, Wallace GA (1976) Calcium ions regulate cyclic AMP and beating in cultured heart cells. Nature 261:60–61

Harary I, Wallace G, Bristol G (1983) A video-computer for the chronotropic and inotropic measurements of the beating of cultured heart cells. Cytometry 3:367–375

Hardie EL, Jones SB, Euler DE, Fishman DL, Randal WC (1981) Sinoatrial node artery distribution and its relation to hierarchy of cardiac automaticity. Am J Physiol 241:H45–H53

Harh JY, Paul MH (1975) Experimental cardiac morphogenesis. 1. Development of the ventricular septum in the chick. J Embryol Exp Morphol 33:13–28

Harris AJ, Kuffler SE, Dennis MJ (1971) Differential chemosensitivity of synaptic and extrasynaptic areas on the neuronal surface membrane in parasympathetic neurones of the frog, tested by microapplication of acetylcholine. Proc Roy Soc Ser B (London) 177:541–553

Hartzell HC (1980) Distribution of muscarinic acetylcholine receptors and presynaptic nerve terminals in amphibian heart. J Cell Biol 86:6–20

Hasin Y, Shimoni Y, Stein O, Stein Y (1980) Effect of cholesterol depletion on the electrical activity of rat heart myocytes in culture. J Mol Cell Cardiol 12:675–683

Hasin Y, Sapoznikov D, Stein O, Stein Y (1982) Effect of fatty acid composition of rat heart myocytes on their electrical activity. J Mol Cell Cardiol 14:163–171

Hatt PY, Jouannot P, Moravec J, Perennec J, Laplace M (1978) Development and reversal of pressure-induced cardiac hypertrophy. Basic Res Cardiol 73:405–421

Hatt PY, Rakusan K, Gatineau P, Laplace M (1979) Morphometry and ultrastructure of heart hypertrophy induced by chronic volume overload (aorto-caval fistula in the rat). J Mol Cell Cardiol 11:989–998

Hay OA, Low FN (1972) The fine structure of progressive stages of myocardial mitosis in chick embryos. Am J Anat 134:175–201

Hayashi K (1962) An electron microscopic study on the conduction system of the cow heart. Jap Circ J 26(10) 765–841

Hayashi S (1971) Electron microscopy of the heart conduction system of the dog. Arch Histol Jap 33:67–86

Hayashi S, Oga K, Otsuka N (1970) The fine structure of nerve endings in the sinus node of the canine heart. J Electron Microsc 19:176–181

Hearse DJ, Humphrey SM, Nayler WG, Slade A, Border D (1975) Ultrastructural damage associated with re-oxygenation of the anoxic myocardium. J Mol Cell Cardiol 7:315–324

Hecht HH, Kossmann CE, Childers RW, Langendorf R, Lev M, Rosen KM, Pruitt RD, Truex RC, Uhley HN, Watt TB (1973) Atrioventricular and intraventricular conduction. Revised nomenclature and concepts. Am J Cardiol 31:232–244

Heidenhain M (1901) Über die Struktur des menschlichen Herzmuskels. Anat Anz 20:33–42

Heintzberger CFM (1974) Development of the sinuatrial node in the mouse. Acta Morphol Neerl Scand 12:317–330

Heitz U (1982) Immunocytochemistry – theory and application. Acta Histochem (Suppl) 25:17–35

Henderson D, Eibl H, Weber K (1979) Structure and biochemistry of mouse hepatic gap junctions. J Mol Biol 132:193–218

Henry CG, Lowry OH (1983) Quantitative histochemistry of canine cardiac Purkinje fibers. Am J Physiol 245(5):H824–H829

Hermsmeyer K (1979) High shortening velocity of isolated single arterial muscle cells. Experientia 35:1599–1602

Hermsmeyer K, Robinson RB (1977) High sensivity of cultured cardiac muscle cells to autonomic agents. Am J Physiol 233:C172–C179

Hertzberg EL, Lawrence TS, Gilula NB (1981) Gap junctional communication. Ann Rev Physiol 43:479–491

Heuser LH, Corner GW (1957) Developmental horizons in human embryos. Description of age group X, four to twelve somites. Contrib Embryol Carnegie Inst 36:29–39

Heuson-Stiennon JA (1965) Morphogenèse de la cellule musculaire striée au microscope électronique. Formation des structures fibrillaires. J Microsc (Paris) 4:657–678

Hewett K, Legato MJ, Danilo P Jr, Robinson RB (1983) Isolated myocytes from adult canine left ventricle: Ca tolerance, electrophysiology, and ultrastructure. Am J Physiol 245:H830–H839

Heymann MA, Creasy RK, Rudolph AM (1973) Quantitation of blood flow patterns in the foetal lamb in vitro. In: Proc Sir Joseph Barcroft Cent Symp. Cambridge University Press, London, pp 129–135

Hiatt EP (1943) The action of adrenaline, acetylcholine and potassium in relation to the innervation of the isolated auricle of the spiny dogfish (Squalus acanthias). Am J Physiol 139:45–48

Hiatt EP, Garrey WE (1943) Drug actions of the spontaneously beating turtle ventricle indicating lack of innervation. Am J Physiol 138:758–762

Hibbs RG (1956) Electron microscopy of developing cardiac muscle in chick embryos. Am J Anat 99:17–52

Hibbs RG, Ellison JP (1973) The atrioventricular valves of the guinea-pig. II. An ultrastructural study. Am J Anat 138:347–365

Hibbs RG, Ferrans VJ (1968) An ultrastructural and histochemical study of rat atrial myocardium. Am J Anat 124:251–280

Higgins TJC, Allsopp D, Bailey PJ (1979) The effect of beta-adrenergic blocking drugs on the intrinsic beating rate of cultured rat myocytes. J Mol Cell Cardiol 11:101–107

Higgins TJC, Bailey PJ, Allsopp D, Imhof DA (1981) Cultured neonate rat myocytes as a model for the study of myocardial ischaemic necrosis. J Pharm Pharmacol 33:644–649

Higuchi M, Nishi K, Takenaka F (1979) A comparison of enzyme activity for energy production in the myocardium and conduction system. Jap Heart J 20:667–673

Hill CE, Hendry IA (1977) Development of neurons synthesizing noradrenaline and acetylcholine in superior cervical ganglion of the rat in vivo and in vitro. Neuroscience 2:741–749

Hirakow R (1966) Fine structure of Purkinje fibres in the chick heart. Arch Histol Jap 27:485–499

Hirakow R (1970) Ultrastructural characteristics of the mammalian and sauropsidian heart. Am J Cardiol 25:195–203

Hirakow R (1983) Development of the cardiac blood vessels in staged human embryos. Acta Anat (Basel) 115:220–230

Hirakow R, Gotoh T (1975) A quantitative ultrastructural study on developing rat heart. In: Lieberman M, Sano T (eds) Developmental and physiological correlates of cardiac muscle. Raven Press, New York, pp 37–49

Hirakow R, Gotoh T (1980) Quantitative studies on the ultrastructural differentiation and growth of mammalian cardiac muscle. II. The atria and ventricles of the guinea pig. Acta Anat (Basel) 108:230–237

Hirakow R, Krause WJ (1980) Postnatal differentiation of ventricular myocardial cells of the opossum (*Didelphis virginiana* Kerr) and T-tubule formation. Cell Tissue Res 210:95–100

Hirakow R, Gotoh T, Watanabe T (1980) Quantitative studies on the ultrastructural differentiation and growth of mammalian cardiac muscle cells. I. The atria and ventricles of the rat. Acta Anat (Basel) 108:144–152

Hirano H, Ogawa K (1967) Ultrastructural localisation of cholinesterase activity in nerve endings in the guinea pig heart. J Electron Microsc 16:313–321

Hirota A, Sakai T, Fujii S, Kamino K (1983) Initial development of conduction pattern of spontaneous action potential in early embryonic precontractile chick heart. Dev Biol 99:517–523

Hirsch EF, Borghard-Erdle AM (1961) The innervation of the human heart. I. The coronary arteries and the myocardium. Arch Pathol 71:384–407

Hoerter J, Mazet F, Vassort G (1981) Perinatal growth of the rabbit cardiac cell: possible implications for the mechanism of relaxation. J Mol Cell Cardiol 13:725–740

Hoff EC, Kramer IC, Du Bois D, Patten BM (1939) The development of the electrocardiogram of the embryonic heart. Am Heart J 17:470–488

Hoffman BF (1979) Fine structure of internodal pathways (editorial). Am J Cardiol 44:385–386

Hoffmeister H, Lickfeld K, Ruska H, Rybak B (1961) Sécretions granulaires dans le cœur branchial de *Myxine glutinosa* L. Z Zellforsch Mikrosk Anat 55:810–817

Hogue MJ (1937) Studies of heart muscle in tissue cultures. Anat Rec 67:521–535

Hogue MJ (1947) Intercalated disks in tissue cultures. Anat Rec 99:157

Hoh JFY, McGrath PA, Hale PT (1977) Electrophoretic analysis of multiple forms of rat cardiac myosin: Effects of hypophysectomy and thyroxine replacement. J Mol Cell Cardiol 10:1053–1076

Hoh JFY, Yeoh G, Thomas M, Higginbottom L (1979) Structural differences in the heavy chains of rat ventricular myosin isoenzymes. FEBS Lett 97:330–334

Holmes AH (1923) Purkinje fibres in the auricles of birds. J Physiol Proc 58:3P–4P

Homles RL (1957) Structures in the atrial endocardium of dog, which stain with methylene blue and the effects of unilateral vagotomy. J Anat 91:259–266

Holmgren E (1907) Über die Trophospongien der quergestreiften Muskelfasern nebst Bemerkungen über den allgemeinen Bau dieser Fasern. Arch Mikrosk Anat 71:165–247

Holmgren S (1977) Regulation of the heart of a teleost, *Gadus morhua*, by autonomic nerves and circulating catecholamines. Acta Physiol Scand 99:62–74

Holsinger JW, Wallace AG, Sealy WC (1968) The identification and surgical significance of the atrial internodal conduction tracts. Ann Surg 167:447–453

Holubarsch CH, Goulette RP, Litten RZ, Martin BJ, Mulieri LA, Alpert NR (1985) The economy of isometric force development, myosin isoenzyme pattern and myofibrillar ATPase activity in normal and hypertrophied rat myocardium. Circ Res 56:78–86

Horres CR, Lieberman M, Purdy JE (1977) Growth orientation of heart cells on nylon monofilament: Determination of the volume-to-surface area ratio and intracellular potassium concentration. J Memb Biol 34:313–329

Hoshino K (1983) How useful are microcomputers in histochemical studies? J Histochem Cytochem 31:244–247

Hoshino T, Fujiwara H, Kawai C, Hamashima Y (1983) Myocardial fibre diameter and regional distribution in the ventricular wall of normal adult hearts, hypertensive hearts and hearts with hypertrophic cardiomyopathy. Circulation 67:1109–1116

Howse HP, Ferrans VJ, Hibbs RG (1970) A comparative histochemical and electron microscopic study of the surface coatings of cardiac muscle cells. J Mol Cell Cardiol 1:157–168

Hoyle G (1983) Muscles and their neural control. Wiley & Sons, New York

Huang TF (1973) The action potential of the myocardial cell of the golden carp. Jap J Physiol 23(5):529–580

Hung K-S, Loosli CG (1977) Electron-microscopic studies of the innervation of the pulmonary veins of the mouse. Acta Anat 97:97–102

Hurle JM, Ojeda JL (1979) Cell death during the development of the truncus and conus of the chick embryo heart. J Anat 129:427–439

Hurle JM, Lafarga M, Ojeda JL (1977) Cytological and cytochemical studies of the necrotic area of the bulbus of the chick embryo heart: phagocytosis by developing myocardial cells. J Embryol Exp Morphol 41:161–173

Hurle JM, Lafarga M, Ojeda JL (1978) In vivo phagocytosis by developing myocardial cells: An ultrastructural study. J Cell Sci 33:363–369

Hutchins GM, Liebman L, Moore GU, Ghara Gozloo F (1979) Atrioventricular canal malformations interpreted as secondary to reduced compression upon the developing heart. Am J Pathol 95:579–595

Huxley AF, Niedergerke R (1954) Interference microscopy of living muscle fibres. Nature 173:971–973

Huxley HE (1969) The mechanism of muscular contraction. Science 164:1356–1360

Huxley HE, Hanson J (1954) Changes in the cross-striations of muscle during contraction and stretch and their structural interpretation. Nature 173:973–976

Hyde A, Blondel B, Matter A, Cheneval JP, Filloux B, Girardier L (1969) Homo- and heterocellular junctions in cell cultures: An electrophysiological and morphological study. Prog Brain Res 31:283–311

Icardo JM (1984) The growing heart: An anatomical perspective. In: Zak R (ed) Growth of the heart in health and disease. Raven Press, New York, pp 41–79

Imamura K (1978) Ultrastructural aspect of left ventricular hypertrophy in spontaneously hypertensive rats: A qualitative and quantitative study. Jap Circ J 42:979–1002

Imanaga I (1974) Cell-to-cell diffusion of Procion Yellow in sheep and calf Purkinje fibres. J Memb Biol 16:381–388

Ingwall JS (1980) Metabolism of proteins and nucleic acids during muscle differentiation: Regulation of gene expression. In: Wildenthal K (ed) Degradative processes in heart and skeletal muscle. Elsevier/North Holland, Amsterdam, pp 131–159

Inoue F (1959) Slow potential and conduction delay at the atrioventricular region in frog's heart. J Cell Comp Physiol 54:231–236

Irisawa A (1978) Fine structure of the small sino atrial node specimen used for the voltage clamp experiment. In: Bonke FIM (ed) The sinus node, chap 26. Nijhoff, The Hague, pp 311–319

Irisawa H (1978) Fine structure of the sino atrial node of rabbit heart. In: Kobayashi T, Ito Y, Rona G (eds) Recent advances in studies on cardiac structure and metabolism, vol 12, Cardiac adaptation. University Park Press, Baltimore, pp 77–80

Ishikawa H (1968) Formation of elaborate networks of T-system tubules in cultured skeletal muscle with special reference to the T-system formation. J Cell Biol 38:51–66

Ishikawa H, Yamada E (1975) Differentiation of the sarcoplasmic reticulum and T-system in developing mouse cardiac muscle. In: Lieberman M, Sano T (eds) Developmental and physiological correlates of cardiac muscle. Raven Press, New York, pp 21–35

Ishikawa H, Bischoff R, Holtzer H (1968) Mitosis and intermediate-sized filaments in developing skeletal muscle. J Cell Biol 38:538–555

Iwayama T (1971) Nexuses between areas of the surface membrane of the same arterial smooth muscle cell. J Cell Biol 49:521–524

Izquierdo JJ (1930) On the influence of the extra-cardiac nerves upon sino-auricular conduction in the heart of Scyllium. J Physiol 69:29

Izumi T, Yamazoe M, Shibata A (1984) Three-demensional characteristics of the intramyocardial microvasculature of hypertrophied human hearts. J Mol Cell Cardiol 16:449–457

Jacobowitz D, Cooper T, Barner HB (1967) Histochemical and chemical studies of the localisation of adrenergic and cholinergic nerves in normal and denervated cat hearts. Circ Res 20:289–298

Jacobson SL (1977) Culture of spontaneously contracting myocardial cells from adult rats. Cell Struct Funct 2:1–9

James TN (1961a) Morphology of the human atrioventricular node, with remarks pertinent to its electrophysiology. Am Heart J 62:756–771

James TN (1961b) Anatomy of the human sinus node. Anat Rec 141:109–139

James TN (1962) Anatomy of the sinus node of the dog. Anat Rec 143:251–265

James TN (1963) The connecting pathways between the sinus node and AV node and between the right and left atrium in the human heart. Am Heart J 66:498–508

James TN (1967) Anatomy of the cardiac conduction system in the rabbit. Circ Res 20:638–648

James TN (1968) Sudden death in babies: New observations in the heart. Am J Cardiol 22:479–506

James TN (1970) Cardiac conduction system: Fetal and postnatal development. Am J Cardiol 25:213–226

James TN (1977) The sinus node. Am J Cardiol 40:965–986

James TN (1978) Anatomy of coronary arteries and veins. In: Hurst JW, Logue RB, Schlant RC, Werger NK (eds) The heart. McGraw Hill, New York

James TN (1980) Neural control of the heart in health and disease. Adv Int Med 26:317–345

James TN (1983) Structure and function of the AV junction. Jap Circ J 47(1):1–48

James TN, Marshal TK (1976) De Subitaneis Mortibus. XVIII. Persistent fetal dispersion of the atrioventricular node and His bundle within the central fibrous body. Circulation 53:1026–1034

James TN, Sherf L (1968) Ultrastructure of the human atrioventricular node. Circulation 37:1049–1070

James TN, Spence CA (1966) Distribution of cholinesterase within the sinus node and AV node of the human heart. Anat Rec 155:151–161

James TN, Sherf L, Fine G, Morales AR (1966) Comparative ultrastructure of the sinus node in man and dog. Circulation 34:139–163

James TN, McKone RC, Hudspeth AS (1975) De subitaneis mortibus. X. Familial congenital heart block. Circulation 51:379–388

James TN, Spencer MS, Klopfer JC (1976) De subitaneis mortibus. XXI. Adult onset syncope, with comments on the nature of congenital heart block and the morphogenesis of the human atrioventricular septal junction. Circulation 54:1001–1009

Jamieson JD, Palade GE (1964) Specific granules in atrial muscle cells. J Cell Biol 23:151–172

Janse MJ, Anderson RH (1974) Specialised internodal atrial pathways – fact or fiction? Eur J Cardiol 2:117–136

Janse MJ, Van Capelle FJL, Anderson RH, Touboul P, Billette J (1976) Electrophysiology and structure of the atrioventricular node of the isolated rabbit heart. In: Wellens HJJ, Lie KI, Janse MJ (eds) The conduction system of the heart. Stenfert Kroese, Leiden, pp 296–315

Janse MJ, Tranum-Jense J, Keber AG, Van Capelle FJL (1978) Techniques and problems in correlating cellular electrophysiology and morphology in cardiac nodal tissues. In: Bonke FIM (ed) The sinus node, chapter 15. Martinus Nijhoff, The Hague, pp 183–194

Januszewicz P, Gutowska J, Den Lean A, Thibault G, Garcia R, Genest J, Cantin M (1985) Synthetic atrial natriuretic factor induces release (possibly receptor-mediated) of vasopressin from rat posterior pituitary. Proc Soc Exp Biol Med 178:321–325

Jedeikin LA (1964) Regional distribution of glycogen and phosphorylase in the ventricles of the heart. Circ Res 14:202–211

Jennings RB, Ganote CE (1976) Mitochondrial structure and function in acute myocardial ischemic injury. Circ Res 38(Suppl 1):80–91

Jensen H, Holtet L, Hoen R (1978) Nerve-Purkinje fiber relationship in the moderator band of bovine and caprine heart. Cell Tissue Res 188:49–55

Jewett PH, Sommer JR, Johnson EA (1971) Cardiac muscle: Its ultrastructure in the finch and hummingbird with special reference to the sarcoplasmic reticulum. J Cell Biol 49:50–65

Jewett PH, Leonard SD and Sommer JR (1973) Chicken cardiac muscle: Its elusive extended junctional sarcoplasmic reticulum and sarcoplasmic reticulum fenestrations. J Cell Biol 56:595–600

Johansen K, Franklin DL, Van Citters RL (1966) Aortic blood flow in free-swimming elasmobranchs. Comp Biochem Physiol 19:151–160

Johansen M (1965) Comparative aspects of cardiovascular function in vertebrates, In: Hamilton WF (ed) Handbook of physiology, section 2, Circulation, vol 3. American Physiological Society, pp 2583–2614

Johnson EA, Sommer JR (1967) A strand of cardiac muscle: Its ultrastructure and the electrophysiological implications of its geometry. J Cell Biol 33:103–129

Johnson M, Ross D, Meyer M, Rees R, Bunge R, Wakshull E, Burton H (1976) Synaptic vesicle cytochemistry changes when cultured sympathetic neurones develop cholinergic interactions. Nature 262:308–310

Johnson MI, Ross CD, Meyers M, Spitznagel EL, Bunge RP (1980a) Morphological and biochemical studies on the development of cholinergic properties in cultured sympathetic neurons. I. Correlative changes in choline acetyltransferase and synaptic vesicle cytochemistry. J Cell Biol 84:680–691

Johnson MI, Ross CD, Bunge RP (1980b) Morphological and biochemical studies on the development of cholinergic properties in cultured sympathetic neurons. II. Dependence on postnatal age. J Cell Biol 84:692–704

Jones CE, Cannon MS (1980) The myocardium and its vasculature: A histochemical comparison of the normal and chronically sympathectomised dog heart. Histochem J 12:9–22

Jongsma HJ, Masson-Pévet M, Hollander CC, De Bruyne J (1975) Synchronization of the beating frequency of cultured rat heart cells. In: Lieberman M, Sano T (eds) Developmental and physiological correlates of cardiac muscle. Perspectives in cardiovascular research, vol 1. Raven Press, New York, pp 185–196

Jorgensen AO, Basher R (1984) Temporal appearance and distribution of the $CA_{2+} + Mg_{2+}$ ATPase of the sarcoplasmic reticulum in developing chick myocardium as determined by immunofluorescence labelling. Dev Biol 106:156–165

Jorgensen AO, Campbell KP (1984) Evidence of the presence of calsequestrin in two structurally different regions of myocardial sarcoplasmic reticulum. J Cell Biol 98:1597–1602

Jorgensen AO, McLeod AG, Campbell KP, Denney GH (1984) Evidence for the presence of calsequestrin in both peripheral and interior regions of sheep Purkinje fibres. Circ Res 55:267–270

Jourdan P (1980) Ultrastructure and electrical activity of newborn rat heart aggregates. Biol Cell 37:149–154

Jourdan P, Sperelakis N (1980) Electrical properties of cultured heart cell reaggregates from newborn rat ventricles: Comparison with intact non-cultured ventricles. J Mol Cell Cardiol 12:1441–1458

Julian FJ, Morgan DL, Moss RL, Gonzalez M, Dwivedi P (1981) Myocyte growth without physiological impairment in gradually induced rat cardiac hypertrophy. Circ Res 49:1300–1310

Junker J, Sommer JR (1977) Anchorfibers and the topography of the junctional SR. In: Bailey GW (ed) The 35th Annual Proceedings of the Electron Microscopy Society of America, Boston, p 582

Kakiuchi S, Yamazaki R (1970) Calcium dependent phosphodiesterase activity and its activating factor from brain. Studies on cyclic 3', 5' nucleotide phosphodiesterase (III), Biochem Biophys Res Commun 41:1104–1110

Kamenskaya VN, Samonina GE, Udel'nov MG (1976) Characteristics of afferent activity of cardiac nerves in the frog *Rana temporaria*. Zh Evol Biokhim Fiziol (USSR) 12:294–296

Kamenskaya VN, Samonina GE, Udel'nov MG (1977a) Characteristics of afferent activity of cardiac nerves in the tortoise *Agrionemys horsfieldi*. Zh Evol Biokhim Fiziol (USSR) 13:24–30

Kamenskaya VN, Samonina GE, Udel'nov MG (1977a) Characteristics of the autoregulation mechanisms of the steppe turtle heart: II. Analysis of the nature and nerve mechanisms of cardio-cardiac reflexes. Biol Nauki (Moscow) 20:67–73

Kamino K, Hirota A, Fujii S (1981) Localization of pacemaking activity in early embryonic heart monitored using voltage sensitive dye. Nature (London) 290:595–597

Kanno T (1963) Electrical activity of the atrioventricular conducting tissue of the toad, studied by a minute suction electrode. Jap J Physiol 13:97–111

Kappagoda CT, Linden RJ, Snow HM (1972) The effect of stretching the superior vena caval-right atrial junction on right atrial receptors in the dog. J Physiol 227:875–887

Karlsson U, Andersson-Cedergren E (1968) Small leptomeric organelles in intrafusal muscle fibers of the frog as revealed by electron microscopy. J Ultrastruct Res 23:417–426

Karrer HE (1959) The striated musculature of blood vessels. I. General cell morphology. J Biophys Biochem Cytol 6:383–403

Karrer HE (1960) The striated musculature of blood vessels. II. Cell interconnections and cell surface. J Biophys Biochem Cytol 8:135–150

Karsten U, Wallukat G, Wollenberger A (1977) α- and β-adrenergic receptor stimulation in cultures of beating rat heart cells and its effects on automaticity and cyclic AMP levels. Acta Biol Med Germ 36:63–69

Kartenbeck J, Franke WW, Moser JG, Stoffels U (1983) Specific attachment of desmin to desmosomal plaques in cardiac myocytes. EMBO J 2:735–742

Kasten FH (1972) Rat myocardial cells *in vitro*: mitosis and differentiated properties. In Vitro 8:128–150

Kasuga M, Karlsson FA, Kahn CR (1982) Insulin stimulates the phosphorylation of 95000-dalton subunit of its own receptor. Science 215:185–187

Katz AM, Messineo FC (1981) Lipid-membrane interactions and the pathogenesis of ischemic damage in the myocardium. Circ Res 48:1–16

Katz B (1948) The electrical properties of the muscle fiber membrane. Proc R Soc, Ser B (London) 135:506–534

Kaufman MH, Navaratnam V (1981) Early differentiation of the heart in mouse embryos. J Anat 133:235–245

Kawamura K (1961a) Electron microscope study of the cardiac conduction system of the dog. I. The Purkinje fibres. Jap Circ J 25:594–616

Kawamura K (1961b) Electron microscope studies on the cardiac conduction system of the dog. II. The sino atrial and atrioventricular nodes. Jpn Circ J 25:973–1013

Kawamura K (1982) Cardiac hypertrophy – scanned architecture, ultrastructure and cytochemistry of myocardial cells. Jap Circ J 46:1012–1030

Kawamura K, Hayashi K (1966) Electron microscope study of the cardiac conduction system. Jap Circ J 30:49–51

Kawamura K, James TN (1971) Comparative ultrastructure of cellular junctions in working myocardium and the conduction system under normal and pathologic conditions. J Mol Cell Cardiol 3:31–60

Kawamura K, James TN (1978) Electron microscopic cytochemistry of cytochrome oxidase activity in the conduction system of the canine heart. In: Kobayashi T, Ito Y, Rano G (eds) Recent advances in studies on cardiac structure and metabolism, vol 12, Cardiac adaptation. University Park Press, Baltimore, pp 93–101

Kawamura K, Urthaler F, James TN (1978) Fine structure of the conduction system and working myocardium in the little brown bat, *Myotis lucifugus*. In: Kobayashi T, Ito Y, Rano G (eds) Recent advances in studies on cardiac structure and metabolism, vol 12, Cardiac adaptation. University Park Press, Baltimore, pp 81–91

Kawamura K, Imamura K, Uehara H, Nakayama Y, Sawada K, Yamamoto S (1984) Architecture of hypertrophied myocardium: Scanning and transmission electron microscopy. In: Abe H, Ito Y, Tada M, Opie LH (eds) Regulation of cardiac function, Japan Scientific Societies Press, Tokyo/VNU Science Press, Utrecht, pp 81–105

Kebabian JW, Zatz M, Romero JA, Axelrod J (1975) Rapid changes in rat pineal β-adrenergic receptor: alterations in l-[^{3}H] alprenolol binding and adenylate cyclase. Proc Natl Acad Sci USA 72:3735–3739

Kefalides NA, Alper R, Clark CC (1979) Biochemistry and metabolism of basement membranes. Int Rev Cytol 61:167–228

Keith A, Flack MW (1906) The auriculo-ventricular bundle of the human heart. Lancet 2:359–364

Keith A, Flack M (1907) The form and nature of the muscular connections between the primary divisions of the vertebrate heart. J Anat Physiol 41:172–189

Kelly AM, Chacko S (1976) Myofibril organisation and mitosis in cultured cardiac muscle cells. Dev Biol 48:421–430

Kelly AM, Chacko S (1977) Internal membrane junctions and unusual T-tubules in cultured chick cardiac muscle cells. J Mol Cell Cardiol 9:419–424

Kelly DE (1969) The fine structure of skeletal muscle triad junctions. J Ultrastruct Res 29:37–49

Kensler RW, Goodenough DA (1980) Isolation of mouse myocardial gap junctions. J Cell Biol 86:755–764

Kent AFS (1913) The structure of the cardiac tissues at the auriculo-ventricular junction. J Physiol 47:17–18

Kent AFS (1914) The right lateral auriculo-ventricular junction of the heart. J Physiol 48:22–24

Kent KM, Epstein SE, Cooper T, Jacobowitz DM (1974) Cholinergic innervation of the canine and human ventricular conducting system. Circulation 50:948–955

Kidd DM, Jones AZ, Lemanski LF, Rudolph A, Allen L (1981) Histological and electron microscopic stereological study of the myocardium of newborn genetically cardiomyopathic hamsters. J Ultrastruct Res 76:107–119

Kikuchi S (1976) The structure and innervation of the sinuatrial node of the mole heart. Cell Tissue Res 172:345

Kilbinger H (1973) Gas chromatographic estimation of acetylcholine in the rabbit heart using a nitrogen selective detector. J Neurochem 21:421–429

Kim D, Marsh JD, Barry WH, Smith TW (1984) Effects of growth in low potassium medium or ouabain on membrane Na, K-ATPase, cation transport and contractility in cultured chick heart cells. Circ Res 55:39–48

Kim S, Baba N (1971) Atrioventricular node and Purkinje fibres of the guinea pig heart. Am J Anat 132:339–354

Kim Y, Yasuda M (1980) Development of the cardiac conducting system in the chick embryo. Anat Histol Embryol 9:7–20

Kirby S, Burnstock G (1969) Pharmacological studies of the cardiovascular system in the anaesthetised sleepy lizard (*Tiliqua rugosa*) and toad (*Bufo marinus*). Comp Biochem Physiol 28:321–331

Kisch B (1948) Electrographic investigations of the heart of fish. Exp Med Surg 6:31–62

Kisch B (1960) Nervendigungen am Herzkammermuskel. Z Kreislaufforsch 49:762–768

Kjellen L, Pettersson I, Hook M (1981) Cell-surface heparan sulfate: An intercalated membrane proteoglycan. Proc Natl Acad Sci USA 78:5371–5375

Kjörell U, Thornell L-E (1982) Identification of a complex between α-actinin and the intermediate filament subunit skeletin in bovine heart Purkinje fibres. Eur J Cell Biol 28:139–144

Klein I (1983) Colchicine stimulates the rate of contraction of heart cells in culture. Cardiovasc Res 17:459–465

Kontani H, Koshiura R (1981) Action of divalent cations of the mechanoreceptor of isolated frog (*Rana catesbeiana*) heart. Folia Pharmacol Jap 77:177–186

Kordylewski L, Karrison T, Page E (1983) P-face particle density of freeze-fractured vertebrate cardiac plasma membrane. Am J Physiol 245:H992–H997

Korecky B, Rakusan K (1978) Normal and hypertrophic growth of the rat heart: changes in cell dimensions and numbers. Am J Physiol 234:H123–H128

Korner PI (1979) Central nervous control of autonomic cardiovascular function. In: Berne RM, Sperelakis N, Geiger SR (eds) Handbook of physiology, sect 2, The cardiovascular system, vol 1. The heart, chapter 20, Williams & Wilkins, Baltimore, pp 691–740

Koteliansky VE, Glukhova MA, Shirinsky VP, Babaev VR, Kondalenko VF, Rukosuv VS, Smirnov VN (1982) A filamin-like protein – a new component of heart myofibrils. J Mol Cell Cardiol 14(Suppl 3):85–89

Kramer AWJN, Marks LS (1965) The occurrence of cardiac muscle in the pulmonary veins of Rodentia. J Morphol 117:135–150

Krause EG, Halle W, Kallabis E, Wolenberger A (1970) Positive chronotropic response of cultured isolated rat heart cells to N6, 2′-0-dibutryl-3′,5′-adenosine mono-phosphate. J Mol Cell Cardiol 1:1–10

Kreutziger GO (1968) Specimen surface contamination and the loss of structural detail in freeze-etch preparations. In: Proceedings of the 26th Annual Meeting of the Electron Microscope Society of America. Claitor's Publishing, Baton Rouge, Louisiana, pp 138–139

Kruger P, Duspira F, Furlinger F (1933) Tetanus und Tonus der Skeletmuskeln des Frosches. Eine histologische, reizphysiologische und chemische Untersuchung. Pflüg Arch 231:750–786

Kuffler SW, Dennis MJ, Harris AJ (1971) The development of chemosensitivity in extrasynaptic areas of the neuronal surface after denervation of parasympathetic ganglion cells in the heart of the frog. Proc R Soc, Ser B (London) 17:555–563

Kulikowski RR (1981) Myofibrillogenesis *in vitro*: Implications for early cardiac morphogenesis. In: Pexieder T (ed) Perspectives in cardiovascular research, vol 5, Mechanisms of cardiac morphogenesis and teratogenesis. Raven Press, New York, p 367

Kumar S (1973) The heart and its conducting system in *Amblystoma tigrinum* (Amphibia, Urodela). J Anat Soc India 22:29–34

Kumar S (1974) Neurohistological studies on the heart of *Labeo rohita* (Ham) with special reference to the conducting system. Geobios (Jodhpur) 1:127–130

Kunos G (1978) Adrenoceptors. Ann Rev Pharmacol Toxicol 18:291–311

Kunos G, Nickerson M (1976) Temperature-induced interconversion of α- and β-adrenoceptors in the frog heart. J Physiol (London) 25:23–40

Kurkiewicz T (1909) O Histogenezie Męśnia Sercowego Zwierząt Kręgowich, Bulletin l'Academie Science Cracovie 148–191

Kuroda M, Masaki T (1980) On the 42000 dalton proteins in vertebrate skeletal muscle: actin and eu-actinin. In: Ebashi S, Maruyama K, Endo M (eds) Muscle contraction: Its regulatory mechanisms, Springer, Berlin Heidelberg New York, pp 507–514

Kushida H (1962) A study of cellular swelling and shrinkage during fixation, dehydration and embedding in various standard media. J Electron Microsc (Tokyo) 11:135–138

Laguens R (1971) Morphometric study of myocardial mitochondria in the rat. J Cell Biol 48:673–676

Laks MM, Nisenson MJ, Swan HJC (1967) Myocardial cell and sarcomere lengths in the normal dog heart. Circ Res 21:671–678

Laks MM, Morady F, Garner D, Swan HJC (1974) Temporal changes in canine right ventricular volume, mass, cell size and sarcomere length after banding the pulmonary artery. Cardiovasc Res 8:106–111

Landis SC (1976) Rat sympathetic neurons and cardiac myocytes developing in microcultures: correlation of the fine structure of endings with neurotransmitter function in single neurons. Proc Natl Acad Sci USA 73:4220–4224

Landis SC (1980) Developmental changes in the neurotransmitter properties of dissociated sympathetic neurons: a cytochemical study of the effects of medium. Dev Biol 77:349

Lane M-A, Sastre A, Salpeter MM (1976) Innervation of heart cells in culture by an endogenous source of cholinergic neurons. Proc Natl Acad Sci USA 73:4506–4510

Lang RE, Tholken H, Ganten D, Luft FC, Ruskoaho H, Unger Th (1985) Atrial natriuretic factor – a circulating hormone stimulated by volume loading. Nature 314:264–266

Langer GA (1984) Calcium at the sarcolemma. J Mol Cell Cardiol 16:147–153

Langer GA, Frank JS, Nudd LM, Seraydarian K (1976) Sialic acid: effect of removal on calcium exchangeability of cultured heart cells. Science 193:1013–1015

Langer GA, Frank JS, Philipson KD (1982) Ultrastructure and calcium exchange of the sarcolemma, sarcoplasmic reticulum and mitochondria of the myocardium. Pharmacol Ther 16:331–376

Larno S, Lhoste F, Auclair M-C, Lechat P (1980) Interaction between parathyroid hormone and the beta-adrenoceptor system in cultured rat myocardial cells. J Mol Cell Cardiol 12:955–964

Laskowski MB, D'Agrosa LS (1983) The ultrastructure of the sinu-atrial node of the bat. Acta Anat 117:85–101

Lathrop DA, Bailey J-C (1977) Lack of electrical interaction between proximal bundle branches and subjacent muscle. J Applied Physiol 42(2):235–239

Lau YH, Robinson RB, Rosen MR, Bilezikian JP (1980) Subclassification of β-adrenergic receptors in cultured rat cardiac myoblasts and fibroblasts. Circ Res 47:41–48

Laurent P, Holmgren S, Nilsson S (1983) Nervous and humoral control of the fish heart: Structure and function. Comp Biochem Physiol 76A:525–542

Lazzara R, Yeh BK, Samet P (1974) Functional anatomy of the canine left bundle branch. Am J Cardiol 33:623–632

Lazarides E (1978) The distribution of desmin (100 Å) filaments in primary cultures of embryonic chick cardiac cells. Exp Cell Res 112:265–273

Lazarides E (1980) Intermediate filaments as mechanical integrators of cellular space. Nature 283:249–255

Lazarides E, Hubbard BD (1976) Immunological characterization of the subunit of the 100 Å filaments from muscle cells. Proc Natl Acad Sci USA 73:4344–4348

Lazarides E, Granger BL, Gard DL, O'Connor CM, Breckler J, Price M, Danto SI (1982) Desmin- and vimentin-containing filaments and their role in the assembly of the Z disc in muscle cells. Cold Spring Harbor Symposium on Quantitative Biology 46:351–378

Lazarus M, Colgan J, Sachs HG (1976) Quantitative light and electron microscopic comparison of the normal and cardiomyopathic Syrian hamster heart. J Mol Cell Cardiol 8:431–441

Leak LV (1967) The ultrastructure of myofibers in a reptilian heart: the Boa constrictor. Am J Anat 120:553–559

Leak LV (1969) Electron microscopy of cardiac tissue in a primitive vertebrate Myxine glutinosa. J Morphol 128:131–157

Leak LV, Burke JF (1964) The ultrastructure of human embryonic myocardium. Anat Rec 9:623–650

Leeson TS (1978) The transverse tubular (T) system of rat cardiac muscle fibers as demonstrated by tannic acid mordanting. Can J Zool 56:1906–1916

Leeson TS (1980) T-tubules, couplings and myofibrillar arrangements in rat atrial myocardium. Acta Anat 108:374–388

Leeson TS (1981) The fine structure of snake myocardium. Acta Anat 109:252–269

Le Furgey A, Ingram P, Henry SC, Murphy E, Lieberman M (1983) Three-dimensional configuration of the mitochondria in cultured heart cells. In: Scanning electron microscopy. AMF O'Hare, Chicago, p 293

Legato MJ (1972) Ultrastructural characteristics of the rat ventricular cell grown in tissue culture, with special reference to sarcomerogenesis. J Mol Cell Cardiol 4:299–317

Legato MJ (1973) The myocardial cell for the clinical cardiologist. Futura Publishing, New York

Legato MJ (1975) Ultrastructural changes during normal growth in the dog and rat ventricular myofiber. In: Lieberman M, Sano T (eds) Developmental and physiological correlates of cardiac muscle. Raven Press, New York, pp 249–274

Legato MJ (1979a) Cellular mechanisms of normal growth in the mammalian heart. I. Qualitative and quantitative features of ventricular architecture in the dog from birth to 5 months of age. Circ Res 44:250–262

Legato MJ (1979b) Cellular mechanisms of normal growth in the mammalian heart. II. Quantitative and qualitative comparison between the right and left ventricular myocytes in the dog from birth to 5 months of age. Circ Res 44:263–279

Legato MJ (1985) The developing heart, clinical implications of its molecular biology and physiology. Martinus Nijhoff publishing, Boston

Legato MJ, Spiro D, Langer GA (1968) Ultrastructural alterations produced in mammalian myocardium by variation in perfusate ionic composition. J Cell Biol 37:1–12

Lehr D (1966) The role of certain electrolytes and hormones in disseminated myocardial necrosis. In: Bajusz E (ed) Electrolyte and cardiovascular disease. Karger, Basel, p 248

Leknes IL (1981) Ultrastructure of myocardial widened Z bands and endocardial cells in two teleostean species. Cell Tissue Res 218:23–28

Lev M (1968) The conduction system. In: Gould SE (ed) Pathology of the heart and blood vessels. Thomas, Springfield, Ill, pp 185–220

Lev M (1972) Pathogenesis of congenital atrio ventricular block. Prog Cardiovasc Dis 15:145–157

Lev M, Thaemert JC (1973) The conduction system of the mouse heart. Acta Anat 85:342–352

Levi-Montalcini R, Meyer H, Hamburger V (1954) In vitro experiments on the effects of mouse sarcomas 180 and 37 on the spinal and sympathetic ganglia of the chick embryo. Cancer Res 14:49–57

Levin KR, Page E (1980) Quantitative studies on plasmalemmal folds and caveolae of rabbit ventricular myocardial cells. Circ Res 46:244–255

Levy MN, Martin PJ (1979) Neural control of the heart. In: Berne RM, Sperelakis N, Geiger SR (eds) Handbook of physiology, sect 2, The cardiovascular system, vol 1, The heart, chapter 16. Williams & Wilkins, Baltimore, pp 581–620

Levy MN, Martin PJ, Stuesse SL (1981) Neural regulation of the heart beat. Ann Rev Physiol 43:443–453

Lewartowski B, Bielecki (1963) The influence of hemicholinium No.3 and vagal stimulation on acetylcholine content of rabbit atria. J Pharmacol Exp Ther 142:24–30

Lewis SEM, Kelly FJ, Goldspink DF (1984) Pre- and post-natal growth and protein turnover in smooth muscle, heart and slow- and fast-twitch skeletal muscles of the rat. Biochem J 217:517–526

Lewis TF (1904) The questions of sinusoids. Anat Anz 25:261–279

Lewis WH (1926) Cultivation of embryonic heart-muscle. Carnegie Inst Wash Contrib Embryol No 90, XVIII:1–21

Liberthson RR, Szidon JP, Bharati S, Lev M, Fishman AP (1975) Haemodynamic consequences or decayed ventriculoconal conduction in the frog (Rana catesbeiana). Am J Physiol 229:1085–1093

Libby P (1984) Long-term culture of contractile mammalian heart cells in a defined serum-free medium that limits non-muscle cell proliferation. J Mol Cell Cardiol 16:803–811

Licata RH (1954) The embryonic heart in the ninth week. Am J Anat 94:73–125

Licata RH (1956) A continuation study of the development of the blood supply of the human heart. Anat Rec 124:362

Lichnovsky V, Obrucnik M, Kraus J (1978) A quantitative morphometric study of capillary length and ventricular volume and surface area in the human embryonic and foetal heart. Folia Morphol (Praha) 26:187–193

Lichnovsky V, Obrucnik M, Machan B (1982a) Ultrastructure of cellular elements of internodal connections of the stimulus conducting system in human embryonal and fetal hearts. Acta Universitatis Pakackianae Olomucensis Facultatis Medicae 102:29–38

Lichnovsky V, Machan B (1982b) Development of the conduction system of human embryonic and fetal heart: Differentiation of internodal connection. Acta Universitatis Palackianae Olomucensis Facultatis Medicae 102:39–46

Lieberman M, Roggeveen AE, Purdy JE, Johnson EA (1972) Synthetic strands of cardiac muscle: growth and physiological implication. Science 175:909–911

Lieberman M, Sawanobori T, Shigeto N, Johnson EA (1975) Physiologic implications of heart muscle in tissue culture. In: Lieberman M, Sano T (eds) Developmental and physiological correlates of cardiac muscle. Raven Press, New York, pp 139–154

Lieberman M, Horres CR, Shigeto N, Ebihara L, Aiton JF, Johnson EA (1981) Cardiac muscle with controlled geometry, Application to electrophysiological and ion transport studies. In: Nelson PG, Lieberman M (eds) Excitable cells in tissue culture. Plenum Press, New York, pp 379–408

Lignon JM (1979) Responses to sympathetic drugs in the ammocoete heart: Probable influence of the small intensely fluorescent (SIF) cells. J Mol Cell Cardiol 11:447–465

Lignon JM, Le Douarin G (1978) Small intensely fluorescent (SIF) cells and myocardial cells in the amocoete heart: A correlative histofluorescence, light and electrion microscopic study with special reference to the action of reserpine. Biol Cell 31:169–176

Lin HL, Katele KV, Grim AF (1977) Functional morphology of the pressure and the volume-hypertrophied rat heart. Circ Res 41:830–836

Lindner E, Schaumburg G (1968) Zytoplasmatische Filamente in den quergestreiften Muskelzellen des kaudalen Lymphherzens von *Rana temporaria* L. Untersuchungen am Lymphherzen. Z Zellforsch Mikrosk Anat 84:549–562

Linzbach AJ (1960) Heart failure from the point of view of quantitative anatomy. Am J Cardiol C5:370–382

Lipsius SL, Gibbons WR (1980) Acetylcholine lengthens action potential of sheep cardiac Purkinje fibres. Am J Physiol 238:H237–H243

Litten RZ, Martin BJ, Low RB, Alpert NR (1982) Altered myosin isozyme patterns from pressure-overloaded and thyrotoxic hypertrophied rabbit hearts. Circ Res 50:856–864

Lompré AM, Poggioli J, Vassort G (1979a) Maintenance of fast Na – channels during primary culture of embryonic chick heart cells. J Mol Cell Cardiol 11:813–825

Lompré AM, Schwartz K, D'Albis A, Lacombe G, Van Theim N, Swynghedauw B (1979b) Myosin isoenzyme redistribution in chronic heart overload. Nature 282:105–107

Lompré AM, Mescadier JJ, Wisnewsky C, Bouveret P, Pantalmi C, D'Albis A, Schwartz K (1981) Species and age-dependent changes in the relative amounts of cardiac myosin isoenzymes in mammals. Dev Biol 84:286–290

Long L, Fabian F, Mason DT, Wikman-Coeffelt J (1977) A new cardiac myosin characterized from the canine atria. Biochem Biophys Res Commun 76:626–635

Loud AV, Anversa P (1984) Biology of disease. Morphometric analysis of biologic processes. Lab Invest 50:250–261

Loud AV, Borany WC, Pack BA (1965) Quantitative evaluation of cytoplasmic structures in electron micrographs. Lab Invest 14:996–1008

Loud AV, Anversa P, Giacomelli F, Wiener J (1978) Absolute morphometric study of myocardial hypertrophy in experimental hypertension. I. Determination of myocyte size. Lab Invest 38:586–596

Loud AV, Olivetti G, Anversa P (1983) Methods in laboratory investigation. Morphometric measurement of cellular hypertrophy. Lab Invest 49:230–234

Loud AV, Behi C, Olivetti G, Anversa P (1984) Morphometry of left and right ventricular myocardium after strenuous exercise in preconditioned rats. Lab Invest 51:104–111

Lowenstein WR, Kanno Y, Socolar SJ (1978) The cell-to-cell channel. Fed Proc 37:2645–2650

Ludatscher RM (1968) Fine structure of the muscular wall of rat pulmonary vein. J Anat 103:345–357

Lund DD, Tomanek RJ (1978) Myocardial morphology in spontaneously hypertensive and aortic-constricted rats. Am J Anat 152:141–152

Lund DD, Schmid PG, Joannsen UJ, Roskoski R Jr (1982) Biochemical indices of cholinergic and adrenergic autonomic innervation in dog heart. Disparate alterations in chronic right heart failure. J Mol Cell Cardiol 14:419–426

Lund DD, Twietmeyer TA, Schmid PG, Tomanek RJ (1979) Independent changes in cardiac muscle fibres and connective tissue in rats with spontaneous hypertension, aortic constriction and hypoxia. Cardiovasc Res 13:39–44

Lundsgaard-Hansen P, Meyer C, Riedwyl H (1967) Transmural gradients of glycolytic enzyme activities in left ventricular myocardium. Pflüg Arch 297:89–106

Lutz BR (1930) The effect of adrenalin on the auricle of elasmobranch fishes. Am J Physiol 94:135–139

Macey DJ, Epple A, Potter IC, Hilliard RW (1984) The effect of catecholamines on branchial and cardiac electrical recordings in adult lampreys (*Geotria australia*). Comp Biochem Physiol 79C:295–300

Mackenzie I (1910) The nodal tissue of the vertebrate heart. J Pathol Bact 14:404–405

Mackenzie I (1913) The excitation and connecting muscular system of the heart. Transcripts of the 17th International Congress of Medicine, Section 1:121–150

MacLennan DH, Campbell KP, Reithmeier RA (1983) Calsequestrin. In: Cheung WY (ed) Calcium and cell function. Academic Press, New York, vol 4, pp 151–173

Maekawa M, Nohara Y, Kawamura K, Hayashi K (1967) Electron microscope study of the conduction system in mammalian hearts. In: Sano T, Mitzuhira V, Matsuda K (eds) Electrophysiology and ultrastructure of the heart. Bunkodo, Tokyo, pp 41–54

Mains RE, Patterson PH (1973) Primary cultures of dissociated sympathetic neurons. I. Establishment of long-term growth in culture and studies of differentiated properties. J Cell Biol 59:329–345

Mair WG, Tome FM (1972) The ultrastructure of the adult and developing human myotendinous junction. Acta Neuropathol (Berlin) 21:239–252

Maki M, Takayanagi R, Misono KS, Pandey KN, Tibbetts C, Inagami T (1984) Structure of rat atrial natriuretic factor precursor deduced for cDNA sequence. Nature 309:722–724

Makowski L, Caspar DLD, Phillips WC, Goodenough DA (1977) Gap junction structures. II. Analysis of the X-ray diffraction data. J Cell Biol 74:629–645

Malkoff DB, Strehler BL (1963) The ultrastructure of isolated and *in situ* human cardiac age pigment. J Cell Biol 16:611–616

Mall FP (1912) On the development of the human heart. Am J Anat 13:249–298

Mall G, Reinhard H, Stopp D, Rossner A (1980) Morphometric observations on the rat heart after high-dose treatment with cortisol. Virchows Arch (Pathol Anat) 385:169–180

Mall G, Mattfelot T, Rieger P, Volk B, Frolov VA (1982) Morphometric analysis of the rabbit myocardium after chronic ethanol feeding – early capillary changes. Basic Res Cardiol 77:57–67

Mallorga P, Tallman JF, Henneberry RC, Hirata F, Strittmatter WT, Axelrod J (1980) Mepacrine blocks β-adrenergic agonist-induced desensitization in astrocytoma cells. Proc Natl Acad Sci USA 77:1341–1345

Mallov S, McKibbin JM, Ross JS (1953) The distribution of some of the essential lipids in beef heart muscle and conducting tissue. J Biol Chem 201:825–838

Malouf NN, Meissner G (1980) Cytochemical localization of a basic ATPase to canine myocardial surface membrane. J Histochem Cytochem 28:1286–1294

Manasek FJ (1968a) Embryonic development of the heart. I. A light and electron microscope study of myocardial development in the early chick embryo. J Morph 125:329–366

Manasek FJ (1968b) Mitosis in developing cardiac muscle. J Cell Biol 37:191–6

Manasek FJ (1969a) Myocardial cell death in the embryonic chick ventricle. J Embryol Exp Morphol 21:271–284

Manasek FJ (1969b) Embryonic development of the heart. II. Formation of the epicardium. J Embryol Exp Morphol 22:333–348

Manasek FJ (1970) Histogenesis of the embryonic myocardium. Am J Cardiol 25:149–168

Manasek FJ (1971) The ultrastructure of embryonic myocardial blood vessels. Dev Biol 26:42–54

Manasek FJ (1975) The extracellular matrix: A dynamic component of the developing embryo. Curr Top Dev Biol 10:35–102

Manasek FJ (1976) Macromolecules of the extracellular compartment of embryonic and mature heart. Circ Res 38:331–337

Manasek FJ (1980) Organisation, interactions and environment of heart cells during myocardial ontogeny. In: Berne RM, Sperelakis N, Geiger SR (eds) Handbook of physiology, sect 2, The cardiovascular system, vol 1, The heart. Williams & Wilkins, Baltimore, pp 29–42

Manasek FJ (1981) Determinants of heart shape in early embryos. Fed Proc 40:2011–2016

Manasek FJ, Monroe RG (1972) Early cardiac morphogenesis is independent of function. Dev Biol 27(4):584–588

Manasek FJ, Kulikowski RR, Fitzpatrick L (1978) Cytodifferentiation: A causal antecedent of looping. Birth defects: Original articles series, vol XIV:161–178

Manasek FJ, Kulikowski RR, Nakamura A, Nguyenphuc Q, Lacktis JW (1984) Early heart development: A new model of cardiac morphogenesis. In: Zak R (ed) Growth of the heart in health and disease. Raven Press, New York, pp 105–130

Manger WM (1982) Catecholamines in normal and abnormal cardiac function. In: Kellermann JJ (ed) Advances in cardiology, vol 30. Karger, Basel München, pp 1–151

Manjunath CK, Goings GE, Page E (1984) Cytoplasmic surface and intramembrane components of rat heart gap junctional proteins. Am J Physiol 246: H865–H875

Marie JP, Giullemot H, Hatt PY (1976) Le degré de granulation des cardiocytes auriculaires. Etude planimétrique au cours de différents apports d'eau et de sodium chez le rat. Path Biol 24: 549–554

Marino TA (1979) The atrioventricular node and bundle in the ferret heart. A light and quantitative electron microscopic study. Am J Anat 154: 365–392

Marino TA, Severdia JB (1982) Fine structural examination of the single heart tube in the fifteen day ferret embryo. Cell Tissue Res 221: 597–605

Marino TA, Severdia JB (1983) The early development of the AV node and bundle in the ferret heart. Am J Anat 167: 299–312

Marino TA, Truex RC, Marino DR (1979) The development of the atrioventricular node and bundle in the ferret heart. Am J Anat 154: 135–150

Marino TA, Biberstein D, Severdia JB (1981) The ultrastructure of the atrioventricular junctional tissues in the newborn ferret heart. Am J Anat 164(4): 383–392

Marino TA, Houser SR, Martin FG, Freeman AR (1983) An ultrastructural morphometric study of the papillary muscle of the right ventricle of the cat. Cell Tissue Res 230: 543–552

Mark GE, Strasser FF (1966) Pacemaker activity and mitosis in cultures of newborn rat heart ventricle cells. Exp Cell Res 44: 217–233

Mark GE, Chamley JH, Burnstock G (1973) Interactions between autonomic nerves and smooth and cardiac muscle cells in tissue culture. Dev Biol 32: 194–200

Markwald RR (1973) Distribution and relationship of precursor Z-material to organizing myofibrillar bundles in embryonic rat and hamster ventricular myocytes. J Mol Cell Cardiol 5: 341–350

Maron BJ, Ferrans VJ, Roberts WC (1975) Ultrastructural features of degenerated cardiac muscle cells in patients with cardiac hypertrophy. Am J Pathol 79: 387–434

Maron BJ, Ferrans VJ (1978) Ultrastructural features of hypertrophied human ventricular myocardium. Prog Cardiovasc Dis 21: 207–238

Maron BJ, Verter J, Kapur S (1978) Disproportionate ventricular septal thickening in the developing normal human heart. Circulation 57: 520–526

Marshall PN, Galbraith W (1984) The calculation of differences in the colours of histological objects. J Histochem 16: 211–218

Martin AF, Haithcoat JL, Dowell RT (1983) Redistribution of ventricular myosin isoenzymes in the neonatal and adult rat heart in response to a chronic pressure overload, in: Alpert NR (ed) Perspectives in cardiovascular research, vol 17. Myocardial hpyertrophy and failure. Raven Press, New York, pp 359–371

Martin AF, Pagani ED, Solaro RJ (1982) Thyroxine-induced redistribution of isoenzymes of rabbit ventricular myosin. Circ Res 50: 117–124

Martinez TT, Walker MJA (1977) The actions of ouabain, dinitrophenol, anoxia and ionic manipulation on cultured heart cells. Proc West Pharmacol Soc 20: 269–273

Martinez-Palomo A (1970) The surface coats of animal cells. Int Rev Cytol 29: 29–75

Martinez-Palomo A, Alanis J (1980) The amphibian and reptilian hearts: Impulse propagation and ultrastructure. In: Bourne GH (ed) Hearts and heart-like organs, vol 1. Academic Press, New York, pp 171–197

Martinez-Palomo A, Alanis J, Benitez D (1970) Transitional cardiac cells of the conductive system of the dog heart. J Cell Biol 47: 1–17

Martinosi AN (1984) Mechanism of Ca^{2+} release from sarcoplasmic reticulum of skeletal muscle. Physiol Rev 64: 1240–1320

Marvin WJ Jr, Atkins DL, Chittick VL, Lund DD, Hermsmeyer K (1984a) *In Vitro* adrenergic and cholinergic innervation of the developing rat myocyte. Circ Res 55: 49–58

Marvin WJ Jr, Chittick VL, Rosenthal JK, Sandra A, Atkins DL, Hermsmeyer K (1984b) The isolated sinoatrial node cell in primary culture from the newborn rat. Circ Res 55: 253–260

Masson-Pévet M (1979) The fine structure of cardiac pacemaker cells in the sinus node and in tissue culture. Thèse de doctorat en Physiologie, Université d'Amsterdam

Masson-Pévet M, Bleeker WK, Mackaay AJC, Gros D, Bouman LN (1978) Ultrastructural and functional aspects of the rabbit sinoatrial node. In: Bonke FIM (ed) The sinus node, chapter 16. Martinus Nijhoff, The Hague, pp 195–211

Masson-Pévet M, Bleeker WK, Mackaay AJC, Bouman LN, Houtkooper JM (1979a) Sinus node and atrium cells from the rabbit heart: A quantitative EM description after electrophysiological localisation. J Molec Cell Cardiol 11:555–568

Masson-Pévet M, Bleeker WK, Gros D (1979b) The plasma membrane of leading pacemaker cells in the rabbit sinus node. Circ Res 45:621–629

Masson-Pévet M, Gros D, Besselsen E (1980) The caveolae in rabbit sinus node and atrium. Cell Tissue Res 208:183–196

Mathur PN, Hurkat PC (1973) The heart and the conducting system of *Gavialis gangeticus* (Gamelin). Acta Morphol Acad Sci Hung 21:137–148

Matsubara I (1980) X-ray diffraction studies of the heart. Ann Rev Biophys Bioeng 9:81–105

Maylie JG (1982) Excitation-contraction coupling in neonatal and adult myocardium of cat. Am J Physiol 242:H834–843

Mazet F (1977) Freeze-fracture studies of gap junctions in the developing and adult amphibian cardiac muscle. Dev Biol 60:139–152

Mazet F, Cartaud J (1976) Freeze-fracture studies of frog atrial fibres. J Cell Sci 22:427–434

McBride RE, Moore GW, Hutchins GM (1981) Development of the outflow tract and closure of the interventricular septum in the normal human heart. Am J Anat 160:309–332

McCall D (1979) Cation exchange and glycoside binding in cultured rat heart cells. Am J Physiol 236:C87–C95

McCallion DJ, Wong WT (1956) A study of the localisation and distribution of glycogen in early stages of the chick embryo. Can J Zool 34:63

McCallister LP, Daiello DC, Tyers GFO (1978) Morphometric observations of the effects of normothermic ischemic arrest on dog myocardial ultrastructure. J Mol Cell Cardiol 10:67–80

McCarl RL, Hartzell CR, Schroedl N, Kunze E, Ross PD (1980) Mammalian heart cells in culture. In: Bourne GH (ed) Hearts and heart-like organs, vol 3. Pathology and surgery of the heart. Academic Press, New York London Toronto Sydney San Francisco, pp 1–44

McDonnell TJ, Oberpriller JO (1983a) The ultrastructure of the atrium in the adult newt *Notophthalmus viridescens* (Amphibia, Salamandridae). J Morph 175:235–251

McDonnell TJ, Oberpriller JO (1983b) The atrial proliferative response following partial ventricular amputation in the heart of the adult newt. A light and electron microscopic autoradiographic sudy. Tissue Cell 15:351–363

McDonnell TJ, Oberpriller JO (1984) The response of the atrium to direct mechanical wounding in the heart of the adult newt, *Notophthalmus viridescens*. An electron microscopic and autoradiographic study. Cell Tissue Res 235:583–592

McKenna N, Meigs JB, Wang Y-L (1985) Identical distribution of fluorescently labelled brain and muscle actins in living cardiac fibroblasts and myocytes. J Cell Biol 100:292–296

McLean MJ, Sperelakis N (1974) Rapid loss of sensitivity to tetrodotoxin by chick ventricular myocardial cells after separation from the heart. Exp Cell Res 86:351–364

McLean MJ, Sperelakis N (1976) Retention of fully differentiated electrophysiological properties of chick embryonic heart cells in culture. Dev Biol 50:134–141

McLean MJ, Lapsley RA, Shigenobu K, Murad F, Sperelakis N (1975) High cyclic AMP levels in young chick embryonic hearts. Dev Biol 42:196–201

McMahan UJ, Kuffler SW (1971) Visual identification of synaptic boutons on living ganglion cells and of varicosities in post ganglionic axons in the heart of the frog. Proc R Soc Ser B (Lond) 177:485–508

McMinn (1969) Tissue repair. Academic Press, New York

McNutt NS (1970) Ultrastructure of intercellular junctions in adult and developing cardiac muscle. Am J Cardiol 25:169–183

McNutt NS (1975) Ultrastructure of the myocardial sarcolemma. Circ Res 37:1–13

McNutt NA, Fawcett DW (1969) The ultrastructure of the cat myocardium. 2. Atrial muscle. J Cell Biol 42:46–67

McNutt NS, Fawcett DW (1974) Myocardial ultrastructure. In: Langer GA, Brady AJ (eds) The mammalian myocardium. Wiley, New York, pp 1–49

McNutt NS, Weinstein RS (1970) The ultrastructure of the nexus: A correlated thin-section and freeze-cleave study. J Cell Biol 47:666–688

McWilliam JA (1884) On a number of facts concerning the reflex inhibition of the eel's heart. J Physiol 5:19p–23p

McWilliam JA (1885a) On the structure and rhythm of the heart in fishes, with special reference to the heart of the eel. J Physiol 6:192–244

McWilliam JA (1885b) Cardiac inhibition in the newt. J Physiol 6:16P–17P

Mead RH, Clusin WT (1984) Origin of the background sodium current and effects of sodium removal in cultured embryonic cardiac cells. Circ Res 55:67–77

Meek WJ, Eyster JAE (1914) The effect of vagal stimulation and of cooling on the location of the pacemaker within the sinoauricular node. Am J Physiol 34:368–383

Meersin F (1969) The myocardium in hyperfunction, hypertrophy and heart failure. Circ Res 25 (Suppl 2):1–163

Meerson FZ, Zaletayeva TA, Lagutcher SS, Pshennikova MG (1964) Structure and mass of mitochondria in the process of compensatory hyperfunction and hypertrophy of the heart. Exp Cell Res 36:568–578

Meijer AEFH, De Vries GP (1978) Enzyme histochemical studies on the Purkinje fibres of the atrioventricular system of the bovine and porcine hearts. Histochem J 10:399–408

Meijer AEFH, Stegehuis F (1980) Histochemical technique for the demonstration of phosphofructokinase activity in heart and skeletal muscles. Histochemistry 6:75–81

Melax H, Leeson TS (1969) Fine structure of developing and adult intercalated discs in rat heart. Cardiovasc Res 3:261–267

Melax H, Leeson TS (1972) Electron microscope study of myocardial tissue space contents of the rat heart. Cardiovasc Res 6:89–94

Mercadier J-J, Lompré AM, Wisnewsky C, Samuel JL, Bercovici J, Swyghedauw B, Schwartz K (1981) Myosin isoenzymic changes in several models of rat cardiac hypertrophy. Circ Res 49:525–532

Mercadier J-J, Bouveret P, Gorza L, Schiaffino S, Clark WA, Zak R, Swynghedauw B, Schwartz K (1983) Myosin isoenzymes in normal and hypertrophied human ventricular myocardium. Circ Res 53:52–62

Merrillees NCR (1968) The nervous environment of individual smooth muscle cells of the guinea pig vas deferens. J Cell Biol 37:794–817

Merrillees NCR (1974) The fine structure of the sinus node in the rat. Adv Cardiol 12:34–44

Meyer H, Queiroga LT (1961) Electron microscope study of the developing heart muscle cell in thin sections of chick-embryo tissue cultures. In: Carvalho AP, Dello WC, Hoffman B (eds) The specialized tissues of the heart. Elsevier, Amsterdam, pp 76–79

Michailow S (1908) Die Nerven des Endocardiums. Anat Anz 32:87–101

Miller MR, Kasahara M (1964) Studies on the nerve endings in the heart. Am J Anat 115:217–234

Mills TW (1884) Some observations on the influence of the vagus and accelerators on the heart of the turtle. J Physiol 5:359–361

Mills TW (1885) The innervation of the heart of the slider terrapin (*Pseudemys rugosa*). J Physiol (London) 6:248–286

Minot CS (1900) On a hitherto unrecognized form of blood circulation without capillaries in the organs of vertebrata. Proc Boston Soc Nat Hist 29:185–215

Mizuhira V, Hirakow R, Ozawa H (1967) Fine structure of Purkinje fibers in the avian heart. In: Sano T, Mizuhira V, Matsuda K (eds) Electrophysiology and ultrastructure of the heart. Grune & Stratton, Bunkodo, Tokyo, New York, pp 12–26

Mobley BA, Page E (1972) The surface area of sheep cardiac Purkinje fibres. J Physiol (Lond) 220:547–563

Mochet M, Moravec J, Guillemot H, Hatt PY (1975) The ultrastructure of the rat conductive tissue; An electron microscope study of the atrioventricular node and the bundle of His. J Mol Cell Cardiol 7:879–889

Mohamed SN, Holmes R, Hartzell CR (1983) A serum-free, chemically-defined medium for function and growth of primary neonatal rat heart cell cultures. In Vitro 19:471–478

Mohsen T, Anthonioz PH, Jadoun G (1976) Sur la présence dans le cœur de *Protopterus annectens* (poisson dipneuste) de cellules nodales et conductrices histologiquement distinctes des cellules myocardiques ordinaires. CR Séanc Soc Biol Ouest Africa 170:712–715

Mönkeberg JG (1910) Beiträge zur normalen und pathologischen Anatomie des Herzens. Verh Dtsch Path Ges 14:64–71

Moore GW, Hutchins GM, Bulkley BH, Tseng JJ, Ki PF (1980) Constituents of the human ventricular myocardium: Connective tissue hyperplasia accompanying muscular hypertrophy. Am Heart J 100:610–616

Moravec M, Moravec J (1982) Presence of mechanoreceptors in the atrioventricular junction of the rat heart: Microanatomical and ultrastructural evidences. J Ultrastruct Res 81:47-65

Moravec M, Moravec J (1984) Intrinsic innervation of the atrioventricular junction of the rat heart. Am J Anat 171:307–319

Moravec-Mochet M, Moravec J, Hatt PY (1977) Presence of synaptic and muscular spindle-like structures in the atrioventricular junction of the rat heart: An electron microscopic study. J Ultrastruct Res 58:196–209

Moravec-Mochet M, Moravec J, Hatt PY (1978) Morphological evidence for neural control of the activity of rat heart pacemaker. In: Kobayashi T, Ito Y, Rano G (eds) Recent advances in studies on cardiac structure and metabolism, vol 12, Cardiac Adaptation. University Park Press, Baltimore, pp 71–75

Mori M (1955) The atrioventricular connecting system of the crocodile heart. Kyushu Mem Med Sci 5:199–205

Moses RL, Claycomb WC (1982) Ultrastructure of terminally differentiated adult rat cardiac muscle cells in culture. Am J Anat 164:113–131

Moses RL, Claycomb WC (1984) Ultrastructure of cultured atrial cardiac muscle cells from adult rats. Am J Anat 171:191–206

Motte G, Guillevin L, Dessouter P, Fruchaud J (1978) Nodoventricular accessory pathway unmasked by a block in the trunk of the bundle of His. Arch Mal Cœur 71:277–234

Mowry RW, Bangle R Jr (1951) Histochemically demonstrable glycogen in the human heart, with special reference to glycogen storage disease and diabetes mellitus. Am J Path 27:611–625

Mubagwa K, Carmeliet E (1983) Effects of acetylcholine on electrophysiological properties of rabbit cardiac Purkinje fibers. Circ Res 53:740–751

Muir AR (1951) The development of the sinuatrial node in the heart of the sheep. J Anat 85:430

Muir AR (1954) The development of the ventricular part of the conducting tissue in the heart of the sheep. J Anat (Lond) 88:381–191

Muir AR (1955) The sinuatrial node of the rat heart. Quart J Exp Physiol 40:378–386

Muir AR (1957a) An electron microscope study of the embryology of the intercalated disc in the heart of the rabbit. J Biochem Biophys Cytol 3:193–202

Muir AR (1957b) Observations on the fine structure of the Purkinje fibres in the ventricles of the sheep heart. J Anat 91(2):251–258

Muir AR (1965) Further observations on the cellular structure of cardiac muscle. J Anat 99:27–46

Mukherjee C, Caron MG, Lefkowitz RJ (1975) Catecholamine-induced subsensitivity of adenylate cyclase associated with loss of β-adrenergic receptor binding sites. Proc Natl Acad Sci USA 72:1945–1949

Mukudai T, Wada T, Kanda S, Otsuka N (1978) Histochemical studies in the conduction system of the mammalian heart. Acta Histochem Cytochem 12:191

Mukudai T (1980) Histochemical study of the heart conduction system. Okayama Igakkai Zasshi (in Japanese) 92:635–702

Mukuno K (1966) The fine structure of the human extraocular muscles. (1) A "laminated structure" in the muscle fibers (in Japanese). J Electron Microsc 15:227–236

Munnell JF (1982) Sensory components in the terminal innervation of the ovine cardiac conduction system. Am J Anat 163:337–350

Munnell JF, Getty R (1968a) Canine myocardial Z-disc alterations resembling those of nemaline myopathy. Lab Invest 19:303–308

Munnell JF, Getty R (1968b) Nuclear lobulation and amitotic division associated with increasing cell size in the aging canine myocardium. J Gerontol 23:363–369

Muntz KH, Gonyea WJ, Mitchell JH (1981) Cardiac hypertrophy in response to an isometric training program in the cat. Circ Res 49:1092–1101

Murakami T, Saito I, Mochizuki K (1981) The sinoatrial node of the avian heart. Exp Anim (Tokyo) 30:263–268

Murphy E, Aiton JF, Horres CR, Lieberman M (1983) Calcium elevation in cultured heart cells: its role in cell injury. Am J Physiol 245:C316–C321

Murray JB (1954) Oxygen uptake of atrioventricular conducting tissue of beef heart. AM J Physiol 177:463–466

Murray MR (1965) Muscle. In: Willmer EN (ed) Cells and tissues in culture, vol 2, Methods, biology and physiology. Academic Press, New York, pp 311–372

Myerburg RJ, Nilsson K, Gelband H (1972) Physiology of canine intraventricular conduction and endocardial excitation. Circ Res 30:217–43

Myklebust R, Jensen H (1978) Leptomere fibrils and T-tubule desmosomes in the Z-band region of the mouse heart papillary muscle. Cell Tissue Res 188:205–215

Myklebust R, Dalen H, Sætersdal TS (1975) A comparative study in the transmission and scanning electron microscope of intracellular structures in sheep heart muscle tissue. J Microsc 105:57–65

Myklebust R, Sætersdal TS, Engedal H (1978a) The T tubule system in the myocardia of the sand rat and mouse as demonstrated by horseradish peroxidase. Cell Tissue Res 192:205–213

Myklebust R, Sætersdal TS, Engedal H, Ultstein M, Degarden S (1978b) Ultrastructural studies on the formation of myofilaments and myofibrils in the human embryonic and adult hypertrophied heart. Anat Embryol (Berl) 152:127–140

Nag AC (1980) Study of non-muscle cells of the adult mammalian heart: a fine structural analysis and distribution. Cytobios 28:41–61

Nag AC, Cheng M (1980) Intercellular adhesion: Co-construction of contractile heart tissue by cells of different species. Sience 208:1150–1152

Nag AC, Carey TR, Cheng M (1983a) DNA synthesis in rat heart cells after injury and the regeneration of myocardia. Tissue Cell 15:597–613

Nag AC, Cheng M, Fischman DA, Zak R (1983b) Long-term cell culture of adult mammalian cardiac myocytes: Electron microscopic and immunofluorescent analyses of myofibrillar structure. J Mol Cell Cardiol 15:301–317

Nagao K, Toyama J, Kodama I, Yanada K (1981) Role of the conduction system in endocardial excitation spread in the right ventricle. Am J Cardiol 48:864–870

Nair MGK (1970) The anatomy and embryology of the heart and its conducting system of the dog fish Carcharias sorrah (Mull and Henle). Zool Anz 185:165–273

Nakao T, Susuki S, Saito M (1981) An electron microscopic study of the cardiac innervation in larval lamprey. Anat Rec 199:555–563

Nakata K (1977) Quantitative analysis of ultrastructural changes in developing rat cardiac muscle during normal growth and during acute volume load. Jpn Circ J 41:1238–1250

Nath K, Bollon AP (1978) Effect of dibutyryl cAMP and analogs on the rate of contractions of myocytes in culture. Experientia 34:1282–1283

Nath K, Shay JW, Bollon AP (1978) Relationship between dibutyryl cyclic AMP and microtubule organization in contracting heart muscle cells. Proc Natl Acad Sci USA 75:319–323

Nathan H, Gloobe H (1970) Myocardial atrio-venous junctions and extensions (sleeves) over the pulmonary and caval veins. Thorax 25:317–324

Nathan RD (1981) Aggregates of fetal rat heart cells: electrophysiology and tetrodotoxin sensitivity. J Mol Cell Cardiol 13:241–249

Nathan RD, De Haan RL (1978) In vitro differentiation of a fast Na conductance in embryonic heart cell aggregates. Proc Natl Acad Sci USA 75:2776–2780

Navaratnam V (1980) Anatomy of the mammalian heart. In: Bourne GH (ed) Hearts and heart-like organs, vol 1. Comparative anatomy and development. Academic Press, New York, pp 349–375

Navaratnam V (1965) Development of the nerve supply to the human heart. Br Heart J 27:640–650

Nayler WG, Dresel PE (1984) Ca and the sarcoplasmic reticulum. J Mol Cell Cardiol 16:165–174

Nayler WG, Merrillees NCR (1964) Some observations on the fine structure and metabolic activity of normal and glycerinated ventricular muscle of toad. J Cell Biol 22:533–550

Nayler WG, Poole-Wilson PA, Williams A (1979) Hypoxia and calcium. J Mol Cell Cardiol 11:683–706

Neffgen JF, Korecky B (1972) Cellular hyperplasia and hypertrophy in cardiomegalies induced by anemia in young and adult rats. Circ Res 30:104–113

Nettleship GA (1936) Experimental studies on the afferent innervation of the cat's heart. J Comp Neur 64:115–133

Nickerson M, Nomaguchi GM (1950) Blockade of epinephrine-induced cardioacceleration in the frog. Am J Physiol 163:484–504

Nicol JAC (1952) The autonomic nervous system in lower chordates. Biol Rev (Cambridge) 27:1–49

Nielson KC, Owman CL (1967) Differential amine depletion from cardiac adrenergic nerves by segontin. Experientia 23:203–204

Nielson KC, Owman CL (1968) Difference in cardiac adrenergic innervation between hibernators and non-hibernating mammals. Acta Physiol Scand Suppl 316:1–30

Nilsson JF, Sporrong G (1970) Electron microscopic investigations of adrenergic and non-adrenergic axons in the rabbit SA node. Z Zellforsch Mikrosk Anat 3:404–412

Nonidez JF (1937) Identification of the receptor areas in the venae cavae, and pulmonary veins which initiate reflex cardiac acceleration (Bainbridge's reflex). Am J Anat 61:203–231

Nonidez JF (1941) Studies on the innervation of the heart. II. Afferent nerve endings in the large arteries and veins. Am J Anat 68:151–189

Novi AM (1968) An electron microscopic study of the innervation of papillary muscles in the rat. Anat Rec 160:123–142

Núñez-Durán H (1981) Electron microscopic study of the sarcolemma of Purkinje cells of the goat heart. Acta Anat 109:19–24

Núñez-Durán H, Peón J, Bárcenas L, Ubaldo E (1983) Site of cellular uncoupling in injured cardiac tissue in the dog. Acta Anat 115:204–211

Oberpriller J, Oberpriller JC (1971) Mitosis in adult newt ventricle. J Cell Biol 49:560–563

Oberpriller JO, Oberpriller JC (1974) Response of the adult newt ventricle to injury. J Exp Zool 187:249–260

Obrucnik M, Lichnovsky V (1977) Atrial ultrastructure of the human embryonic and foetal heart. Folia Morphol (Praha) 25:357–359

Obrucnik M, Malinsky J, Lichnovsky V (1972) The early stages of differentiation of the vascular bed in the ventricular wall of the human embryonic heart as seen in the electron microscope. Folia Morphol (Praha) 20:49–51

Oets J (1950) Electrocardiograms of fishes. Physiol Comp Ecol 2:181–186

Ohyumi M (1975) Histochemical and ultrastructural studies of enzymes, particulary in carbohydrate metabolism of the bovine heart conduction system. Acta Histochem Cytochem 8(3):164–174

Oishi H (1967) Manner of stimulus conduction in atria. Jpn Heart J 8:276–290

Okada R (1984) Histology of the human conduction system. In: Abe H, Ito Y, Tada M, Opie LH (eds) Regulation of cardiac function. Japan Science Societies Press, Tokyo/VNU Science Press, Utrecht, pp 117–126

Okamato N, Satow Y (1976) Cell death in bulbar cushion of normal and abnormal developing heart. In: Lieberman M, Sano T (eds) Developmental and physiological correlates of cardiac muscle. Raven Press, New York, pp 51–66

Oliphant LW, Loewen RD (1976) Filament systems in Purkinje cells of the sheep heart: Possible alterations in myofibrillogenesis. J Mol Cell Cardiol 8:679–688

Olivetti G, Anversa P, Loud AV (1980) Morphometric study of early postnatal development in the left and right ventricular myocardium of the rat. II. Tissue composition, capillary growth and sarcoplasmic alterations. Circ Res 46:503–512

Opie LH (1969) Metabolism of the heart in health and disease. Part II. Am Heart J 77:100–122

O'Rahilly R (1971) The timing and sequence of events in human cardiogenesis. Acta Anat 79:70–75

Orts-Llorca F, Fonolla JP, Sobrado J (1982) The formation, septation and fate of the truncus arteriosus in man. J Anat 134:41–56

Oshima T, Currie MG, Geller DM, Needleman P (1984) An atrial peptide is a potent renal vasodilator substance. Circ Res 54:612–616

Ostadal B, Rychter Z, Poupa O (1968) Qualitative development of the terminal vascular bed in the perinatal period of the rat. Folia Morphol (Praha) 16:116–123

Ostadal B, Schiebler TH (1971) Terminal blood bed in the heart of the turtle (Testudo hermanni): Z Anat Entwicklungsgesch 134:111–116

Ostadal B, Schiebler TH, Rychter Z (1975) Relations between development of the capillary wall and myoarchitecture of the rat heart. Adv Exp Med Biol 53:375–388

Ostlund E (1954) The distribution of catechol amines in lower animals and their effect on the heart. Acta Physiol Scand 31(Suppl 112):1–67

Ostlund E, Siegman M, Bloom G, Nordenstam H, Adams-Ray J, Lishajko F, Ritzen M, Van Euler US (1960) Storage and release of catecholamines and the occurrence of a specific submicroscopic granulation in hearts of cyclostomes. Nature 188:324–325

Otsuka N, Tomisawa M (1969) Fluorescence microscopy of the catecholamine-containing nerve fibers in vertebrate hearts (in Japanese). Acta Anat Nippon (Tokyo) 44:1–6

Otsuka N, Hara T, Okamoto H (1967) Histotopochemische Untersuchungen am Reizleitungssystem des Hundeherzens. Histochemie 10:66–73

Otsuka N, Chihara J, Sakurada H, Kanda S (1977) Catecholamine-storing cells in the cyclostome heart. Arch Histol Jap 40(Suppl):241–244

Otsuka N, Wada T, Mukudai T, Kanda S, Sasaki J (1979) Embryologic and histochemical study of the conduction system of the heart in rats. Acta Histochem Cytochem 12:590

Overy HR, Priest RE (1966) Mitotic cell division in postnatal cardiac growth. Lab Invest 15:1100–1103

Owman C, Sjöberg N-O, Swedin G (1971) Histochemical and chemical studies on pre- and postnatal development of the different systems of "short" and "long" adrenergic neurons in peripheral organs of the rat. Z Zellforsch Mikrosk Anat 116:319–341

Padieu P, Frelin C , Pinson A, Charbonne F, Athias P (1978) Effect of environmental factors and tissue culture methodology in producing and studying cultured cardiac cells. In: Kobayashi T, Ito Y, Rona G (eds) Recent advances in studies on cardiac structure and metabolism, vol 12, Cardiac adaptation. University Park Press, Baltimore, pp 609–620

Paes de Almeida O, Bohm GM, De Parla Carvalho M, Paes de Carvalho A (1974) The cardiac muscle in the pulmonary vein of the rat: A morphological and electrophysiological study. J Morph 145:409–434

Paes de Carvalho A, De Mello WC, Hoffman BF (1959) Electrophysiological evidence for specialised fiber types in rabbit atrium. Am J Physiol 196:483–488

Paff GH (1935) Conclusive evidence for sino-atrial dominance in isolated 48-hour embryonic chick hearts cultivated in vitro. Anat Rec 63:203–210

Page E (1967) Tubular systems in Purkinje cells of the cat heart. J Ultrastruct Res 17:72–83

Page E (1978) Quantitative ultrastructural analysis in cardiac membrane physiology. Am J Physiol 235:C147–C158

Page E, Buecker JL (1981) Development of dyadic junctional complexes between sarcoplasmic reticulum and plasmalemma in rabbit left ventricular myocardial cells. Circ Res 48:519–522

Page E, Fozzard HA (1973) Capacitive, resistive and syncytial properties of heart muscle – Ultrastructural and physiological considerations. In: Bourne GH (ed) The structure and function of muscle, Structure, vol II, 2nd edn. Academic Press, New York, pp 91–158

Page E, McCallister LP (1973a) Studies on the intercalated disc of rat left ventricular myocardial cells. J Ultrastruct Res 43:388–411

Page E, McCallister LP (1973b) Quantitative electron microscopic description of heart muscle cells: Application to normal, hypertrophied and thyroxin-stimulated hearts. Am J Cardiol 31:172–181

Page E, Shibata Y (1981) Permeable junctions between cardiac cells. Ann Rev Physiol 43:431–441

Page E, Upshaw-Earley J (1977) Volume changes in sarcoplasmic reticulum of rat hearts perfused with hypertonic solutions. Circ Res 40:355–366

Page E, Power B, Fozzard HA, Meddoff DA (1969) Sarcolemmal evaginations with knob-like or stalked projections in Pukinje fibres of the sheep heart. J Ultrastruct Res 28:288–300

Page E, McCallister LP, Power B (1971) Stereological measurements of cardiac ultrastructures implicated in excitation – contraction coupling sarcotubules and T-system. Proc Natl Acad Sci USA 68:1465–1466

Page E, Karrison T, Upshaw-Earley J (1983) Freeze-fractured cardiac gap junctions: Structural analysis by three methods. Am J Physiol 244:H525–H539

Paiment JM, McMillan DB (1975) The extracardiac chromaffin cells of larval lampreys. Gen Comp Endocrinol 27:495–508

Pannese E (1955) Observazioni istochimiche sull'apparecchio di conduzione cardiaco. Acta Histochem 2:102–201

Pannese E (1969) Unusual membrane-particle complexes within nerve cells of the spinal ganglia. J Ultrastruct Res 29:334–342

Pardo JV, D'Angelo Siliciano J, Craig SW (1983) Vinculin is a component of an extensive network of myofibril-sarcolemma attachment regions in cardiac muscle fibers. J Cell Biol 97:1081–1088

Paris S, Samuel D, Romey G, Ailhaud G (1979) Uptake of fatty acids by cultured cardiac cells from chick embryo: evidence for a facilitation process without energy dependence. Biochimie 61:361–367

Pathak CL, Goyal S (1978) Phosphorylase activity as a histochemical marker of specialised tissue of heart. Histochem J 10(6):633–640

Patten BM (1949) Initiation and early changes in the character of the heart beat in vertebrate embryos. Physiol Rev 29:31–47

Patten BM (1953) The development of the heart. In: Gould SB (ed) The Pathology of the heart, chapter II. Thomas, Springfield, Ill. pp 20–90

Patten BM (1956) The development of the sino-ventricular conduction system. Univ Mich Med Bull 22:1–21

Patten BM, Kramer TC (1933) The initiation of contraction in the embryonic chick heart. Am J Anat 53:349–375

Patterson PH, Chun LLY (1974) The influence of non-neuronal cells on catecholamine and acetycholine synthesis and accumulation in cultures of dissociated sympathetic neurons. Proc Natl Acad Sci USA 71:3607–3610

Patterson PH, Chun LLY (1977) The induction of acetylcholine synthesis in primary cultures of dissociated rat sympathetic neurons. I. Effects of conditioned medium. Dev Biol 56:263–280

Pavlovich ER, Cherova IA (1981) Morphometric identification of specialized internodal pathways of the heart. Byull Eksp Biol Med 92:496–498

Payet MD, Bkaily G, Schanne OF, Ruiz-Ceretti E (1980) Influence of streptomycin on spontaneous activity of clusters of cultured cardiac cells from neonatal rats. Can J Physiol Pharmacol 58:433–435

Peachey LD (1965) The sarcoplasmic reticulum and transverse tubules of the frog's sartorius. J Cell Biol 25:209–231

Peachey LD, Franzini-Armstrong C (1983) Structure and function of membrane systems of skeletal muscle cells. In: Peachey LD (ed) Handbook of physiology, sect 10, Skeletal muscle. American Physiological Society, Bethesda Maryland, pp 23–71

Pearlman ES, Weber KT, Janicki JS, Pietra GG, Fischman AP (1982) Muscle fibre orientation and connective tissue content in the hypertrophied human heart. Lab Invest 46:158–164

Penefsky ZJ (1983) Perinatal development of cardic contractile mechanisms. In: Gootman N, Gootman PM (eds) Perinatal cardiovascular function. Marcel Dekker, New York, pp 110–199

Pepe FA (1975) Structure of muscle filaments from immunochemical and ultrastructural studies. J Histochem Cytochem 23:543–562

Peracchia C (1973) Low resistance junctions in crayfish. I. Two arrays of globules in junctional membranes. J Cell Biol 57:54–65

Peracchia C (1980) Structural correlates of gap junction permeation. Int Rev Cytol 66:81–146

Pexieder T (1975) Cell death in the morphogenesis and teratogenesis of the heart. Adv Anat Embryol Cell Biol 51:6–100

Pexieder T (1982) La solution de deux énigmes de l'organogenèse du cœur. Fondation Suisse de Cardiologie 13:3–14

Petit A (1968) Ultrastructure de la rétine de l'œil pariétal d'un Lacertilien, Anguis fragilis. Z Zellforsch Mikrosk Anat 92:70–93

Pfeffer MA, Pfeffer JM, Dunn FG, Nishiyama K, Tsuchiya M, Frohlich ED (1979) Natural biventricular hypertrophy in normotensive rats. I. Physical and haemodynamic characteristics. Am J Physiol 236:H640–H643

Philipson KD, Bers DM, Nishimoto AY (1980) The role of phospholipids in the Ca binding of isolated cardiac sarcolemma. J Mol Cell Cariol 12:1159–1173

Pinson A, Frelin C, Padieu P (1973) The lipoprotein lipase activity in cultured beating heart cells of the postnatal rat. Biochimie 55:1261–1264

Pitlick PT, Kirkpatrick SE, Friedman WF (1976) Distribution of fetal cardiac output: importance of pacemaker location. Am J Physiol 231:204–208

Piwnica-Worms D, Lieberman M (1983) Microfluorometric monitoring of pH in cultured heart cells: Na-H exchange. Am J Physiol 244:C422–C428

Piwnica-Worms D, Jacob R, Horres CR, Lieberman M (1983) Transmembrane chloride flux in tissue-cultured chick heart cells. J Gen Physiol 81:731–748

Pollack GH (1976) Intercellular coupling in the atrioventricular node and other tissues of the heart. J Physiol 255:275–298

Porte A, Stoeckel M-E, Sacrez A, Batzenschlagen A (1980) Unusual familial cardiomyopathy with storage of intermediate filaments in the cardiac muscular cells. Virchows Arch A Path Anat Histol 386:43–58

Porter KR, Palade GE (1957) Studies on the endoplasmic reticulum. III. Its form and distribution in striated muscle cells. J Biophys Biochem Cytol 3:269–300

Prakash R (1954) The heart and its conducting system in the tadpoles of the frog (Rana tigrina Daudin). Proc Zool Soc Bengal 7:23–37

Prakash R (1956) The heart and its conducting system in the common Indian fowl. Proc Natl Inst India 22:22–23

Prakash R (1960) The heart and its conducting system in the lizard *Calotes versicolor* (Daudin). Anat Rec 136:469–476

Pressler ML, Elharrar U, Bailey JC (1982) Effects of extracellular calcium ions, verapamil, and lanthanum on active and passive properties of canine cardiac Purkinje fibers. Circ Res 51:637–651

Price KM, Littler WA, Cummins P (1980) Human atrial and ventricular myosin light chains subunits in the adult and during development. Biochem J 191(2):571–580

Price MG (1984) Molecular analysis of intermediate filament cytoskeleton – a putative load-bearing structure. AM J Physiol 246:H560–H572

Priola DV (1980) Intrinsic innervation of the canine heart – effects on conduction in the atrium, AV node, and proximal bundle branch. Circ Res 47:74–79

Puff A, Langer H (1965) Das Problem der diastolischen Entfaltung der Herzkammer (eine Untersuchung über das elastische Gewebe im Myocard). Gegenbaurs Morphol Jahrb 107:184–212

Punkt K, Punkt J, Krug H, Schippel G (1984) Comparison of histophotometric and biochemical myosin-ATPase estimations. J Histochem 16:385–387

Purves RD, Hill CE, Chamley JH, Mark GE, Fry DM, Burnstock G (1974) Functional autonomic neuromuscular junctions in tissue culture. Pflüg Arch 350:1–7

Rakusan K, Poupa O (1963) Changes in the diffusion distance in the rat heart muscle during development. Physiol Bohemoslov 12:220–227

Rakusan K, Poupa O (1964) Capillaries and muscle fibres in the heart of old rats. Gerontologia (Basel) 9:107–112

Rakusan K, Raman S, Layberry R, Korecky B (1978) The influence of aging and growth in the postnatal development of cardiac muscle in rats. Circ Res 42:212–218

Rakusan K, Moravec J, Hatt P (1980) Regional capillary supply in normal and hypertrophied rat heart. Microvasc Res 20:319–326

Randall DJ (1966) The nervous control of cardiac activity in the tench *(Tinca tinca)* and the goldfish *(Carassius auratus)*. Physiol Zool 39:185–192

Randall DJ (1970) The circulatory system, In: Hoar WS, Randall DJ (eds) Fish physiology, vol IV. Academic Press, New York

Randall WC (1977) Neural regulation of the heart. Oxford University Press, New York

Randall WC, Armour JA (1977) Gross and microscopic anatomy of the cardiac innervation. In: Randall NC (ed) Neural regulation of the heart, chapter II. Oxford University Press, New York, pp 13–41

Rao GH (1958) First appearance of coronary arteries in the human embryo. J Anat Soc India 7:55

Rardon DP, Bayley JC (1983) Direct effects of cholinergic stimulation on ventricular automaticity in guinea pig myocardium. Circ Res 52:105–110

Rash JE, Shay JW, Biesele JJ (1968) Urea extraction of Z bands, intercalated disks, and desmosomes. J Ultrastruct Res 24:181–189

Rasch JE, Biesele JJ, Gey GO (1970) Three classes of filaments in cardiac differentiation. J Ultrastruct Res 33:408–435

Rawles ME (1943) The heart-forming areas of the early chick blastoderm. Physiol Zool 16:22–42

Rayns DG, Simpson FO, Bertaud WS (1967) Transverse tubule apertures in mammalian myocardial cells: surface array. Science 156:656–657

Rayns DG, Simpson FO, Bertaud WS (1968) Surface features of striated muscle. I. Guinea-pig cardiac muscle. J Cell Sci 3:467–474

Rayns DG, Simpson FO, Ledingham JM (1969) Ultrastructure of desmosomes in mammalian intercalated disc; appearances after lanthanum treatment. J Cell Biol 42:322–326

Rayns DG, Devine CE, Sutherland CL (1975) Freeze fracture studies of membrane systems in vertebrate muscle. I. Striated muscle. J Ultrastruct Res 50:306–321

Raz A, Isakson PC, Minkes MS, Needleman P (1977) Characterization of a novel metabolic pathway of arachidonate in coronary arteries which generates a potent endogenous coronary vasodilator. J Biol Chem 252:1123–1126

Retzer R (1909) The moderator band and its relation to the papillary muscles with observations on the development and structure of the right ventricle. Bull Johns Hopkins Hosp 20:108–176

Retzer R (1920) The sino-ventricular bundle; A functional interpretation of morphological findings. Contrib Embryol 9:143–156

Reuter H (1974) Exchange of calcium ions in the mammalian myocardium. Mechanisms and physiological significance. Circ Res 34:598–605

Reuter H, Seitz N (1968) The dependence of calcium efflux from cardiac muscle on temperature and external ion composition. J Physiol (Lond) 195:451–470

Revel JP, Karnovsky MJ (1967) Hexagonal array of subunits in intercellular junctions of the mouse heart and liver. J Cell Biol 33:C7–C12

Rhodin JAG, Del Missier P, Reid LC (1961) The structure of the specialised impulse-conducting system of the steer heart. Circulation 24:349–367

Richardson KC (1964) The fine structure of the albino rabbit iris with special reference to the identification of adrenergic and cholinergic nerves and nerve endings in its intrinsic muscles. Am J Anat 114:173–205

Richardson KC (1966) Electron microscopic identification of autonomic nerve endings. Nature 210:756

Rigby WFC, Graboys TB (1981) Current concepts and management of the pre-excitation syndromes. J Cardiovasc Med 277–293

Risnik VV, Verin AD, Gusev NB (1985) Comparison of the structure of two cardiac troponin T isoforms. Biochem J 225:549–552

Robb JS (1953) Specialised (conducting) tissue in the turtle heart. Am J Physiol 172:7–13

Robb JS (1965) Comparative basic cardiology. Grune & Stratton, New York

Robb JS, Petri R (1961) Expansions of the atrioventricular system. In: De Carvalho AP, De Mello WG, Hoffman BF (eds) The specialised tissues of the heart. Elsevier, New York, pp 1–21

Roberts JT, Wearn JT (1941) Quantitative changes in the capillary-muscle relationship in human hearts during normal growth and hypertrophy. Am Heart J 21:617–633

Robertson JD (1957) New observations on the ultrastructure of the membranes of frog peripheral nerve fibers. J Biophys Biochem Cytol 3:1043–1048

Robertson JD (1958) The cell membrane concept. J Physiol (Lond) 140:58P–59P

Robinson PM (1969) A cholinergic component in the innervation of the longitudinal smooth muscle of the guinea pig vas deferens. The fine structural localisation of acetylcholinesterase. J Cell Biol 41:462–476

Robinson RB (1982) Action potential characteristics of rat cardiac cells do not change with time in culture. J Mol Cell Cardiol 14:367–370

Robinson RB (1985) Models of cardiac development: Transplants, organ culture, cell dispersion, and cell culture. In: Legato MJ (ed) The developing heart. Martinus Nijhoff Publishing, Boston, pp 69–94

Robinson RB, Legato MJ (1980) Maintained differentiation in rat cardiac monolayer cultures: tetrodotoxin sensitivity and ultrastructure. J Mol Cell Cardiol 12:493–498

Robinson RB, Reder RF, Danilo PJr (1984) Preparation and electrophysiological characterization of cardiac cell cultures derived from the fetal canine ventricle. Dev Pharmacol Ther 7:307–318

Robinson TF (1980) Lateral connections between heart muscle cells as revealed by conventional and high voltage transmission electron microscopy. Cell Tissue Res 211:353–359

Robinson TF, Winegrad S (1979) The measurement and dynamic implication of thin filament lengths in heart muscle. J Physiol (London) 286:607–619

Robinson TF, Winegrad S (1981) A variety of intercellular connections in heart muscle. J Mol Cell Cardiol 13:185–195

Robinson TF, Cohen-Gould L, Factor SM (1983) Skeletal framework of mammalian heart muscle. Arrangement of inter and pericellular connective tissue structures. Lab Invest 49:482–498

Romanoff AL (1960) The avian embryo: Structural and functional development. Macmillan, New York, pp 681–780

Rosen KM, Bauernfeind RA, Swiryn S, Dhingra RC, Wyndham CR (1980) The significance of normal and anomalous atrioventricular conducting pathway in cardiac arrhythmias. Adv Intern Med 25:277–302

Rosenbluth J (1978) Glial membrane specialization in extra paranodal regions. J Neurocytol 7:709–719

Rosenquist GC, De Haan RL (1966) Migration of precardiac cells in the embryonic chick heart. Contrib Embryol 38:113–121

Ross D, Johnson M, Bunge R (1977) Development of cholinergic characteristics in adrenergic neurones is age dependent. Nature 267:536–539

Roy PE, Morin PJ (1971) Variations of the Z-band in human auricular appendages. Lab Invest 25:422–426

Rubio R, Sperelakis N (1971) Entrance of colloidal ThO tracer into the T-tubules and longitudinal tubules of the guinea pig heart. Z Zellforsch Mikrosk Anat 116:20–36

Rudolph AM (1974) Congenital diseases of the heart: Clinical, physiologic considerations in diagnosis and management. Year Book Publishers Chicago, pp 1–16

Rudolph AM (1976) Cardiac output in the mammalian fetus. Rev Perinat Med 1:35–47

Rumyantsev PP (1972) Electron microscope study of the myofibril partial disintegration and recovery in the mitotically dividing cardiac muscle cells. Z Zellforsch Mikrosk Anat 129:471–499

Rumyantsev PP (1974) Ultrastructural reorganization, DNA synthesis and mitotic division of myocytes of atria of rats with left ventricular infarction. Virchows Arch B Zellpath 15:357–378

Rumyantsev PP (1977) Interrelations of the proliferation and differentiation processes during cardiac myogenesis and regeneration. Int Rev Cytol 51:187–273

Rumyantsev PP (1979) Some comparative aspects of myocardial regeneration. In: Mauro A (ed) Muscle regeneration. Raven Press, New York, pp 335–355

Rumyantsev PP (1981) New comparative aspects of myocardial regeneration with special reference to cardiomyocyte proliferative behaviour. In: Becker RO (ed) Mechanisms of growth control. Thomas, Springfield, Ill, pp 311–342

Rumyantsev PP (1982) DNA synthesis in atrial myocytes of rats with aortic stenosis. In: Chazov E, Saks V, Rona G (eds) Advances in myocardiology, vol 4. Plenum Press, New York, pp 147–162

Ruska H (1965) Electron microscopy of the heart. In: Taccardi B, Marchetti G (eds) Electrophysiology of the heart. Pergamon Press, Oxford, pp 1–19

Rybicka K (1978a) Microtubules in the ventricular specialized conducting fibers of the dog heart. J Mol Cell Cardiol 10:409–414

Rybicka K (1978b) Ultrastructural study on the extrusion of multivesicular bodies from cardiac cells. Virchows Arch B Cell Path 28:119–133

Rybicka K (1979) Glycosomes (protein-glycogen complex) in the canine heart. Ultrastructure, histochemistry and changes induced by acidic treatment. Virchows Arch B Cell Pathol 30:335

Rybicka K (1981) Binding of glycosomes to endoplasmic reticulum and to intermediate filaments in cardiac conduction fibers. J Histochem Cytochem 29:553–560

Rybicka K (1981) Simultaneous demonstration of glycogen and protein in glycosomes of cardiac tissue. J Histochem Cytochem 29:4–8

Rychter Z, Ostadal B (1968) Periodization of the blood supply development of the myocardium in chick embryo. Physiol Bohemoslov 17:485

Rychter Z, Ostadal B (1971) Mechanisms of the development of coronary arteries in chick embryo. Folia Morphol (Praha) 19:113–124

Rychter Z, Jelinek K, Marhan O (1971) Progress of vascularization of the ventricular myocardium in the rat embryo. Physiol Bohemoslov 20:527–532

Rychter Z, Jelinek R, Marhan O (1972) Shape and location of non vascularized area of ventricular myocardium in rat embryo during terminal phase of heart vascularization. Folia Morphol (Praha) 20:21–28

Rychter Z, Jirasek JE, Rychterova V, Uher J (1975) Vascularization of heart in human embryo: location and shape on non-vascularized part of cardiac wall. Folia Morphol (Praha) 23:88–96

Sachs HG, De Haan RL (1973) Embryonic myocardial cell aggregates: Volume and pulsation rate. Dev Biol 30:233–240

Sætersdal TS, Myklebust R (1975) Ultrastructure of the pigeon papillary muscle with special reference to the sarcoplasmic reticulum. J Mol Cell Cardiol 7:543–551

Sætersdal TS, Justesen N-P, Kroustad AW (1974) Ultrastructure and innervation of the teleostean atrium. J Mol Cell Cardiol 6:415–437

Sætersdal TS, Sorensen E, Myklebust R, Helle KB (1975) Granule containing cells and fibers in the sinus venosus of elasmobranchs. Cell Tissue Res 163:471–490

Saito T (1973) Effects of vagal stimulation on the pacemaker action potentials of carp heart. Comp Biochem Physiol 44A:191–199

Saito K, Tanura Y, Saito M, Matsumura K, Niki T, Mori H (1981) Comparison of the subunit compositions and ATPase activities of myosin in the myocardium and conduction system. J Mol Cell Cardiol 13:311–322

Sahai R, Chawla DS (1967) Specialised conducting (connecting) muscles and innervations of the heart of the house shrew, *Suncus murinus* Linnaeus, J Zool Soc India 24(1):61–67

Sakai T, Hirota A, Fujii S, Kamino K (1983) Flexibility of regional pacemaking priority in early embryonic heart monitored by simultaneous optical recording of action potentials from multiple sites. Jap J Physiol 33:337–350

Samuel D, Paris S, Ailhaud G (1976) Uptake and metabolism of fatty acids and analogues by cultured cardiac cells from chick embryo. Eur J Biochem 64:583–595

Samuel JL, Bertier B (1984) Different distributions of microtubules, desmin and filaments and isomyosins during the onset of cardiac hypertrophy in the rat. Eur J Cell Biol 34:300–306

Samuel JL, Rappaport L, Mercadier J-J, Lompre A-M, Sartore S, Triban C, Schiaffino S, Schwartz K (1983) Distribution of myosin isozymes within single cardiac cells. An immunohistochemical study. Circ Res 52:200–209

Sandborn EB, Cote MG, Roberge J, Bois P (1967) Microtubules et filaments cytoplasmiques dans le muscle de mammifères. J Microsc (Paris) 6:169–178

Sandri C, Van Buren JM, Akert K (1977) Membrane morphology of the vertebrate nervous system. A study with freeze-etch technique. Prog Brain Res 46:1–384

Sano T, Iida Y (1968) The sinoatrial connection and wandering pacemaker. J Electrocardiol 1:147–154

Sano T, Yamagishi S (1965) Spread of excitation from the sinus node. Circ Res 16:423–430

Santer RM (1977) Monoaminergic nerves in the central and peripheral nervous system of fishes. Gen Pharmacol 8:157–172

Santer RM, Cobb JLS (1972) Ultrastructural and histochemical studies on the innervation of the heart of a teleost, *Pleuronectes platessa* L. Z Zellforsch 131:1–14

Santini M (1969) New fibers of sympathetic nature in the inner core region of Pacinian corpuscles. Brain Res 16:535–538

Sartore S, Pierobon-Bormioli S, Schiaffino S (1978) Immunohistochemical evidence for myosin polymorphisin in the chicken heart. Nature 274:82–83

Sartore S, Gorza L, Pierobon Bormioli S, Dalla Libera, Schiaffino S (1981) Myosin types and fiber types in cardiac muscle. I. Ventricular myocardium. J Cell Biol 88:226–233

Sashida H, Uchida K, Abiko Y (1984) Changes in cardiac ultrastructure and myofibrillar proteins during ishaemia in dogs, with special reference to changes in Z lines. J Mol Cell Cardiol 16:1161–1172

Sato S, Ashraf M, Millard RW, Fujiwara H, Schwartz A (1983) Connective tissue changes in early ischemia of porcine myocardium: An ultrastructural study. J Mol Cell Cardiol 15:261–275

Scales DJ (1981) Aspects of the mammalian cardiac sarcotubular system revealed by freeze fracture electron microscopy. J Mol Cell Cardiol 13:373–380

Scales DJ (1983) III. Three-dimensional electron microscopy of mammalian cardiac sarcoplasmic reticulum at 80 KV. J Ultrastruct Res 83:1–9

Schaper J, Hehrlein F, Schlepper M, Thiedemann K-U (1977) Ultrastructural alterations during ischemia and reperfusion in human hearts during cardiac surgery. J Mol Cell Cardiol 9:175–189

Scher AM, Spach MS (1979) Cardiac depolarisation and repolarisation and the electrocardiogram. In: Berne RM, Sperelakis N, Geiger SR (eds) Handbook of physiology, sect 2, The cardiovascular system, vol 1, The heart. Williams & Wilkins, Baltimore, pp 357–392

Sherf L, James TN (1979) Fine structure of cells and their histologic organisation within internodal pathways of the hearts: Clinical and electrocardiographic implications. Am J Cardiol 44:345–369

Scheuer J, Bhan AK (1979) Cardiac contractile proteins. Adenosine triphosphatase activity and physiological function. Cir Res 45:1–12

Schiaffino S, Gorza L, Pierobon-Bormioli S, Sartore S (1980) Myosin polymorphisin cellular heterogeneity and plasticity of cardiac muscle. In: Pette D (ed) Plasticity of muscle. Walter de Gruyter, Berlin, pp 559–568

Schier JJ, Adelstein RS (1982) Structural and enzymatic comparison of human cardiac muscle myosins isolated from infants, adults, and patients with hypertrophic cardiomyopathy. J Clin Invest 69:816–825

Schiebler TH (1953) Histochemische Untersuchung der Purkinje-Fasern von Säugern. Z Zellforsch Mikrosk Anat 39:152–167

Schiebler TH (1955) Herzstudie. II. Mitteilung. Histologische, histochemische und experimentelle Untersuchungen am Atrioventrikularsystem von Huf- und Nagetieren. Z Zellforsch Mikrosk Anat 43:243–306

Schlafer M, Kane PF, Kirsch MM (1982) Superoxide dismutase plus catalase enhances the efficacy of hypothermic cardioplegia to protect the globally ischaemic reperfused heart. J Thorac Cardiovasc Surg 83:830–839

Schmid PG, Dykstoa RH, Mayer HE, Oda RP, Donnell JJ (1979) Evidence of non uniform sympathetic neural activity to heart regions in guinea pigs. Am J Physiol 237:H606–H611

Schmid PG, Grief BG, Lund DD, Roskoski R Jr (1978) Regional choline acetyltransferase in guinea-pig heart. Circ Res 42:657–660

Schneider MF (1981) Membrane charge movement and depolarization-contraction coupling. Annu Rev Physiol 43:507–517

Schollmeyer JE, Goll DE, Stromer MH, Dayton WR, Singh I, Robson RM (1974) Studies on the composition of the Z disc. J Cell Biol 63:303a

Schoultz TW, Swett JE (1972) The fine structure of the Golgi tendon organ. J Neurocytol 1:1–26

Schroedl NA, Hartzell CR (1983) Myocytes and fibroblasts exhibit functional synergism in mixed cultures of neonatal rat heart cells. J Cell Physiol 117:326–332

Schulze W (1982) Cytochemistry of adenylate cyclase quantitative analysis of the effect of cytochemical procedures on adenylate cyclase in heart tissue homogenates. Histochemistry 75:133–143

Schulze W, Krause EG (1983) Cytochemical demonstration of guanylate cyclase activity in cardiac muscle. Preferential localisation at sarcolemma and junctional sarcoplasmic reticulum. Histochemistry 77:243–254

Schwartz K, Lecarpentier Y, Martin JL, Lompre AM, Mercadier JJ, Swynghedauw B (1981) Myosin isoenzymic distribution correlates with speed of myocardial contraction. J Mol Cell Cardiol 13:1071–1075

Schwartz K, Lompre AM, Bouveret P, Wisnewsky C, Swynghedauw B (1980) Use of antibodies against dodecylsulfate denatured heavy meromyosins to probe structural differences between muscular myosin isoenzymes. Eur J Biochem 104:341–346

Schwartz K, Lompre AM, Bouveret P, Wisnewsky C, Whaler RG (1982) Comparisons of rat cardiac myosins at fetal stages in young animals and in hypothyroid adults. J Biol Chem 257:14412–14418

Schwartz K, Lompre AM, Lacombe G, Bouveret P, Wisnewsky C, Whaler RG, D'Albis A, Swynghedauw B (1983) Cardiac myosin isoenzymic transitions in mammals. In: Alpert NR (ed) Myocardial hypertrophy and failure. Perspectives in cardiovascular research, vol 7. Raven Press, New York, pp 345–358

Schwarzfeld TA, Jacobson SL (1981) Isolation and development in cell culture of myocardial cells of the adult rat. J Mol Cell Cardiol 13:563–575

Scott JA, Khaw BA, Locke E, Haber E, Homcy CH (1985) The role of free radical-mediated processes in oxygen-ralated damage in cultured murine myocardial cells. Circ Res 56:72–77

Scott TM (1971) The ultrastructure of ordinary and Purkinje cells of the fowl heart. J Anat 110:259–273

Seidl W, Schulze M, Steding G, Kluth D (1981) A few remarks on the physiology of the chick Gallus gallus embryo heart. Folia Morphol (Prague) 29:237–242

Seraydarian MW, Artaza L (1976) Regulation of energy metabolism by creatine in cardiac and skeletal muscle cells in culture. J Mol Cell Cardiol 8:669–678

Shaner RF (1929) The development of the atrio-ventricular node, bundle of His, and sinuatrial node in the calf, with a description of a third embryonic node-like structure. Anat Rec 44:85–99

Shelanski ML, Albert S, DeVries GH, Norton WT (1971) Isolation of filaments from brain. Science 174:1242–1245

Shelanski ML, Liem RKH (1979) Neurofilaments. J Neurochem 33:5–13

Sheldon CA, Friedman WF, Sybers HD (1976a) Scanning electron microscopy of developing cardiac myocytes. In: Johari O, Becker RP (eds) Scanning electron microscopy Part V. IIT, Chicago, pp 631–663

Sheldon CA, Friedman WF, Sybers HD (1976b) SEM of fetal and neonatal lamb cardiac cells. J Mol Cell Cardiol 8:853–862

Shen AC, Jennings RB (1972) Myocardial calcium and magnesium in acute ischemic injury. Am J Pathol 67:417–440

Sheridan DJ, Cullen MJ, Tynan MJ (1977) Postnatal ultrastructural changes in the cat myocardium: a morphometric study. Cardiovasc Res 11:536–540

Sheridan DJ, Cullen MJ, Tynan MJ (1979) Qualitative and quantitative observations on ultrastructural changes during postal development in the cat myocardium. J Mol Cell Cardiol 11:1173–1181

Shibata Y (1977) Comparative ultrastructure of cell membrane specializations in vertebrate cardiac muscles. Arch Histol Jap 40:391–406

Shibata Y, Yamamoto T (1979) Freeze-fracture studies of gap junctions in vertebrate cardiac muscle cells. J Ultrastruct Res 67:79–88

Shibata Y, Page E (1981) Gap junctional structure in intact and cut sheep cardiac Purkinje fibers: A freeze-fracture study of Ca-induced resealing. J Ultrastruct Res 75:195–204

Shibata Y, Nakata K, Page E (1980) Ultrastructural changes during development of gap junctions in rabbit left ventricular myocardial cells. J Ultrastruct Res 71:258–271

Shimada T, Nakamura M, Kitahara Y, Sachi M (1983) Surface morphology of chemically digested Purkinje fibers of the goat heart. J Electron Microsc 32:187–196

Shimada T, Horita K, Murakami M, Ogura R (1984) Morphological studies of different mitochondrial populations in monkey myocardial cells. Cell Tissue Res 238:577–582

Shindler R, Harakal C, Sevy RW (1968) Catecholamine content of the sino-atrial node and common right atrial tissue. Proc Soc Exp Biol Med 128:789–800

Shipley RA, Shipley LJ, Wearn JT (1937) The capillary supply in normal and hypertrophied hearts of rabbits. J Exp Med 65:29–42

Shore PA, Cohn VH, Highman B, Haling HM (1958) Distribution of norepinephrine in the heart. Nature 181:848–849

Simionescu N, Simionescu M, Palade GE (1978) Structural basis of permeability in sequential segments of the microvasculature of the diaphragm. II. Pathways followed by microperoxidase across the endothelium. Microvasc Res 15:17–36

Simpson P, McGrath A (1983) Norepinephrine-stimulated hypertrophy of cultured rat myocardial cells is an alpha, adrenergic response. J Clin Invest 72:732–738

Simpson P, Savion S (1982) Differentiation of rat myocytes in single cell cultures with and without proliferating non myocardial cells, Cross-striations, ultrastructure, and chronotropic response to isoproterenol. Circ Res 50:101–116

Simpson FO, Rayns DG (1968) The relationship between the transverse tubular system and other tubules at the Z-disc levels of myocardial cells in the ferret. Am J Anat 122:193–208

Simpson FO, Rayns DG, Ledingham JM (1973) The ultrastructure of ventricular and atrial myocardium. In: Challice CE, Viragh S (eds) Ultrastructure of the mammalian heart. Academic Press, New York, pp 1–41

Simpson FO, Rayns DG, Ledingham JM (1974) Fine structure of mammalian myocardial cells. Adv Cardiol 12:15–33

Simpson P, McGrath A, Savion S (1982) Myocyte hypertrophy in neonatal rat heart cultures and its regulation by serum and catecholamines. Circ Res 51:787–801

Sineva IV, Borisova EM, Kelareva NA, Udelnov MG (1976) Bioelectrical characteristics of the structures of the atrioventricular (AV) junction and formation of atrioventricular delay. Vestn Mosk Uni Ser VI Biol Pochvoved 31(2):13–21

Singh S, White FC, Bloor CM (1981) Myocardial morphometric characteristics in swine. Circ Res 49:434–441

Sissman NJ (1970) Developmental landmarks in cardiac morphogenesis: Comparative chronology. Am J Cardiol 25:141–149

Sjöstrand FS, Andersson CE, Dewey MM (1958) Ultrastructure of the intercalated disc of frog, mouse and guinea-pig cardiac muscle. J Ultrastruct Res 1:271–287

Slezák J, Geller SA (1979) Cytochemical demonstration of adenylate cyclase in cardiac muscle effect of dimethyl sulfoxide. J Histochem Cytochem 27:774–781

Slezák J, Tribulová N (1984) Subcellular localisation of adenylate cyclase and guanylate cyclase in heart muscle cells. Histochem J 16:380–382

Small JV, Sobieszek A (1977) Studies on the function and composition of the 10 nm (100Å) filaments of vertebrate smooth muscle. J Cell Sci 23:243–268

Smith HE, Page E (1976) Morphometry of rat heart mitochondrial subcompartments and membranes: Application to myocardial cell atrophy after hypophysectomy. J Ultrastruct Res 55:31–41

Smith HE, Page E (1977) Ultrastructural changes in rabbit heart mitochondria during the perinatal period. Neonatal transition to aerobic metabolism. Dev Biol 57:109–117

Smith P, Clark DR (1979) Myocardial capillary density and muscle fibre size in rats born at simulated high altitude. Br J Exp Pathol 60:225–230

Smith TW, Barry WH, Marsh JD, Lorell B (1982) Hypoxia, calcium fluxes and inotropic state: studies in cultured heart cells. Am Heart J 103:716–723

Snijder J, Meijer AEFH (1970) Enzyme histochemical studies on the Purkinje fibres of canine, bovine, and procine hearts. Histochem J 2:395–409

Sobel H (1972) Effect of age on cardiac metabolism. In: Bajusz E, Rona G (eds) Recent advances in studies on cardiac structure and metabolism, vol 1, Myocardiology. University Park Press, Baltimore, pp 101–111

Somlyo AV (1979) Bridging structures spanning the junctional gap at the triad of skeletal muscle. J Cell Biol 80:743–750

Sommer JR (1982) Ultrastructural considerations concerning cardiac muscle. J Mol Cell Cardiol 14(Suppl 3):77–83

Sommer JR, Johnson EA (1968a) Cardiac muscle. A comparative study of Purkinje fibers and ventricular fibers. J Cell Biol 36:497–526

Sommer JR, Johnson EA (1968b) Purkinje fibres of the heart examined with the peroxidase reaction. J Cell Biol 37:570–574

Sommer JR, Johnson EA (1969) Cardiac muscle. A comparative ultrastructural study with special reference to frog and chicken hearts. Z Zellforsch Mikrosk Anat 98:437–468

Sommer JR, Johnson EA (1970) Comparative ultrastructure of cardial cell membrane specialisations. A review. Am J Cardiol 25:184–194

Sommer JR, Johnson EA (1979) Ultrastructure of cardiac muscle. In: Berne RM, Sperelakis N, Geiger SR (eds) Handbook of physiology, sect 2, The cardiovascular system, vol 1, chapter 5. American Physiological Society, Bethesda, pp 113–186

Sommer JR, Waugh RA (1976) The ultrastructure of the mammalian cardiac muscle cell – with special emphasis on the tubular membrane systems. Am J Pathol 82:191–232

Somogyi E, Sótonyi P, Balogh I, Sreter F (1982) Diagnostic electron microscopy and biochemistry of human post-mortem sudden death cardiac biopsy. Acta Med Leg Soc (Liege) 32:17–22

Sonnenblick EH, Napolitano LM, Dogett WM, Cooper T (1967) An intrinsic neuromuscular basis for mitral valve motion in the dog. Circ Res 21:9–15

Sótonyi P, Somogyi E (1983) Cytochemical demonstration of the molecular forms of cardiac glycosides in the heart muscle. Acta Histochem 72:117–122

Sótonyi P, Kerenmin A, Somogyi E (1982) A new method for cytochemical demonstration of calcium in heart muscle. Histochemistry 75:425–436

Spach MS (1982) The electrical representation of cardiac muscle based on discontinuities of axial resistivity at a microscopic and macroscopic level. A basis for saltatory propagation in cardiac muscle. In: De Carvalho AP, Hoffman BF, Lieberman M (eds) Normal and abnormal conduction in the heart. Futura Publishing, Mt Kisco, New York, pp 145–179

Spach MS, Kootsey JM (1983) The nature of electrical propagation in cardiac muscle. Am J Physiol 244:H3–H22

Spach MS, King TD, Barr RC, Boaz DE, Morrow MN, Herman-Giddens S (1969) Electrical potential distribution surrounding the atria during depolarization and repolarization in the dog. Circ Res 24:857–873

Spach MS, Lieberman M, Scott JG, Barr RC, Johnson EA, Kootsey JM (1971) Excitation sequences of the atrial septum and AV node in isolated hearts of the dog and rabbit. Circ Res 29:156–172

Spach MS, Miller WT, III, Dolber PC, Kootsey JM, Sommer JR, Mosher CE Jr (1982a) The functional role of structural complexities in the propagation of depolarization in the atrium of the dog. Cardiac conduction disturbances due to discontinuities of effective axial resistivity. Circ Res 50:175–191

Spach MS, Kootsey JM, Sloan JD (1982b) Active modulation of electrical coupling between cardiac cells of the dog, A mechanism for transient and steady state variations in conduction velocity. Circ Res 51:347–362

Sperelakis N, Lee EC (1971) Characterization of (Na, K)-ATPase isolated from embryonic chick hearts and cultured chick heart cells. Biochim Biophys Acta 233:562–579

Sperelakis N, Lehmkuhl D (1967) Effects of temperature and metabolic poisons on membrane potentials of cultured heart cells. Am J Physiol 213:719–724

Sperelakis N, McLean MJ (1978) Electrical properties of cultured heart cells. In: Kobayashi T, Ito Y, Rona G (eds) Recent advances in studies on cardiac structure and metabolism, vol 12, Cardiac adaptation, University Park Press, Baltimore, pp 645–666

Sperelakis N, Rubio R (1971) Ultrastructural changes produced by hypertonicity in cat cardiac muscle. J Mol Cell Cardiol 3:139–156

Sperelakis N, Mayer G, McDonald R (1970) Velocity of propagation in vertebrate cardiac muscles as a function of tonicity and $[K^+]_o$. Am J Physiol 219:952–963

Sperelakis N, Forbes MS, Rubio R (1974) The tubular systems of myocardial cells: Ultrastructure and possible function. In: Dhalla NS (ed) Myocardial biology. Recent advances in studies on cardiac structure and metabolism, vol 4. University Park Press, Baltimore, pp 163–194

Spira ME (1971) The nexus in the intercalated disc of the canine heart; quantitative data for an estimation of its resistance. J Ultrastruct Res 34:409–425

Spiro D, Hagopian M (1967) On the assemblage of myofibrils, In: Warren KB (ed) Formation and fate of cell organelles. Academic Press, New York, pp 71–98

Squire JM (1981) The structural basis of muscular contraction. Plenum Press, New York

Sreter FA, Faris R, Balogh I, Somogyi E, Sótonyi P (1982) Changes in myosin isozyme distribution induced by low doses of isoproterenol. Arch Int Pharmacodyn Ther 260:159–164

Stabrovsky EM (1967) The distribution of adrenaline and noradrenaline in the organs of the Baltic lamprey, *Lampetra fluviatilis* at rest and during various functional stresses. J Evol Biochem Physiol 3:216–221

Staley NA, Benson ES (1968) The ultrastructure of frog ventricular cardiac muscle and its relationship to mechanisms of excitation-contraction coupling. J Cell Biol 38:99–114

Stalsberg H (1970) Mechanism of dextral looping of the embryonic heart. Am J Cardiol 25:265–271

Stalsberg H, De Haan RL (1969) The pre-cardiac areas and formation of the tubular heart in the chick embryo. Dev Biol 19:128–159

Steding G, Seidl W (1980) Contribution to the development of the heart. Part 1. Normal development. Thorac Cardiovasc Surgeon 28:386–409

Steding G, Seidl W (1981) Contribution to the development of the heart. Part 2: Morphogenesis of congenital heart diseases. Thorac Cardiovasc Surgeon 29:1–16

Steding G, Seidl W, Kluth D, Schulze M (1981) On the morphogenesis of the normal and the malformed heart. Folia Morphol (Praha) 29:243–246

Stein O, Stein Y (1968) Lipid synthesis, intracellular transport and storage. 3. Electron microscopic radioautographic study of the rat heart perfused with tritiated oleic acid. J Cell Biol 36:63–77

Steinbeck G, Bonke FIM, Allessie MA, Lammers WJEP (1978) Cardiac glycosides and pacemaker activity of the sinus node – a microelectrode study on the isolated right atrium of the rabbit. In: Bonke FIM (ed) The sinus node, chapter 21. Martinus Nijhoff, The Hague, pp 258–269

Stene-Larsen G, Helle KB (1978a) Cardiac β_2-adrenoceptor in the frog. Comp Biochem Physiol C60:165–173

Stene-Larsen G, Helle KB (1978b) Evidence against a transformation of the β_2-adrenoceptor in the frog heart by changes in temperature or metabolic state. Life Sci 23(27/28):2681–2688

Stene-Larsen G, Helle KB (1979) Temperature effects on the inotropic and chronotropic responses to adrenaline in the frog (*Rana temporaria*) heart. J Comp Physiol B 132:313–318

Stoker ME, Gerdes AM, May JF (1982) Regional differences in capillary density and myocyte size in the normal human heart. Anat Rec 202:187–191

Streeter DD Jr (1979) Gross morphology and fiber geometry of the heart. In: Berne R, Sperelakis N (eds) Handbook of physiology, sect II, The cardiovascular system, vol 1, The heart. American Physiological Society, Bethesda, pp 61–112

Streeter GL (1942) Developmental horizons in human embryos. Description of age group XI, 13 to 20 somites and age group XII, 21 to 29 somites. Contrib Embryol Carnegie Inst 30:211–245

Streeter GL (1945) Developmental horizons in human embryos. Description of age group XIII, embryos about 4 or 5 millimeters long, and age group XIV, period of indentation of lens vesicle. Contrib Embryol Carnegie Inst 31:27–63

Streeter GL (1948) Developmental horizons in human embryos. Description of age groups XV, XVII and XVIII. Contrib Embryol Carnegie Inst 32:133–203

Streeter GL (1951) Developmental horizons in human embryos, Description of age group XIX, XX, XXI, XXII, XXIII, being the issue of a survey of the carnegie collection. Prepared for publication by CH Heuser and GW Corner. Contrib Embryol Carnegie Inst 34:165–196

Sung RJ, Styperek JL (1979) Electrophysiologic identification of dual atrioventricular nodal pathway conduction in patients with reciprocating tachycardia using anomalous by-pass tracts. Circulation 60:1464

Suenson M (1984) Ephaptic impulse transmission between ventricular myocardial cell *in vitro*. Acta Physiol Scand 120:445–455

Sweeney LJ, Clark EB, Rosenquist GC (1980) Left heart hypoplasia in the chick embryo, an experimental study. Anat Rec 196(3):186A

Sweeney LJ, Clark WA Jr, Umeda PK, Zak KR, Manasek FJ (1984) Immunofluorescence analysis of the primordial myosin detectable in embryonic striated muscle. Proc Natl Acad Sci USA 81:797–800

Swett FH (1923) The connecting systems of the reptile heart; alligator. Anat Rec 26:129–140

Swiryn S, Bauernfeind RA, Palileo EA, Strasberg B, Duffy CE, Rosen KM (1982) Electrophysiologic study demonstrating triple antegrade AV nodal pathways in patients with spontaneous and/or induced supraventricular tachycardia. Am Heart J 103:168–176

Sybers HD, Gann MK (1975) Non-specificity of a "selective" stain for sarcoplasmic reticulum in cardiac cells. In: Arceneaux CJ (ed) Thirty-third Annu, EMSA Meeting, Claitor's Publishing, Baton Rouge, pp 539–540

Syrovy I (1979) Polymorphism and specificity of myosin. Int J Biochem 10:383–389

Syrovy I, Delcayre C, Swynghedauw B (1979) Comparison of ATPase activity and light subunits in myosins from left and right ventricles and atria in seven mammalian species. J Mol Cell Cardiol 11:1129–1135

Szentivanyi M, Juhasz-Nagy A (1959) A new aspect of the nervous control of the coronary blood vessels. Q J Exp Physiol 44:67–79

Szentivanyi M, Kunos G, Juhsz-Nagy A (1976) Unmasking of sympathetic vasoconstriction of the coronary vessels after acute administration of reserpine to dogs. Can J Physiol Pharmacol 54:116–169

Taegtmeyer H, Peterson MB, Ragavan VV, Fergussan AG, Lesch M (1977) De novo alanine synthesis in isolated oxygen deprived rabbit myocardium. J Biol Chem 252:5010–5018

Takayasu M, Tomita T, Hirakawa S, Takegami S, Konishi T, Nohara Y (1969) Conduction pathways of excitation in the right atrium of the rabbit under hyperpotassemia. Israel J Med Sci 5:485–487

Tandler J (1912) Die Entwicklungsgeschichte des Herzens. In: Keibel F, Mall EP (eds) Manual of human embryology, Lippincott, Philadelphia

Taylor IM, Anderson RH (1973) Cholinesterase and the atrioventricular node and bundle in the human fetus up to midterm. J Histochem Cytochem 21:464–468

Taylor SM, Jones PA (1980) Histochemical demonstration of myosin CA^{2+}-ATPase accumulation in primary cultures of skeletal and heart muscle cells. Histochem J 12:169–181

Taxi J (1976) Morphology of the autonomic nervous system, In: Llinas R, Precht W (eds) Frog neurobiology. A handbook, Springer, New York, pp 137–139

Tebecis AK (1967) A study of electrograms recorded from the conus arteriosus of an elasmobranch heart. Aust J Biol Sci 20:843–841

Thaemert JC (1966) Ultrastructure of cardiac muscle and nerve contiguities. J Cell Biol 29:156–162

Thaemert JC (1969) Fine structure of neuromuscular relationships in mouse heart. Anat Rec 163:575–586

Thaemert JC (1970) Atrioventricular node innervation in ultrastructural three dimensions. Am J Anat 128:239–263

Thaemert JC (1973) Fine structure of the atrio ventricular node as viewed in serial sections. Am J Anat 136:43–66

Thaemert JC, Emmett SS (1968) The ultrastructure of neuromuscular relationships within the atrioventricular node of the heart. Anat Rec 160:518 (Abstract)

Theiler K (1972) The house mouse, Springer, New York

Theron JJ, Biagio R, Meyer AC, Boekkooi S (1978) Ultrastructural observations on the maturation and secretion of granules in atrial myocardium. J Mol Cell Cardiol 10:567–572

Thiedemann KU, Ferrans VJ (1976) Ultrastructure of sarcoplasmic reticulum in atrial myocardium of patients with mitral valvular disease. Am J Pathol 83:1–38

Thiedemann KU, Ferrans VJ (1977) Left atrial ultrastructure in mitral valvular disease. Am J Pathol 89:575–604

Thoenes W, Ruska H (1960) Über "Leptomere Myofibrillen" in der Herzmuskelzelle. Z Zellforsch Mikrosk Anat 51:560–570

Thomas EI (1976) The minute anatomy of the heart of Anabas testudineus. Zool Anz 196:397–404

Thompson EJ, Wilson SH, Schuette WH, Whitehouse WC, Nirenberg MW (1973) Measurement of the rate and velocity of movement by single heart cells in culture. Am J Cardiol 32:162–166

Thornell L-E (1972) Myofilament-polyribosome complexes in the conducting system of hearts from cow, rabbit, cat. J Ultrastruct Res 41:579–596

Thornell L-E (1973a) Evidence of an imbalance in synthesis and degradation of myofibrillar proteins in rabbit Purkinje fibres. J Ultrastruct Res 44:85–95

Thornell L-E (1973b) Ultrastructural variations of Z bands in cow Purkinje fibres. J Mol Cell Cardiol 5:409–417

Thornell L-E (1974a) An ultrahistochemical study on glycogen in cow Purkinje fibres. J Mol Cell Cardiol 6:439–448

Thornell L-E (1974b) Distinction of glycogen and ribosome particles in cow Purkinje fibres by enzymatic digestion en bloc and in sections. J Ultrastruct Res 47:153–168

Thornell L-E (1974c) The fine structure of Purkinje fibre glycogen. A comparative study of negatively stained and cytochemically stained particles. J Ultrastruct Res 49:157–166

Thornell L-E (1975) Morphological characteristics of Purkinje fibre bundles separated from their connective tissue sheath. J Mol Cell Cardiol 7:191–194

Thornell L-E, Eriksson A (1981) Filament systems in the Purkinje fibres of the heart. Am J Physiol 241:H291–H305

Thornell L-E, Forsgren S (1982) Myocardial cell heterogeneity in the human heart with respect to myosin ATPase activity. Histochem J 14:479–490

Thornell L-E, Sjöström M, Anderson KE (1976) The relationship between mechanical stress and myofibrillar organization in heart Purkinje fibres. J Mol Cell Cardiol 8:689–695

Thornell L-E, Eriksson A, Stigbrand T, Sjöström M (1978) Structural proteins in cow Purkinje and ordinary ventricular fibres – A marked difference. J Mol Cell Cardiol 10:605–616

Thornell L-E, Johanssen B, Eriksson A, Lehto V-P, Virtanen I (1984) Intermediate filament and associated proteins in the human heart: An immunofluorescence of normal and pathological hearts. Eur Heart J 5(Suppl):F231/F241

Thornell L-E, Eriksson A, Johansson B, Kjörell U, Franke WW, Virtanen I, Lehto V-P (1985) Intermediate filament and associated proteins in heart Purkinje fibers. A membrane-myofibril anchored cytoskeletal system. Ann NY Acad Sci 455 (in press)

Timpl R, Rohde H, Robey PG, Rennard SI, Foidart J-M, Martin GR (1979) Laminin – a glycoprotein from basement membranes. J Biol Chem 254:9933–9937

Tirri R, Harri MNE, Laitinen L (1974) Lowered chronotropic sensitivity of rat and frog hearts to sympathomimetic amines following cold acclimation. Acta Physiol Scand 90:260–266

Tokuyasu KT (1983) Visualization of longitudinally-oriented intermediate filaments in frozen sections of chicken cardiac muscle by a new staining method. J Cell Biol 97:562–565

Tokuyasu KT, Dutton AH, Geiger B, Singer SJ (1981) Ultrastructure of chicken cardiac muscle as studied by double immunolabelling in electron microscopy. Proc Natl Acad Sci USA 78:7619–7623

Tokuyasu KT, Dutton AH, Singer SJ (1983) Immunoelectron microscopic studies of desmin (skeletin) localization and intermediate filament organization in chicken cardiac muscle. J Cell Biol 96:1736–1742

Tomanek RJ (1979) The role of prevention or relief of pressure overload on the myocardial cell of the spontaneously hypertensive rat. A morphometric and stereologic study. Lab Invest 40:83–91

Tomanek RJ, Hovanec JM (1981) The effects of long-term pressure-overloaded and aging on the myocardium. J Mol Cell Cardiol 13:471–488

Tomanek RJ, Karlsson UL (1973) Myocardial ultrastructure of young and senescent rat. J Ultrastruct Res 42:201–220

Tomanek RJ, Davis JW, Anderson SC (1979) The effect of methyldopa on cardiac hypertrophy in spontaneously hypertensive rats: Ultrastructural, stereological and morphometric analysis. Cardiovasc Res 13:173–182

Torii H (1962) Electron microscope obversations of the SA and AV nodes and Purkinje fibres of the rabbit. Jap Circ J 26:39–77

Touboul P, Gressard A, Vexler RM, Chatelain MT (1977) Unusual re-entrant tachycardias associated with accessory pathways. In: Kulbertus HE (ed) Re-entrant arrhythmias. MTP Press, Lancaster, pp 132–143

Touboul P, Vexler RM, Chatelain MT (1978) Re-entry via mahaim fibers as a possible basis for tachycardia. Br Heart J 40:806–811

Toyoshima H, Park YD, Ishikawa Y, Nagata S, Hirata Y, Sakakibara H, Shimomura K, Nakayama R (1982) Effect of ventricular hypertrophy on conduction velocity of activation front in the ventricular myocardium. Am J Cardiol 49:1938–1945

Tranum-Jensen J (1975) The ultrastructure of the sensory end-organs (baroreceptors) in the atrial endocardium of young mini-pigs. J Anat 119:255–275

Tranum-Jensen J (1976) The fine structure of the atrial and atrio-ventricular (AV) junctional specialised tissues of the rabbit heart. In: Wellens HJJ, Lie KI, Janse MJ (eds) The conduction system of the heart-structure, function and clinical implications. Lea & Febiger, New York, pp 55–81

Tranum-Jensen J (1978) The fine structure of the sinus node: A survey. In: Bonke FIM (ed) The sinus node, chapter 13. Martinus Nijhoff, The Hague, pp 149–165

Tranum-Jensen J (1979) Ultrastructural studies on atrial nerve-end formations in mini-pigs. In: Hainsworth R, Kidd C, Linden RF (eds) Cardiac receptors, chapter 2. Cambridge University Press, Oxford, pp 27–50

Tranzer JP, Thoenen H, Snipes RL, Richards JG (1969) Recent developments on the ultrastructural aspect of adrenergic nerve endings in various experimental conditions. Prog Brain Res 31: 33–46

Trautwein W, Uchizono K (1963) Electron microscopic and electrophysiologic study of the pacemaker in the sinoatrial node of the rabbit heart. Z Zellforsch Mikrosk Anat 61:96–109

Tribulova N, Slezák J, Gabauer I, Holec VI (1984) Effect of chlorpromazine on the extent of ischemic changes of the myocardium. A histochemical study. Histochem J 383:384

Trillo A, Bencosme SA (1965) La ultrastructura de las celulas musculares cardiacas con referencia especial a los granulos especificos. I. El miocardio de las rama. Archiv Inst Cardiol Mek 35:803–817

Truex RC (1961) Comparative anatomy and functional considerations of the cardiac conduction system. In: De Carvalho AP, De Mello WC, Hoffman BF (eds) The specialised tissues of the heart. Elsevier, Amsterdam, pp 22–43

Truex RC (1972) Myocardial cell diameters in primate hearts. Am J Anat 135:269–280

Truex RC (1976) The sinoatrial node and its connections with the atrial tissues. In: Wellens HJJ, Lie KI, Janse MJ (eds) The conduction system of the heart-structure, function and clinical implications. Lea & Febiger, New York, pp 209–226

Truex RC, Smythe MQ (1965a) Comparative morphology of the cardiac conduction tissue in animals. Ann NY Acad Sci 127:19–33

Truex RC, Smythe MQ (1965b) Recent observations on the human cardiac conduction system, with special considerations of the atrioventricular node and bundle. In: Taccardi B, Marchetti G (eds) Electrophysiology of the heart. Pergamon Press, Oxford, pp 177–198

Truex RC, Smythe MQ (1967) Reconstruction of the human atrioventricular node. Anat Rec 158:11–19

Truex RC, Belej R, Ginsberg LM, Hartman RL (1974) Anatomy of the ferret heart: An animal model for cardiac research. Anat Rec 179:411–422

Truex RC, Marino TA, Marino DR (1978) Observations of the development of the human atrioventricular node and bundle. Anat Rec 192:337–350

Tsien R, Weingart R (1976) Inotropic effect of cyclic AMP in calf ventricular muscle studied by a cut-end method. J Physiol 260:117–141

Tsuchimochi H, Sugi M, Kuro-o M, Ueda S, Takaku F, Furuta S, Shirai T, Yazaki Y (1984) Isozymic changes in myosin of human atrial myocardium induced by overload. J Clin Invest 74:662–665

Tsuyuguchi N, Matsumara K, Mikawa K, Niki T, Mori H, Aki K (1978) Characteristics of energy metabolism in specialised muscle of bovine heart. In: Kobayashi T, Sano T, Dhalla NS (eds) Recent advances in studies on cardiac structure and metabolism, vol 2. University Park Press, Baltimore, pp 343–347

Uehara Y, Campbell GR, Burnstock G (1976) Muscle and its innervation. An atlas of fine structure. Arnold, London

Urthaler F, Walker AA, Hefner LL, James TN (1975) Comparison of contractile performance of canine atrial and ventricular muscles. Circ Res 37:762–771

Urthaler F, Walker AA, Kawamura K, Hefner LL, James TN (1978) Canine atrial and ventricular muscle mechanics studied as a function of age. Circ Res 42:703–713

Uyeda CT, Eng LF, Bignami A (1972) Immunological study of the glial fibrillary acidic protein. Brain Res 37:81–89

Van Capelle FJL, Janse MJ (1976) Influence of geometry on the shape of the propagated action potential. In: Wellens HJJ, Lie KI, Janse MJ (eds) The conduction system of the heart, chapter 18. Lea & Febiger, Philadelphia, pp 316–335

Van der Laarse A, Hollaar L, Van der Valk LJM, Witteveen SAGJ (1978) Enzyme release from and enzyme depletion in rat heart cell cultures during anoxia. J Mol Med 3:123–131

Van der Laarse A, Hollaar L, Kokshoorn JM (1979a) The acitivity of cardio-specific isoenzymes of creatine phosphokinase and lactate dehydrogenase in monolayer cultures of neonatal rat heart cells. J Mol Cell Cardiol 11:501–510

Van der Laarse A, Hollaar L, Van der Valk LJM (1979b) Release of alpha hydroxybutyrate from neonatal rat heart cell cultures exposed to anoxia and reoxygenation: comparison with impairment of structure and function of damaged cardiac cells. Cardiovasc Res 13:345–353

Van Mierop LHS (1967) Location of pacemaker in chick embryo heart at the time of initiation of heartbeat. Am J Physiol 212:407–415

Van Mierop LHS (1979) Morphological development of the heart. In: Berne RM, Sperelakis N, Geiger SR (eds) Handbook of physiology, sect II, The cardiovascular system, vol 1, The heart. Williams & Wilkins, Baltimore, pp 1–28

Van Mierop LHS, Gessner IH (1970) The morphological development of the sinoatrial node in the mouse. Am J Cardiol 25:204–212

Van Mierop LHS, Patterson PR, Reynolds RW (1964) Two cases of congenital asplenia with isomerism of the cardiac atria and the sinoatrial nodes. Am J Cardiol 13:407–414

Van Winkle (1977) The fenestrated collar of mammalian cardiac sarcoplasmic reticulum: a freeze-fracture study. Am J Anat 149:277–282

Vassall-Adams PR (1982a) The development of the atrioventricular bundle and its branches in the avian heart. J Anat 134:169–183

Vassall-Adams PR (1982b) The developing atrioventricular bundle and its branches in relation to small induced ventricular septal defects in the heart of chick embryos. J Anat 134:209–214

Vernall DG (1962) The human embryonic heart in the seventh week. Am J Anat 111:17–24

Vieweg WVR, Alpert JS, Hagan AD (1975) Origin of the sinoatrial node and atrioventricular node arteries in right, mixed and left inferior emphasis systems. Catheter Cardiovasc Diagn 1:361–373

Virágh S, Challice CE (1969) Variations in filaments and fibrillar organization, and associated sarco-lemmal structures in cells of the normal mammalian heart. J Ultrastruct Res 28:321–334

Virágh S, Challice CE (1973a) The impulse generation and conduction system of the heart. In: Ultrastructure in biological systems, vol 6, Ultrastructure of the mammalian heart. Academic Press, New York London, pp 43–89

Virágh S, Challice CE (1973b) Origin and differentiation of cardiac muscle cells in the mouse. J Ultrastruct Res 42:1–24

Virágh S, Challice CE (1977a) The development of the conduction system in the mouse embryo heart. I. The first embryonic AV conduction pathway. Dev Biol 56:382–396

Virágh S, Challice CE (1977b) The development of the conduction system in the mouse embryo heart. II. Histogenesis of the atrioventricular node and bundle. Dev Biol 56:397–411

Virágh S, Challice CE (1980) The development of the conduction system in the mouse embryo heart. III. The development of sinus muscle and sino atrial node. Dev Biol 80:28–45

Virágh S, Challice CE (1981) The origin of the epicardium and the embryonic myocardial circulation in the mouse. Anat Rec 201:157–168

Virágh S, Challice CE (1982) The development of the conduction system in the mouse embryo heart. IV. Differentiation of the atrioventricular conduction system. Dev Biol 89:25–40

Virágh S, Challice CE (1983) Development of the early atrioventricular conduction system in the embryonic heart. Can J Physiol Pharmacol 61:775–792

Virágh S, Porte A (1961) Elements nerveux intracardiaques et innervation du myocarde. Etude au microscope électronique dans le cœur de rat. Z Zellforsch Mikrosk Anat 55:282–296

Virágh S, Porte A (1973a) The fine structure of the conducting system of the monkey heart (*Macaca mulatta*). I. The sino-atrial node and internodal connections. Z Zellforsch Mikrosk Anat 145:191–211

Virágh S, Porte A (1973b) On the impulse conducting system of the monkey heart (*Macaca mulatta*). II. The atrio-ventricular node and bundle. Z Zellforsch Mikrosk Anat 145:363–388

Virágh S, Szabo E, Challice CE (1982) Glycogen-containing lysosomes and glycogen loss in the cardiocytes of embryonic and neonatal mice. In: Chazov E, Smirnov V, Dhalla NS (eds) Advances in myocardiology, vol 3. Plenum Press, New York, pp 553–561

Vitali-Mazza L, Anversa P, Tedeschi F, Mastandrea R, Mavilla V, Visioli O (1972) Ultrastructural basis of acute left ventricular failure from severe acute aortic stenosis in the rabbit. J Mol Cell Cardiol 4:661–671

Vlk J (1958) Über den Acetylcholingehalt im rechten und linken Herzvorhof bei Kaninchen und Ratten. Naunyn-Schmiedebergs Arch Exp Pathol Pharmakol 235:19–22

Vogel WOP (1985) The caudal heart of fish: not a lymph heart. Acta Anat 121:41–45

Vrensen GFJM (1970) Further observations concerning the involvement of rough endoplasmic reticulum and ribosomes in early stages of glycogen repletion in rat liver. J Microsc (Paris) 9:517–534

Vrensen GFJM, Kuyper CLMA (1969) Involvement of rough endoplasmic reticulum and ribosomes in early stages of glycogen repletion in rat liver. J Microsc (Paris) 8:599–614

Vye MV, Fischman DA (1970) The morphological alteration of particulate glycogen by en bloc staining with uranyl acetate. J Ultrastruct Res 33:278–291

Wachtlova M, Rakusan K, Poupa O (1965) The coronary terminal vascular bed in heart of the hare *Lepus europeus* and the rabbit *Oryctolagus cuniculus* dom. Physiologia Bohemoslov 14:328–331

Wachtlova M, Rakusan K, Roth Z, Poupa O (1967) The terminal vascular bed of the myocardium in the wild rat *Rattus norvegicus* and the laboratory rat *Rattus norvegicus* lab. Physiologia Bohemoslov 16:548–554

Wahab NS, Zucker IH, Gilmore JP (1975) Lack of a direct effect of efferent cardiac vagal nerve activity on atrial receptor activity. Am J Physiol 229:314–317

Wainrach S, Sotello JR (1961) Electron microscope study of the developing chick embryo heart. Z Zellforsch Mikrosk Anat 55:622–634

Wakabayashi S, Goshima K (1981) Kinetic studies on sodium-dependent calcium uptake by myocardial cells and neuroblastoma cells in culture. Biochim Biophys Acta 642:158–172

Walker SM, Schrodt GR, Edge MB (1970) Electron-dense material within sarcoplasmic reticulum apposed to transverse tubules and to the sarcolemma in dog papillary muscle fibres. Am J Anat 128:33–44

Walker SM, Schrodt GR, Edge MB (1971) The density attached to the inside surface of the apposed sarcoplasmic reticular membrane in the vertebrate cardiac and skeletal muscle fibres. J Anat 108:217–230

Walker SM, Schrodt GR, Currier GT (1975) Evidence for a structural relationship between successive parallel tubules in the SR network and supernumerary striations of Z line material in Purkinje fibers of the chicken, sheep, dog and Rhesus monkey heart. J Morphol 147:459–474

Walls EW (1942) Specialised conducting tissue in the heart of the golden hamster (*Mesocricetus auratus*). J Anat 76:359–369

Walls EW (1947) The development of the specialised conducting tissue of the human heart. J Anat 81:93–110

Ward DE, Camm AJ, Spurrell RAJ (1979) Ventricular pre-excitation due to anomalous nodo-ventricular pathways: report of 3 patients. Eur J Cardiol 9:111–127

Wearn JT (1928) The extent of the capillary bed of the heart. J Exp Med 47:273–292

Wearn JT (1941) Alterations in the heart accompanying growth and hypertrophy. Bull Johns Hopkins Hosp 68:363–374

Weibel ER (1979) Stereological methods for biological morphometry, vol 1. Academic Press, London

Weidmann S (1966) The diffusion of radiopotassium across intercalated discs of mammalian cardiac muscle. J Physiol (Lond) 187:323–342

Weiner J, Giacomelli F, Loud AV, Anversa P (1979) Morphometry of cardiac hypertrophy induced by experimental renal hypertension. Am J Cardiol 44:919–929

Weingart R (1974) The permeability to tetraethylammonium ions of the surface membrane and the intercalated discs of sheep and calf myocardium. J Physiol (London) 240:741–762

Weisberg A, Winegrad S, Tucker M, McClellan G (1982) Histochemical detection of specific isozymes of myosin in rat ventricular cells. Circ Res 51:802–809

Wendt-Gallitelli MF, Ebrecht G, Jacob R (1979) Morphological alterations and their functional interpretation in the hypertrophied myocardium of Goldblatt hypertensive rats. J Mol Cell Cardiol 11:275–287

Wenink ACG (1976) Development of the human cardiac conducting system. J Anat 121:617–631

Whalen RG, Sell SM, Butler-Browne G, Schwartz K, Bouveret P, Pinset-Harstrom I (1981) Three myosin heavy-chain isozymes appear sequentially in rat muscle development. Nature 292:805–809

Whalen RG, Sell SM, Eriksson A, Thornell LE (1982) Myosin subunit types in skeletal and cardiac tissues and their developmental distribution. Dev Biol 91(2):478–484

Wheat MW (1965) Ultrastructure autoradiography and lysosome studies in myocardium. J Mount Sinai Hosp 32:107–121

Wheeler DM, Horres CR, Lieberman M (1982) Sodium tracer kinetics and transmembrane flux in tissue-cultured chick heart cells. Am J Physiol 243:C169–C176

Widran J, Lev M (1951) Dissection of atrioventricular node, bundle and bundle branches in human heart. Circulation 4:863–867

Wikman-Coffelt J, Srivastava S (1979) Differences in atrial and ventricular myosin light chain LC. FEBS Lett 106:207–212

Wikman-Coffelt J, Parmley WW, Mason DT (1979) The cardiac hypertrophy process, Analyses of factors determining pathological vs physiological developmesch. Circ Res 45:697–707

Williams EH, De Haan RL (1981) Electrical coupling among heart cells in the absence of ultrastructurally defined gap junctions. J Memb Biol 60:237–248

Williams MA (1977) Quantitative methods in biology. In: Glauert AM (ed) Practical methods in electron microscopy, vol 6, Elsevier/North-Holland Biomedical Press, Amsterdam

Winegart S (1984) Regulation of cardiac contractile proteins, correlations between physiology and biochemistry. Circ Res 55:565–574

Winegrad S (1979) Electromechanical coupling in heart muscle. In: Berne RM, Sperelakis N (eds) Handbook of Physiology, sect 2. The cardiovascular system, vol 1, The heart. American Physiological Society, Bethesda, pp 393–428

Winegrad S (1982) Calcium release from cardiac sarcoplasmic reticulum. Ann Rev Physiol 44:451–462

Winkler B, Schaper J, Thiedemann KU (1976) Hypertrophy due to chronic volume overloading in the dog heart. A morphometric study. Basic Res Cardiol 172:222–227

Winquist RJ, Faison EP, Waldman SA, Schwartz K, Murad F, Rapoport RM (1984) Atrial natriuretic factor elicits an endothelium-independent relaxation and activates particulate guanylate cyclase in vascular smooth muscle. Proc Natl Acad Sci USA 81:7661–7664

Witschi E (1972) Characterisation of developmental stages, Part I man, Part II rat. In: Altman PL, Dittmer DS (eds) Biology data book, 2nd edn, vol 1. Federation of American Societies for Experimental Biology, Bethesda, p 176

Wisenbaugh T, Allen P, Cooper G, Holzgrefe H, Beuer G, Capabello B (1983) Contractile function, myosin ATPase activity and isozymes in the hypertrophied pig left ventricle after a chronic progressive pressure overload. Circ Res 53:332–341

Wollenberger A (1984) The cultured myocardial cell as a model in heart research, In: Abe H, Ito Y, Tada M, Opie LH (eds) Regulation of heart function. Japan Science Societies Press, Tokyo/VNU Science Press, Utrecht, pp 269–296

Woods RI (1970a) The innervation of the frog's heart. I. An examination of the autonomic postganglionic nerve fibres and a comparison of autonomic and sensory ganglion cells. Proc Roy Soc Ser B (London) 176:43–54

Woods RI (1970b) The innervation of the frog's heart. III. Electronmicroscopy of the autonomic nerve fibres and their vesicles. Proc Roy Soc Ser B (London) 176:63–68

Woods RI (1977) Changes in sizes of the adrenaline-containing vesicles and their cores in frog cardiac sympathetic nerves after pharmacological treatments. J Neurocytol 6(4):375–396

Woods WT, Sherf L, James TN (1982) Structure and function of specific regions in the canine atrioventricular node. Am J Physiol 243:H41–H50

Woods WT, Urthaler F, James TN (1976) Spontaneous action potentials of cells in the canine sinus node. Circ Res 39(1):76–82

Yamada E (1960) The fine structure of the paraboloid in the turtle retina as revealed by electron microscopy. Anat Rec 137:172

Yamaguchi M, Robson RM, Stromer MH (1983) Evidence for actin involvement in cardiac Z-lines and Z-line analogues. J Cell Biol 96:435–442

Yamamoto TJ (1965) Fine structure of the atrial muscle in snake heart. J Electron Microsc 14:134

Yamanaka M, Greenberg B, Johnson L, Seilhamer J, Brewer M, Friedman T, Miller J, Atlas S, Laragh J, Lewicki J, Fiddes J (1984) Cloning and sequence analysis of the cDNA for the rat atrial natriuretic factor precursor. Nature 309:719–722

Yamauchi A (1965) Electron microscopic observations on the development of SA nad AV nodal tissues in the human embryonic heart. Z Anat Entwicklungsgesch 124:562–587

Yamauchi A (1969) Innervation of the vertebrate heart as studied with the electron microscope. Arch Histol Jap 31:83–117

Yamauchi A (1973) Ultrastructure of the innervation of the mammalian heart. In: Challice CE, Virágh S (eds) The ultrastructure of the mammalian heart. Academic Press, New York, pp 127–178

Yamauchi A (1976) Ultrastructure of chromaffin-like adrenergic interneurons in the autonomic ganglia. In: Coupland RE, Fujita T (eds) Functional morphology of chromaffin, enterochromaffin and related cells. Elsevier, Amsterdam, pp 128–130

Yamauchi A (1979) Comparisons of the fine structure of receptor end-organs in the heart and aorta. In: Hainsworth R, Kidd C, Linden RJ (eds) Cardiac receptors, chapter 3. Cambridge University Press, Oxford, pp 51–70

Yamauchi A (1980) Fine structure of the fish heart. In: Bourne GH (ed) Hearts and heart-like organs, vol 1. Academic Press, New York, pp 119–148

Yamauchi A, Burnstock G (1968a) On the fine structure of the trout heart. J Anat 103:207–208

Yamauchi A, Burnstock G (1968b) An electron microscopic study on the innervation of the trout heart. J Comp Neurol 132:567–588

Yamauchi A, Fujimaki Y, Yokota R (1973) Fine structure studies of the sino-auricular nodal tissue in the heart of a teleost fish, *Misgurnus*, with particular reference to the cardiac internuncial cell. Am J Anat 138:407–430

Yamauchi A, Fujimaki Y, Kumagai K (1974) Direct contact between a neuron and some myocardial cells: An electron microscopic finding in the sinus venosus of the turtle heart. Anat Rec 179:491–496

Yamauchi A, Fujimaki Y, Yokota R (1975) Reciprocal synapses between cholinergic post ganglionic axon and adrenergic interneuron in the cardiac ganglion of the turtle. J Ultrastruct Res 50(1):47–57

Yamazaki K (1929) Biochemical studies on the atrioventricular junctional system of heart I. J Biochem 10:481–490

Yamazaki K (1930a) Biochemical studies on the atrioventricular junctional system of heart II. J Biochem 12:223–234

Yamazaki K (1930b) Biochemical studies on the atrioventricular junctional system of heart III. J Biochem 12:241–246

Yazaki Y, Ueda S, Nagai R, Shimada K (1979) Cardiac atrial myosin adenosine triphosphatase of animals and humans. Distinctive enzymatic properties compared with cardiac ventricular myosin. Circ Res 45:522–527

Yazaki Y, Ueda S, Tsuchimochi H, Nagai R (1984) Cardiac myosin isozymes: A molecular and cellular study. In: Abe H, Ito Y, Tada M, Opie LH (eds) Regulation of cardiac function. Japan Science Societies Press, Tokyo/VNU Science Press, Utrecht, pp 51–58

Yokota R, Ide C, Mitatori T, Onodera S (1983) Electron microscopic histochemisty of cholinesterase activity in the baroreceptor of the cat atrial endocardium. Acta Histochem Cytochem 16:129–137

Yokota S (1982) Immunoelectron microscopic localization of albumin in smooth and striated muscle tissues of rat. Histochemistry 74:379–386

Young JZ (1931) On the autonomic nervous system of the teleostean fish *Uranoscopus scaber*. Quart J Microsc Sci 74:491–535

Young JZ (1933) The autonomic nervous system of selachians. Quart J Microsc Sci 75:571–624

Young JZ (1936) The innervation and reactions to drugs of the viscera of teleostean fish. Proc Roy Soc Ser B (London) 12:303–318

Yoshinaga T (1921) A contribution to the early development of the heart in mammalia with special reference to the guinea-pig. Anat Rec 21:239–308

Yunge L, Benchimol S, Cantin M (1979) Ultrastructural cytochemistry of atrial muscle cells. VII. Radioautographic study of synthesis and migration of glycoproteins. J Mol Cell Cardiol 11:375–388

Zacchei AM, Caravita S (1972) Observations on the ultrastructure of chick-embryo cardiac myoblasts re-aggregated in long-term cultures. J Embryol Exp Morphol 28:571–589

Zak R (1973) Cell proliferation during cardiac growth. Am J Cardiol 31:211–219

Zak R (1974) Development and proliferative capacity of cardiac muscle cells. Circ Res 34(Suppl II):11–17

Zak R, Kizu A, Bugaisky L (1979) Cardiac hypertrophy: its characteristics as a growth process. Am J Cardiol 44:941–946

Zak R, Chizzonite RA, Everett AW, Clark WA (1982) Study of ventricular isomyosins during normal and thyroid hormone induced cardiac growth. J Mol Cell Cardiol 14(Suppl 3):111–117

Zucker IH, Gilmore JP (1974) Evidence of an indirect sympathetic control of atrial stretch receptor discharge in the dog. Circ Res 34:441–446

Author Index

Page numbers in *italics* refer to the bibliography

Subject Index

Handbook of Microscopic Anatomy

Volume II/6

H. Schmalbruch

Skeletal Muscle

1985. 129 figures. XI, 440 pages. ISBN 3-540-15608-9

This volume is a much-needed critical review of 30 years' research into the morphology of normal skeletal muscle in humans and mammals. It covers the field from myogenesis to muscle as a tissue and motor unit differentiation, and reflects the author's interest in normal and diseased human muscle.

Physiological results are discussed if they derive from morphology, are necessary for the understanding of morphology, they help acquire morphological information. Historical developments are presented where they help facilitate the understanding of scientific aberrations or help clarify controversial issues. The integration of normal morphology, physiology, and research in muscle diseases is so far advanced nowadays that it is usual to speak of "Muscle Research". Workers in any of these fields should be interested in the book's appraoch. It also intended as an introduction to the field for young researchers and as a comprehensive reference book for established anatomists, pathologists and physiologists.

Volume VI/8

P. Böck

The Paraganglia

1982. 61 figures. XV, 315 pages. ISBN 3-540-10978-1

The volume is the continuation of a comprehensive description begun in 1943 by M. Watzka. Since that time, the application of modern histologic methods – such as electron microscopy, immunohistochemistry, and fluorescence microscopy – to the study of paraganglionic tissue has resulted in a wide variety of new data. They are summarized in this volume to illustrate selected aspects of paraganglionic cells, among them: a general description of chromaffin cells; paragangionic cells as secretory (endocrine) elements and as interneurons: and the significance of paraganglionic cells in the process of chemoreception as excemplified by the glomus caroticum. The volume covers the literature from 1940 to the end of 1980.

Springer-Verlag
Berlin Heidelberg
New York Tokyo

Springer

Handbook of Microscopic Anatomy

Volume VI/7
L. Vollrath

The Pineal Organ

1981. 190 figures. XVII, 665 pages
ISBN 3-540-10313-9

The pineal body – a cerebral organ whose
function long remained clouded in mystery –
has yielded new insights to research during the
last 20 years. Today, the pineal body is one of
the biological sciences' preferred models for
studying regulatory mechanisms on all levels,
including the molecular. A neuroendocrine
transformer attached directly to the sympathetic nervous system, the pineal body has
characteristics of interest to biologists,
morphologists, endocrinologists, physiologists,
biochemists, pharmacologists, and clinicians
alike.
This volume contains a summary of all functional research into the pineal body conducted
during the last 35 years. Emphasis is placed on
functional morphology (from microscopic,
histochemical, and electronmicroscopic stand-
points) in organisms ranging from mammals to
lampreys. With its inclusion of pharmacological as well as biochemical and physiological
aspects, this volume represents the most up-
to-date and comprehensive work available on
the pineal gland.

Volume I/3

Chromosomes
in Mitosis and Interphase

By H. G. Schwarzacher

1976. 116 figures, 3 tables. VIII, 182 pages
ISBN 3-540-07456-2

A presentation complete in itself of the structure of human chromosomes, as studied under
light and electron microscopy. Emphasis is on
the somatic chromosomes in mitosis and interphase. Results in other species and the particu-
lars of meiosis are considered insofar as they
are necessary for a full understanding. The
following are discussed in detail: the form and
changes of chromosmes in mitosis, the human
karyotype, new banding methods, electron
microscopy showing a model of chromosome
structure in the light of current proven results,
the interphase cellular nucleus, heterochro-
matin, and the position of chromosomes in the
cell. The international literature on the
subject, as well as numerous, original and, in
part, yet unpublished data are presented.
Finally, a comprehensive analysis of genetic
materials is given as it appears in cytologic
preparations.

Volume VII/6
G. Aumüller

Prostate Gland and Seminal Vesicles

1979. 142 figures (some in color) in 181
separate illustratins. X, 380 pages
ISBN 3-540-09191-2

This authoritative reference for residents,
specialists and researchers begins with a
description of normal organ embryology. The
latest embryologic research findings are
dicussed and there is a survey of comparative
anatomy and normal relationships in man.
Emphasis throughout is on the functional
morphology of the human prostate and
seminal vesicle. In addition, a number of
special questions are addressed, regarding,
comparative histology, molecular endocrino-
logy, biochemisty, physiology and pathology.
Current knowledge about organ ultrastructure
is reviewed from the standpoint of functional
production, such as growth and secretion and
the underlying molecular biologic mechanisms
explained.
Replete with original illustrations and citations
from current literature in the field, this handbook allows a deeper insight into the morphology and function of these organs.

Springer-Verlag
Berlin Heidelberg New York Tokyo

Springer